An Introduction to Metalogic

An Introduction to Metalogic

Aladdin M. Yaqub

broadview press

© 2015 Aladdin M. Yaqub

All rights reserved. The use of any part of this publication reproduced, transmitted in any form or by any means, electronic, mechanical, photocopying, recording, or otherwise, or stored in a retrieval system, without prior written consent of the publisher—or in the case of photocopying, a licence from Access Copyright (Canadian Copyright Licensing Agency), One Yonge Street, Suite 1900, Toronto, Ontario M5E 1E5—is an infringement of the copyright law.

Library and Archives Canada Cataloguing in Publication

Yaqub, Aladdin Mahmud, author
    An introduction to metalogic / Aladdin M. Yaqub.

Includes bibliographical references and index.
ISBN 978-1-55481-171-7 (pbk.)

    1. First-order logic. I. Title.

BC128.Y26 2014   511.3   C2014-905949-3

Broadview Press is an independent, international publishing house, incorporated in 1985.

We welcome comments and suggestions regarding any aspect of our publications—please feel free to contact us at the addresses below or at broadview@broadviewpress.com.

North America
PO Box 1243, Peterborough, Ontario K9J 7H5, Canada
555 Riverwalk Parkway, Tonawanda, NY 14150, USA
Tel: (705) 743-8990; Fax: (705) 743-8353
email: customerservice@broadviewpress.com

UK, Europe, Central Asia, Middle East, Africa, India, and Southeast Asia
Eurospan Group, 3 Henrietta St., London WC2E 8LU, United Kingdom
Tel: 44 (0) 1767 604972; Fax: 44 (0) 1767 601640
email: eurospan@turpin-distribution.com

Australia and New Zealand
NewSouth Books
c/o TL Distribution, 15-23 Helles Ave., Moorebank, NSW 2170, Australia
Tel: (02) 8778 9999; Fax: (02) 8778 9944
email: orders@tldistribution.com.au

www.broadviewpress.com

Edited by Robert M. Martin

Broadview Press acknowledges the financial support of the Government of Canada through the Canada Book Fund for our publishing activities.

PRINTED IN CANADA

To the women who made me the man I am:
my mother, Imtithal,
my wife, Connie,
my daughters, Mariam and Ranah,
and my sisters, Basma, Maha, Nisreen, Intesar, and Nagham

Contents

Introduction • xi

Chapter One: First-Order Predicate Logic (PL) • 1

1.1 The Syntax of PL • 1

- 1.1.1 The basic vocabulary of PL • 1
- 1.1.2 PL terms • 5
- 1.1.3 PL formulas • 7
- 1.1.4 Bound and free variables and PL sentences • 8

1.2 The Semantics of PL • 10

- 1.2.1 PL interpretations • 10
- 1.2.2 Examples of PL interpretations • 14
- 1.2.3 Objectual and substitutional quantification • 19
- 1.2.4 The size of a PL interpretation • 21
- 1.2.5 The truth conditions of PL sentences • 22
- 1.2.6 Possible objections to substitutional quantification and replies • 28

1.3 Logical Concepts in PL • 31

- 1.3.1 Definition of a PL argument • 31
- 1.3.2 Validity and logical consequence • 31
- 1.3.3 Definition of a valid argument • 31
- 1.3.4 Definition of an invalid argument • 33
- 1.3.5 Definition of a valid sentence • 34
- 1.3.6 Definition of a contradictory sentence • 35
- 1.3.7 Definition of a contingent sentence • 36
- 1.3.8 Definition of logically equivalent sentences • 37
- 1.3.9 Definition of a satisfiable set • 38
- 1.3.10 Definition of an unsatisfiable set • 39
- 1.3.11 Decidable and semidecidable concepts • 40

1.4 PL Proof Theory • 41

- 1.4.1 The notion of formal derivation • 41
- 1.4.2 The statements of the Soundness and Completeness Theorems for PL • 43
- 1.4.3 Corollaries of the Soundness and Completeness Theorems • 45

1.4.4   The structure and application of inference rules • 47
1.4.5   The Natural Deduction System (NDS) • 52
1.4.6   A justification for the rule Explosion • 59
1.4.7   The Gentzen Deduction System (GDS) • 61

1.5   Exercises • 64

Solutions to the Starred Exercises • 69

Chapter Two: Resources of the Metatheory • 87

2.1   Linguistic and Logical Resources • 87

2.1.1   The metatheory of PL • 87
2.1.2   The language of the metatheory • 88
2.1.3   Logical resources • 89

2.2   Arithmetical Resources • 90

2.2.1   The structure of the natural numbers • 90
2.2.2   The Principle of Mathematical Induction • 91

2.3   Set-Theoretic Resources • 94

2.3.1   Basic concepts and principles • 94
2.3.2   Russell's Paradox • 97
2.3.3   Relations and functions • 98
2.3.4   Cardinalities of sets • 101
2.3.5   Cantor's Diagonal Argument • 105

2.4   An Economical Version of PL • 108

2.4.1   Expressive completeness • 108
2.4.2   The Mini Deduction System (MDS) • 115

2.5   Exercises • 124

Solutions to the Starred Exercises • 127

Chapter Three: The Soundness and Completeness Theorems • 139

3.1   The Soundness Theorem • 139

3.2   The Completeness Theorem • 144

3.2.1   An equivalent formulation of the Completeness Theorem • 144
3.2.2   Lindenbaum's Lemma • 148
3.2.3   Henkin sets • 152
3.2.4   Henkin interpretations of Henkin sets without the identity predicate • 157

3.2.5   Henkin interpretations of Henkin sets with the identity predicate • 161

3.3   **The Compactness Theorem** • **164**

    3.3.1   The Compactness Theorem • 164
    3.3.2   The Finite-Satisfiability Theorem • 165

3.4   **Elementary Equivalence and Isomorphism** • **167**

3.5   **Properties of PL Sets** • **169**

    3.5.1   PL theories • 169
    3.5.2   Axiomatizable PL theories • 172
    3.5.3   Complete and categorical PL sets • 175

3.6   **The Löwenheim-Skolem Theorem** • **177**

    3.6.1   A proof of the theorem • 177
    3.6.2   Skolem's Paradox • 178

3.7   **Exercises** • **180**

Solutions to the Starred Exercises • **182**

Chapter Four: Computability • 189

4.1   **Effective Procedures and Computable Functions** • **189**

4.2   **Turing Computability** • **193**

    4.2.1   Turing machines • 193
    4.2.2   An example • 194
    4.2.3   Instruction lines and diagrams of Turing machines • 195
    4.2.4   The zero, successor, and addition functions • 197
    4.2.5   Turing-computable functions and Church's Thesis • 203
    4.2.6   Decidable and semidecidable sets • 205

4.3   **The Halting Problem** • **209**

4.4   **Partial Recursive Functions** • **213**

4.5   **Exercises** • **218**

Solutions to the Starred Exercises • **219**

Chapter Five: The Incompleteness Theorems • 223

5.1   **Peano Arithmetic** • **223**

5.2   **Representability in Peano Arithmetic** • **228**

5.3	**Arithmetization of the Metatheory** • **231**	
5.4	**Diagonalization and the First Incompleteness Theorem** • **238**	
	5.4.1 The Diagonalization Lemma • 238	
	5.4.2 Gödel's First Incompleteness Theorem • 241	
5.5	**Consequences of Diagonalization and Incompleteness** • **245**	
	5.5.1 The undecidability of consistent extensions of Peano Arithmetic • 245	
	5.5.2 Tarski's Indefinability Theorem • 250	
	5.5.3 Church's Undecidability Theorem • 252	
5.6	**The Incompleteness of Second-Order Predicate Logic** • **257**	
	5.6.1 Second-Order Predicate Logic (PL²) • 257	
	5.6.2 Second-Order Peano Arithmetic • 260	
	5.6.3 PL² and the Compactness Theorem • 265	
	5.6.4 The incompleteness of PL² • 268	
5.7	**Gödel's Second Incompleteness Theorem** • **272**	
	5.7.1 Hilbert's Program • 272	
	5.7.2 The Provability Conditions • 275	
5.8	**Exercises** • **281**	

Solutions to the Starred Exercises • **284**

Index • **291**

Introduction

I

An Introduction to Metalogic is an introductory textbook devoted to the metatheory of First-Order Predicate Logic (PL). The book takes the student from the syntax of PL to a detailed outline of proofs of the incompleteness, undecidability, and indefinability results, passing through the semantics of PL and its natural deduction system, the Soundness and Completeness Theorems, the Compactness Theorem, the Löwenheim-Skolem Theorem, elementary equivalence and isomorphism, PL theories and their properties, Turing-computable functions and partial recursive functions, the Halting Problem, the incompleteness of Second-Order Predicate Logic, brief introductions to set theory and number theory, and other introductory topics of the metatheory of PL. For the material in Chapter One, no more than standard high school mathematical and linguistic competence is required; no specific background in symbolic logic, philosophy, or mathematics is presupposed. For the more advanced material in the book—Chapters Two through Five—a certain degree of what is vaguely and commonly termed "mathematical sophistication" is needed. Students should know the difference between an axiom and a theorem; be familiar with the basic algebraic features of the integers; have encountered rational and real numbers; be able to work with structural symbols; be able to understand and use technical terms introduced by means of sufficient and necessary conditions; be able to follow a demonstrative proof and occasionally to construct one; be able to think abstractly without recourse to visual aids; and, most importantly, not suffer from "math phobia."

An Introduction to Metalogic has three unique features, which are also its primary virtues: it is self-contained; the motivations behind almost all its topics are fully explored; and it avoids becoming mired in excessive technical detail. The reason I say the book is self-contained is that instructors can use it for their logic courses without having to rely on other sources or other courses. The book is designed to provide the foundation for two types of logic courses. The first is an introductory course to PL and its metatheory, in which no background in symbolic logic is presupposed. This course might be attractive to instructors who want to teach, and to students who want to study, First-Order Predicate Logic and the fundamental results of its metatheory without expository accounts of Sentence Logic and, possibly, Term Logic. The first chapter supplies an adequate introduction to PL syntax, semantics, logical concepts, and natural deduction system. The expository discussion, exercises, and solutions to selected exercises of this chapter are intended to build expertise into the basics of First-Order Predicate Logic without presupposing the reader has encountered any system of symbolic logic. Furthermore, the introductory chapter, like the rest of the book, is suitable for self-directed study. Instructors can assign sections of it for homework and devote class time to covering only the more difficult points. They can then move on to the last four chapters, spending most of the semester on metalogic. Should instructors wish to include an introduction to Sentence Logic and cover more of the basics of symbolic logic, they can assign readings and exercises from my introductory textbook *An Introduction to Logical Theory*. For those instructors and students, Broadview Press offers both books as a single bundle. The second course would concern primarily the metatheory of PL, beginning with a quick review of PL, its syntax, semantics, and proof theory. This course would be appropriate for students who have had the standard first course in symbolic logic. I have taught both types of course with great success. I had first-year and humanities students with no background in logic or college mathematics, as well as students with extensive technical background. My statistics show no significant performance biases at all. I had first-year students who scored A's and mathematics and engineering students who scored C's. There were also no significant performance biases between students who took the first and those who took the second type of course.

The motivation behind each of the book's topics is fully explained, with a notable exception: the book does not say explicitly why PL is important. This is because PL serves multiple purposes: instructors will inevitably frame PL's purpose differently depending on whether they are teaching students of philosophy, mathematics, computer science, or one of the other fields, such as linguistics and cognitive science, that make use of logical tools, concepts, and results. A philosophy instructor might view the semantics of PL as intended principally to capture the notion of logical possibility or of logical consequence; a mathematics instructor might introduce PL as a formal development of the background logic of mathematical theories; a cognitive science instructor might see PL as the means for describing a certain type of mental representation and the computational processes that can operate on this type of representation. In order to allow for this kind of flexibility, no single motivation for PL syntax or semantics is identified in the book. Chapter One, however, which deals with PL, contains elaborate explanations of the syntactical categories of PL, and of several of its semantical concepts, including the central concept of logical consequence. Thus even a student who is not guided by an instructor can gather from the discussion and examples in this chapter that PL is a symbolic system that offers precise descriptions of necessary and contingent properties of and relations between propositions. This in itself is sufficient motivation for developing PL syntax and semantics.

Aside from PL syntax and semantics, all other topics raise explicitly or implicitly the specific questions and problems that generate them—questions such as: What is the nature of demonstrative proof? What is the nature of computability? Is it possible to give a precise characterization of the intuitive notion of effective procedure? Are there well-defined functions that are not computable? Are the formal rules of inference sufficiently strong to generate derivations for all logical consequences? Are all truths about a certain field derivable from well-demarcated sets of axioms? Is the concept of logical consequence decidable in PL? Are infinite sets definable in PL? Are uncountable sets definable in PL? Is it ever possible to prove that a mathematical theory is consistent without invoking resources that exceed what is available to the theory itself?

In order to make the book user-friendly, I avoid excessive technicality. On the one hand, I give complete proofs of the Soundness and Completeness Theorems and their corollaries and of many of the theorems in the chapters on computability and the incompleteness results. Furthermore, I go into considerable detail in explaining and illustrating difficult topics. On the other hand, however, I omit the proofs of some important theorems. Although I explain the statements of those theorems and frequently give examples to illustrate their application, I leave the matter at that and accept their truth without proof. The equivalence between partial recursive functions and Turing-computable functions is stated without proof. I also omit the proof for the representability of all recursive functions in Peano Arithmetic. The existence of a recursive diagonalization function is illustrated diagrammatically but not proved (though the Diagonalization Lemma is proved). Two proofs are given for the Second Incompleteness Theorem: the first presupposes that the First Incompleteness Theorem can be formalized in Peano Arithmetic, and the second presupposes that the standard provability predicate satisfies the three familiar provability conditions; both presuppositions are explained but not proved. With a few minor exceptions, those are all the proofs that I omit. Everything else is proved with sufficient attention to details and with as much clarity as I can muster.

The proofs I omit typically involve tedious details that do not advance or enhance students' grasp of the material. For instance, once students are shown a matrix of 1-variable formulas and their diagonal instances, they readily perceive that there must be a computable function that associates the gödel number of any 1-variable formula with the gödel number of its diagonal instance. To prove the existence of this function would take them into recondite details about gödel numbering and the arithmetization of certain syntactical operations, such as substitution. The same can be said about the representability of recursive functions in Peano Arithmetic, and about formalizing the First Incompleteness Theorem in Peano Arithmetic. My experience has shown that including such proofs is actually detrimental. Once these proofs are included, they must be fully explained. The explanations are inevitably tedious and uninteresting and actually discourage students from persevering through the rest of the chapter. No matter how carefully and "enthusiastically" these proofs were presented, they always cost me a number of students. Yet these proofs afford students no deeper insight, illuminating explanation, or further application.

II

Chapter One covers the syntax, semantics, and proof theory of PL. This is the background material whose metatheory will be studied in the next four chapters. As stated above, this material is accessible to any student with high school mathematics, and it hardly requires any degree of mathematical sophistication, other than being comfortable working with symbols. Chapter One begins with the syntax of PL. It presents elaborate discussion of PL syntactical categories and their relation to natural language. After the syntax, the central semantical notions of interpretation and true-on-an-interpretation are introduced and illustrated with several examples. The next section deals with the definitions and examples of the standard logical concepts: valid and invalid arguments; valid, contradictory, and contingent sentences; satisfiable and unsatisfiable sets; and logically equivalent sentences. The last section of Chapter One is devoted to the natural deduction system for PL. Early in the section, PL proof theory and PL semantics are linked via the Soundness and Completeness Theorems, which are stated without proof (their complete proofs will be given in Chapter Three). Several corollaries of the Soundness and Completeness Theorems are proved. Two deduction systems are described: one with many primitive rules, and the standard system, which consists of introduction and elimination rules. In the solutions to the starred exercises, many examples of PL interpretations and derivations are presented.

Chapter Two is devoted to preliminary topics that the next chapters invoke. Linguistic, logical, arithmetical, and set-theoretic resources of metalogic are discussed in an accessible yet rigorous fashion. Among the topics covered are the natural numbers and the Principle of Mathematical Induction, the basic notions of set theory, finite and infinite sets and cardinalities, Cantor's Theorem, and his Diagonal Argument. An economical version of PL is developed to simplify the proofs of metatheorems. The exercises mostly expand on set-theoretic notions and relations, and several of them require the use of mathematical induction.

Chapter Three focuses on the Soundness and Completeness Theorems and other related topics. Both theorems are given their standard proofs. Two versions of the Compactness Theorem are proved as corollaries of the Soundness and Completeness Theorems. The Löwenheim-Skolem Theorem is also proved as a corollary of Henkin's proof of the Completeness Theorem. The chapter covers some basic topics of PL model theory, such as elementary equivalence, isomorphism, PL theories, and some of their properties.

Chapter Four is an introduction to computability theory. The intuitive notions of effective procedure and decidability are introduced. Special care is taken to show that these questions can be formulated as questions about computable numerical functions. Computable function is defined via the most intuitive approach—namely, Turing machines. Turing machines are carefully explained with the aid of diagrams, and illustrated with several examples. The notion of Turing-computable function and other related notions, such as decidable and effectively enumerable sets, are defined rigorously, and a complete proof is given of the Halting Problem. Church's Thesis is also discussed. The chapter ends with a definition of partial recursive function, and the equivalence between partial recursive functions and Turing-computable functions is stated without proof.

Chapter Five contains detailed outlines of proofs of the incompleteness, undecidability, and indefinability results. The most interesting theorems in this book are found in this chapter. The central components of Gödel's proof of the First Incompleteness Theorem are discussed with elaboration: Peano Arithmetic (PA), the arithmetization of the metatheory, the representability of recursive functions in PA, and diagonalization. Several consequences of diagonalization and incompleteness are proved, including the undecidability of PA, the non-axiomatizability of Arithmetic, Church's Undecidability

Theorem, and Tarski's Indefinability Theorem. Next, Second-Order Logic (PL^2) and Second-Order Peano Arithmetic (PA^2) are introduced. Several classical theorems are proved, such as the categoricity of PA^2, the axiomatizability of Second-Order Arithmetic, and the incompleteness of PL^2. The chapter concludes with an outline of two proofs of the Second Incompleteness Theorem and a discussion of Hilbert's Program. The exercises of Chapter Five extend the discussion to a few important topics, such as Löb's Theorem, the Rosser sentences, and nonstandard models of Arithmetic.

III

There are several introductions to metalogic available as textbooks. Some of them are more popular than others. All of the metalogic books I have examined, read, or taught, other than this book, presuppose a high degree of mathematical sophistication. At several universities in which I have taught, including my current institution, some of these books are used in metalogic courses. In my experience, I found that even advanced undergraduate mathematics students find them quite difficult to understand. My book is designed for a mixed audience and it does not favor those with technical background. Last time I offered a metalogic course, in which I used a draft of this book, a student whose only exposure to college mathematics was an introductory statistics course and who never took symbolic logic prior to my course outperformed most of the students in the class and earned a solid A. He was able to produce proofs that were elegant, clear, and rigorous; his examples were imaginative; and his assignments were a pleasure to read.

Many years ago, when I offered my first metalogic course, I used one of the more popular metalogic textbooks. My students were a mix of philosophy, mathematics, computer science, and linguistics students. They found much of the discussion in the book too hard to understand and many of its proofs too difficult to penetrate. At some point, I had to stop using the book and rely on my lecture notes alone. Since then I have taught from my lecture notes. During the summer of 2011, I completed mature drafts of Chapters One and Two. In the fall of 2011 I taught metalogic again. This time I was determined to produce complete drafts of Chapters Three through Five. I wrote the chapters as the course progressed, staying one chapter ahead of the students until the end of the semester. If student performance and evaluations are any measure of the success of a textbook, then I can say with confidence that this book was greatly successful and achieved all its intended objectives.

It is difficult to say how long it took me to complete this book. I wrote its first set of notes in the spring of 1997. I revisited those notes every time I offered the course. I added to them some theorems, definitions, and exercises. As indicated above, I wrote the first draft of this book from June through December 2011. I transformed those fragmented sets of notes into a coherent, self-contained book that is suitable both for class use and for self-directed study. Since it is unusual to be able to write such a book in seven moths while revising the manuscript of another book and teaching two courses, the reader may wonder how I was able to complete all of this. I was for those seven months suffering from a hypomanic episode of a manic-depressive cycle. Therefore I had an extraordinary abundance of energy,

I didn't need to sleep for more than four hours a night, and I was operating at a high level of clarity. After I climbed down and returned to this manuscript in 2012, and worked very carefully through my expositions, proofs, and exercises, I was very surprised to see how often I left matters for the reader as "easy" exercises, when in fact they were quite challenging. Needless to say, I rewrote all of these proofs to make sure that they are all clear and accessible. I share this experience with my readers in hopes of removing the stigma from mental illness, which is no different from any other physical illness. Since I worked on this book with devotion also for the whole of 2012, and since my preexisting notes gave me about a year advantage, it is fair to say that it took me a little more than two and a half years to transform my underdeveloped notes into a book that, I hope, is worthy of an instructor's consideration and a student's effort.

Four anonymous colleagues reviewed a partial manuscript of this book for Broadview Press and recommended its publication. They also offered several suggestions for improvements, which I incorporated for the most part. I am grateful to them for their recommendations and suggestions. I am also grateful to Mariam Yaqub who copyedited the manuscript thoroughly before I submitted it to the publisher. Ranah Yaqub copyedited two sample chapters, which I included with the book proposal I submitted to Broadview Press. I thank them both for their diligent work. I am deeply grateful to Professor Robert Martin for his meticulous copyediting and artful typesetting of the manuscript for Broadview Press. Thanks are also due to the philosophy editor at Broadview Press, Mr. Stephen Latta, with whom I worked on two books. I couldn't possibly have hoped for a better editor. I would like to express my sincere thanks to Ms. Tara Lowes of Broadview Press, who supervised the production of this book and made sure that it appeared in perfect form. I am indebted to the many students who took metalogic from me. Their suggestions greatly enhanced the presentation in this book. As always, I am grateful to my dear friend and colleague Roslyn Weiss, who read the prospectus of this book, which was part of the proposal I submitted to Broadview Press, and made sure it was a winning prospectus, and who was immensely supportive during my recent struggle to regain my mental stability. I am also grateful to my dear friends Michael Mendelson, Sergio Tenenbaum, Jennifer Nagel, Fred and Karen Schueler, James and Elizabeth DeVault, Amy Schmitter, and Donna Wagner, who contributed in many important ways to my well-being during very difficult times. I would like to register my heartfelt thanks to my psychiatrist, Dr. Claudia Baldassano, whose knowledge and acumen are quieting the tribulations of my mind and bringing its affliction into a state of sustained remission. In addition, this goal cannot be met without the skill and wisdom of my psychotherapist, Patricia Fuisz, who is patiently working with me to teach me how to take charge of my depressed mood. I am very grateful to her. Finally, my deepest gratitude is to the woman who makes sure that I remain a decent man in spite of the logical and analytical excesses of my vocation, and who never wavered in her love and support during the many years of my turbulent mental states. I dedicate this work to her, my wife, Connie, and to the other women in my family, without whom I would not be here today to write this book.

Chapter One

First-Order Predicate Logic (PL)

1.1 The Syntax of PL

1.1.1 First-Order Predicate Logic (PL)[1] is the most important and the most studied modern symbolic system. Although it has, when compared to natural languages, severe expressive limitations, its resources for constructing complex, yet extremely precise, declarative sentences go far beyond anything typically available for making similar sentences in natural languages. The **basic vocabulary** of PL consists of the following six categories.

1.1.1a **Names**, which are the following lowercase italic letters: $a, b, c, \ldots, r, s, t$ (excluding 'f', 'g', and 'h'); with numeric subscripts if needed.[2]

1.1.1b **Function symbols**, which are the following lowercase italic letters with numeric superscripts: $f^1, g^1, h^1, f^2, g^2, h^2, f^3, g^3, h^3, \ldots$; with numeric subscripts if needed.

1.1.1c **Predicates**, which are uppercase italic letters with numeric superscripts: $A^1, B^1, C^1, \ldots, X^1, Y^1, Z^1; A^2, B^2, C^2, \ldots, X^2, Y^2, Z^2; A^3, B^3, C^3, \ldots, X^3, Y^3, Z^3; \ldots$; with numeric subscripts if needed.

1.1.1d **Variables**, which are the following lowercase italic letters: u, v, w, x, y, z; with numeric subscripts if needed.

1.1.1e Eight **logical symbols**: $\neg, \land, \lor, \rightarrow, \leftrightarrow, \forall, \exists, =$

1.1.1f **Parentheses**: '(' and ')'

The variables, logical symbols, and parentheses are referred to as the **logical vocabulary** of PL, and the names, function symbols, and predicates are referred to as the **extra-logical vocabulary** of PL.

In the study of language it is important to distinguish between two languages: The **object language** and the **metalanguage**. The object language is the language under study

---

1 First-Order Predicate Logic is also called "First-Order Logic," "Predicate Logic," and "Quantificational Logic."
2 A numeric subscript or superscript is to be understood as a subscript or a superscript n, where n is a natural number, that is, 0, 1, 2, 3, … .

and the metalanguage is the language in which the object language is discussed. In our particular case, the object language is the language of PL, whose syntactical categories are described above, and the metalanguage is English augmented with appropriate symbols. When an expression of the language of PL (i.e., a PL expression) is mentioned in the metalanguage, it is customary to enclose the PL expression between two single-quotation marks, ' '. However, in order to simplify our notation, almost always we will not follow this usage where there is no cause for misunderstanding. When we need to say something about all PL predicates, function symbols, or variables, we use **metalinguistic variables** (i.e., variables in the metalanguage) to range over these PL expressions. We designate the boldfaced letters '**P**', '**Q**', and '**R**', sometimes with numeric superscripts, to be metalinguistic variables ranging over PL predicates; the boldfaced letters '**f**' and '**g**', possibly with numeric superscripts, to be metalinguistic variables ranging over function symbols; and the boldfaced letters '**x**', '**y**', and '**z**' to be metalinguistic variables ranging over PL variables.

As we will discuss in the next section, the names stand for **individuals**, the function symbols for **functions**, and the predicates for **properties** and **relations**. The term 'individual' is used here in its philosophical sense. Individuals can be any kind of entity or object. They can be people, chairs, animals, trees, stars, galaxies, ideas, numbers, geometric figures, linguistic expressions, or what have you. The logical symbols allow us to form complex expressions that designate complex properties, complex relations, or complex states of affairs. The variables have a syntactical role that pertains to the use of quantifiers such as 'every' and 'some'. To illustrate the point, we translate the English sentence 'all whales are mammals' into a hybrid PL-English sentence. First, we reformulate the sentence as 'if something is a whale, then it is a mammal'. The latter could be paraphrased, using variables, as 'for every x, if x is a whale, then x is a mammal'. The variable 'x' here assumes the role of the pronoun 'it' in the English sentence. In general, PL variables in some ways behave like pronouns in English. As we will explain later, in PL sentences variables are placeholders: they indicate which quantifier applies to which place.

PL predicates come with superscripts. A superscript indicates the number of "places" the predicate has. For example, the predicate A^1 is a 1-place PL predicate, meaning that it applies to single individuals. 1-place predicates stand for **properties** because properties apply to single individuals. The English expression 'is a man' is a 1-place English predicate: if we put a singular term, such as 'Socrates' to the left of 'is', we obtain a declarative sentence. You can think of a 1-place predicate as an expression with a "blank"; if you fill the blank with a singular term, the resulting (complete) expression is a declarative sentence. The same analysis applies to PL. However, the "blanks" of any PL predicate, except the identity predicate '=', are always to the right of the predicate. So if the PL name 's' stands for Socrates and the 1-place PL predicate A^1 stands for the property of being a man, then the English sentence 'Socrates is a man' correspond to the PL sentence A^1s. 1-place predicates are also called **unary predicates** or **monadic predicates**.

A predicate that has more than one place stands for a **relation** with the same number of places. Thus, for instance, 2-place predicates stand for 2-place relations, which are commonly called **binary relations**, and 3-place predicates stand for 3-place relations. In general, an n-place predicate (where n is greater than one) stands for an n-place relation. These predicates are called **relational predicates**. In order to improve the readability of n-place English predicates, we use variables instead of blanks. We will refer to the language that results from augmenting English with PL variables as **PL-English**. For example, instead of writing '— shares borders with ... and ---' we will write 'x shares borders with y and z'. This 3-place relational predicate may be used to express the 3-place relation of a country's sharing borders with two other countries or of a region's sharing borders with two other regions, or other similar 3-place relations. We will say that the English sentence 'The USA shares borders with Canada and Mexico' is obtained from the (PL-English) predicate 'x shares borders with y and z' by *substituting* the names 'The USA', 'Canada', and 'Mexico' for the variables 'x', 'y', and 'z', respectively. The language of the metatheory we will study in Chapter Two is PL-English augmented with a large assortment of symbols. It is important to note that English is part of PL-English. Hence every natural English sentence is a PL-English sentence.

We extend the use of variables to PL predicates. In order to simplify the notation and improve the readability of n-place PL predicates, we use variables instead of superscripts to indicate the number of places. For instance, instead or writing B^3 we write $Bxyz$. We can let the PL predicate $Bxyz$ correspond to the PL-English predicate 'x shares borders with y and z', and the PL names s, c, and m correspond to the English names 'The USA', 'Canada', and 'Mexico', in this order. Thus the PL sentence $Bscm$ corresponds to the English sentence 'The USA shares borders with Canada and Mexico'.

Function symbols also have places. The superscript of a function symbol indicates the number of places this function symbol has. As with PL predicates, we will employ the convention of using variables instead of superscripts to mark the number of places a function symbol has. For instance, we might write $gxyz$ instead of g^3 and hvw instead of h^2. We abandon this convention if there is a cause for ambiguity. Function symbols are used to generate complex terms. Later we will define what we mean by 'PL term'. For now it suffices to consider the case of names. If we apply a function symbol to a name, we obtain a complex name that refers to some individual. A complex PL name is similar to a **definite description**. In natural languages definite descriptions are complex expressions that refer to unique individuals, if they refer at all. In English definite descriptions are typically generated with the aid of the definite article 'the'. For example, 'The first president of the United States' is a definite description that refers to one and only one individual—namely George Washington. Definite descriptions can be analyzed syntactically as being composed of **functional descriptions** and **singular terms**. Functional descriptions are similar to predicates in that they have "blanks." The expression 'The first president of ___' is a 1-place functional description. In PL-English we can paraphrase this functional description as 'The first president of x'. Substituting a name for x generates a definite description. Depending on the name we substitute for x, the resulting definite description might or might not have a referent. For instance, if we substitute 'Iraq' for x, the definite description 'The first president of Iraq' has a unique referent: General Qasim (we say in this case that this definite description is **referring**). But if we replace x with the name 'England', the resulting definite description has no referent (we say in this case that this definite description is **non-referring**). It is possible to replace x with another definite description. For instance, instead of using the name 'The USA', we may replace x with the definite description 'The country that dropped a nuclear bomb on Japan'. In this case we generate the definite description 'The first president of the country that dropped a nuclear bomb on Japan'. Again, this definite description refers to George Washington.

We call names and definite descriptions collectively "singular terms." We said above that a definite description is composed of a functional description and a singular term. This assertion is not exactly correct. A functional description might have more than one "blank." A 2-place functional description has two "blanks," a 3-place functional description has three "blanks," and, in general, an n-place functional description has n "blanks." The functional description 'The sum of x, y, and z' is 3-place. If we substitute three numerals for the variables x, y, and z, we obtain a definite description that refers to exactly one number. For example, the definite description 'The sum of 7, 11, and 16' refers to the number 34. Here too it is possible to replace a variable with another definite description. Consider the definite description 'The square of 4'. We can substitute this definite description for the variable z in the original functional description. In this case, we obtain the definite description 'The sum of 7, 11, and the square of 4', which refers to 34. Thus we can assert more accurately that a definite description is composed of a functional description and one or more singular terms. As we have seen, definite descriptions might be referring or non-referring.

PL function symbols correspond to natural-language functional descriptions. For example, the PL function symbols *fx* and *hxyz* might correspond to the English functional descriptions 'The first president of x' and 'The sum of x, y, and z' respectively. If we let the name *e* stand for the USA, the PL term *fe* would correspond to the definite description 'The first president of the USA', and if we let *a*, *b*, and *c* stand for the numbers 7, 11, and 16, respectively, then the PL term *habc* would correspond to the definite description 'The sum of 7, 11, and 16'. Just as the case with an English definite description, a PL term may be substituted for a variable in another PL term. Consider the following example. Say we let the function symbol *gv* correspond to the functional description 'The square of v' and we let the name *s* stand for the number 4. We obtain the PL term *habgs*, which corresponds to the definite description 'The sum of 7, 11, and the square of 4'. There is an important difference between singular terms in natural languages and PL terms that contain no variables. Natural-language singular terms might fail to have referents. For instance, 'Pegasus' and 'The present king of France' have no referents. PL terms that contain no variables can never fail to refer: every such PL term has exactly one referent on each PL interpretation (we will define this notion later).

The logical symbols of PL consist of five sentential connectives, $\neg$, $\wedge$, $\vee$, $\rightarrow$, and $\leftrightarrow$,[3] which represent, respectively, negation and denial expressions such as 'not' and 'it is not the case that', conjunction connectives such as 'and', disjunction connectives such as 'or', conditional connectives such as 'if-then', and biconditional connectives such as 'if and only if'; two quantifier symbols, $\forall$ and $\exists$, which are used to express the quantifiers 'all' and 'some' (in the sense that there is at least one), respectively; and a symbol, =, for the relation of token identity. Since the quantifier symbols are not really sentential connectives, we will refer to the logical symbols of PL (except the symbol =) as 'logical operators'. Thus one of these logical operators is a unary connective, $\neg$, four of them are binary connectives, $\wedge$, $\vee$, $\rightarrow$, and $\leftrightarrow$, and two of them are quantifier symbols, $\forall$ and $\exists$. When we speak of the PL connectives we intend the standard five sentential connectives. When we speak of the logical operators of PL we mean the connectives and the quantifier symbols.

A PL expression of the form '($\forall$z)', where **z** is a PL variable, is called a **universal quantifier**,[4] and one of the form '($\exists$z)' is called an **existential quantifier**. We also refer to these quantifiers as **z-quantifiers**. A PL universal quantifier ($\forall x$) is read in PL-English as 'for all x', 'for every x', or 'for each x'; and a PL existential quantifier ($\exists x$) is read in PL-English as 'for some x' or 'there is (exists) x' (in the sense that there is at least one x). The symbol = is called **the identity predicate**. It is a 2-place predicate that stands for the relation of **token identity**. This relation holds between each individual and itself and never between distinct individuals.

1.1.2   A PL **term** is either a PL name, a PL variable, or a PL expression that is generated from the names and variables by applying the following formation rule any finite number of times.

The Term-Formation Rule: If f^n is an n-place function symbol and $t_1, t_2, ..., $ and t_n are any PL terms, then the PL expression $f^n t_1 t_2 ... t_n$ is a PL term.

We use the boldfaced letters '**r**', '**s**', and '**t**' (with numeric subscripts if needed) as metalinguistic variables ranging over PL terms. If a PL term contains function symbols and variables, we call it a **functional term**; and if it is a name or contains function symbols but no variables, we call it a **singular term**. Note that, by definition, PL variables are neither functional nor singular terms; they are only PL terms. PL functional terms correspond to English functional descriptions and PL singular terms to English singular terms. It is intuitively clear, from our previous discussion, that a PL functional term does not designate any individual (it is an "incomplete" expression) while a PL singular term designates a single individual. We will state these facts with precision when we define the notion of a PL interpretation.

---

3   An alternative list of symbols is: '~' or '–' instead of '$\neg$'; '&' or '•' instead of '$\wedge$'; '⊃' instead of '$\rightarrow$'; and '≡' instead of '$\leftrightarrow$'. Philosophers and philosophical logicians tend to use the alternative list of symbols, and mathematicians and mathematical logicians tend to use the symbols given in this book.

4   It is common in the philosophical literature to express the universal quantifier ($\forall$z) as simply (**z**) without the symbol '$\forall$'.

The use of function symbols could engender ambiguities if we insist on following the convention of dropping the superscripts. To avoid misreading PL terms we need to do two things. First, we should keep in mind that only the letters 'f', 'g', and 'h', with or without numeric subscripts or superscripts, may be used as function symbols, that only the letters 'a', 'b', 'c', 'd', 'e', 'i', 'j', ..., and 't', with or without subscripts, may be used as PL names, and that only the letters 'u', 'v', 'w', 'x', 'y', and 'z', with or without subscripts, may be used as PL variables. Second, if there is any possibility of ambiguity, we do not follow the convention of dropping the superscripts of function symbols. Let us illustrate by means of an example. The expression *Pavcgt* is not ambiguous. It can only be an abbreviation of the expression P^4avcg^1t. However, the expression *Pavgtc* is ambiguous because it can be an abbreviation of one of the following two PL expressions: P^4avg^1tc and P^3avg^2tc. In the first expression we have a 4-place predicate 'P^4' followed by four PL terms, the name 'a', the variable 'v', the singular term 'g^1t', and the name 'c'. In the second expression the 3-place predicate 'P^3' is followed by three PL terms, the name 'a', the variable 'v', and the singular term 'g^2tc'. So we need to be careful with applying the convention of dropping superscripts.

While a PL term might be a single name or a single variable, the general case is that of a PL term that consists of an n-place function symbol followed by n PL terms, many of which could contain several function symbols followed by other PL terms. As an example consider the PL term $f^4xg^2h^1bf^3ang^1esf^1y$. This is not a singular term; rather it is a functional term because it contains function symbols and variables. It consists of the 4-place function symbol f^4 followed by four PL terms: (1) the variable x, (2) the singular term $g^2h^1bf^3ang^1e$, (3) the name s, and (4) the functional term f^1y. The singular term $g^2h^1bf^3ang^1e$ consists of the 2-place function symbol g^2 followed by two singular terms: (1) the singular term h^1b, and (2) the singular term f^3ang^1e. The last singular term consists of the 3-place function symbol f^3 followed by three singular terms: (1) the name a, (2) the name n, and (3) the singular term g^1e. Each of the singular terms h^1b and g^1e consists of a 1-place function symbol followed by a name. The functional term f^1y consists of a 1-place function symbol followed by a variable. We can describe the structure of this complex term diagrammatically as follows.

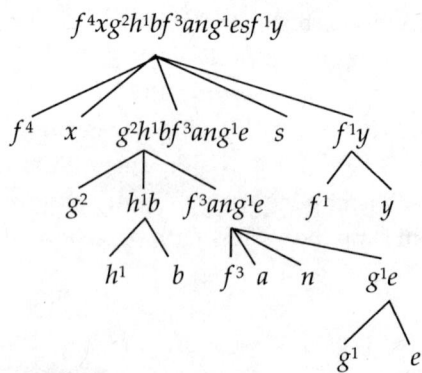

1.1.3   The **formulas** of PL are either atomic or compound. The **atomic formulas** of PL are all expressions of the form **r** = **s** where **r** and **s** are any PL terms, and all expressions of the form $Q^n t_1 t_2 \ldots t_n$ where Q^n is any n-place PL predicate (except the identity predicate =) and t_1, t_2, ..., and t_n are any PL terms. The **compound formulas** of PL are those expressions constructed from the atomic ones by applying, some finite number of times, one or more of the seven **formation rules** listed below. Let **X** and **Y** be any PL formulas. ('R' stands for 'formation rule'.)

R¬:   ¬**X** is a PL formula.

R∧:   (**X**∧**Y**) is a PL formula.

R∨:   (**X**∨**Y**) is a PL formula.

R→:   (**X**→**Y**) is a PL formula.

R↔:   (**X**↔**Y**) is a PL formula.

R∀:   If **X** contains occurrences of the PL variable **z** but no **z**-quantifiers, then (∀**z**)**X** is a PL formula.

R∃:   If **X** contains occurrences of the PL variable **z** but no **z**-quantifiers, then (∃**z**)**X** is a PL formula.

In the next subsection we will explain the difference between PL formulas and PL sentences. The only thing that concerns us now is to understand how to construct PL formulas. We begin with atomic formulas. If we put a PL term to the left of the identity predicate and a PL term to its right, we obtain an atomic formula. The metalinguistic variable Q^n stands for any n-place PL predicate, other than the identity predicate. The symbols t_1, t_2, ..., and t_n are metalinguistic variables that stand for at most n PL terms. These terms could be variables, functional terms, or singular terms, or any combination of them. Since some of these metalinguistic variables may stand for the same term, the number of the distinct terms may be less than n. Consider, as examples, two PL predicates L^3 and E^5. The following expressions are atomic formulas of PL: L^3xya, L^3uuf^1e, L^3cg^1vc, E^5uvwwl, E^5axbg^2xda, and E^5acg^1zbd.

Once we have atomic formulas, we can construct compound ones from these formulas by using one or more of the seven formation rules any finite number of times. Here are a few examples of compound formulas produced by using the first five rules. The formulas are listed on the left and the rules used in constructing the formulas are listed on the right in the order of their application.

$(\neg A^1z \land B^3xay)$	R¬ R∧
$((\neg A^1z \land B^3xay) \to K^2zw)$	R¬ R∧ R→
$\neg(((\neg A^1z \lor B^3xay) \leftrightarrow K^2zw) \land D^1y)$	R¬ R∨ R↔ R∧ R¬
$((A^1z \land B^3xay) \lor (K^2zw \land D^1y))$	R∧ (twice) R∨

There are two conditions that must be met in order to apply the quantifier rules: (1) the variable of the quantification must occur in the formula, and (2) no quantifier of *that* variable occurs in the formula. For example, the variable x occurs in the formula $(\neg A^1 z \land B^3 xay)$ and no x-quantifier occurs in the formula. Thus we can apply either R$\forall$ or R$\exists$. Say, we apply R$\forall$; we obtain the compound formula $(\forall x)(\neg A^1 z \land B^3 xay)$. We say that we have *quantified over x*. Once we quantify over a variable, we cannot quantify over it again. The second condition prevents us from quantifying over the same variable twice.[5] Since an x-quantifier occurs in $(\forall x)(\neg A^1 z \land B^3 xay)$, the second condition cannot be satisfied. So the expression $(\exists x)(\forall x)(\neg A^1 z \land B^3 xay)$ is ungrammatical, according to our rules. However, the variables z and y in the formula $(\forall x)(\neg A^1 z \land B^3 xay)$ are available for quantification. Applying R$\exists$ to z and y, we get $(\exists z)(\exists y)(\forall x)(\neg A^1 z \land B^3 xay)$. No more quantification may be applied to the last formula. We cannot quantify over x, y, or z since the formula already contains an x-quantifier, a y-quantifier, and a z-quantifier, and we cannot quantify over any variable that does not occur in the formula (the first condition). For instance, the expression $(\exists v)(\forall x)(\neg A^1 z \land B^3 xay)$ is ungrammatical because the formula $(\forall x)(\neg A^1 z \land B^3 xay)$ contains no occurrences of the variable v.

The **immediate component** of a PL formula of the form ¬**X** is **X** and its **main operator** is the unary connective ¬. The immediate components of PL formulas of the forms (**X**∧**Y**), (**X**∨**Y**), (**X**→**Y**), and (**X**↔**Y**) are **X** and **Y** and their main operators are the binary connectives ∧, ∨, →, and ↔, respectively. The immediate components of (∀**z**)**X** and (∃**z**)**Y** are **X** and **Y** and their main operators are the quantifiers (∀**z**) and (∃**z**), respectively. If a formula **X** occurs in a formula **Y**, we say that **X** is a **subformula** of **Y**. Technically speaking, every formula is a subformula of itself, because every formula "occurs" in itself. We say that **X** is a **proper subformula** of **Y** just in case **X** is a subformula of **Y** and it is not identical with **Y**. An **atomic component** of a PL formula **X** is a subformula of **X** that is an atomic formula. ¬**X** is called the **negation** of **X**. (**X**∧**Y**) is called the **conjunction** of **X** and **Y**, and they are the **conjuncts** of (**X**∧**Y**). (**X**∨**Y**) is called the **disjunction** of **X** and **Y**, and they are the **disjuncts** of (**X**∨**Y**). (**X**→**Y**) is called the **conditional** of **X** and **Y**, and **X** is the **antecedent** and **Y** the **consequent** of (**X**→**Y**). (**X**↔**Y**) is the called the **biconditional** of **X** and **Y**.

1.1.4   The **scope** of a quantifier is the PL formula that immediately follows the quantifier. An occurrence of a variable **z** in a PL formula is **bound** if it is an occurrence inside a quantifier in the formula or inside the scope of a **z**-quantifier in the formula. An occurrence of a variable in a PL formula is **free** if it is not bound. The **sentences** of PL are precisely those PL formulas that contain no free occurrences of any variable.

---

[5] There are approaches to PL syntax that allow for multiple quantifications over the same variable and for "vacuous quantification," which occurs when a **z**-quantifier is applied to a formula that contains no occurrences of the variable **z**; for instance, the quantification in '$(\forall x)Ray$' is vacuous since the variable x does not occur in *Ray*. There are some technical reasons for preferring such approaches. However, whatever the technical advantages might be for permitting vacuous quantification or multiple quantifications over the same variable, these forms of quantification are counter-intuitive and cumbersome. We prefer to forgo those advantages and retain a more natural and intuitive syntax.

It is important to note that there is only one formula that immediately follows any quantifier. Here is an example that illustrates the definitions above. Both occurrences of the variable y in the formula $\neg(\exists z)(z = y \rightarrow (\exists v)D^3evy)$ are free, since neither is an occurrence inside a quantifier or inside the scope of a y-quantifier (there are no y-quantifiers in this formula). On the other hand, the two occurrences of the variable z in the formula $\neg(\exists z)(z = y \rightarrow (\exists v)D^3evy)$ are bound because the first is inside the quantifier $(\exists z)$ and the second is inside the scope of a z-quantifier. The occurrences of the variable y in the previous formula are inside the scope of a quantifier—namely, $(\exists z)$. However, since the quantifier is a z-quantifier and not a y-quantifier, these occurrences are free and not bound. An occurrence of a variable x inside the scope of a quantifier whose variable is not x need not be bound. A bound occurrence of x must be inside a quantifier or inside the scope of an x-quantifier. The first occurrence of the variable v in the formula $\neg(\exists z)(z = y \rightarrow (\exists v)D^3evy)$ is bound because it is inside the quantifier $(\exists v)$, and the second occurrence of v is bound not because it is inside the scope of $(\exists z)$ but because it is inside the scope of the v-quantifier $(\exists v)$.

The notion of a free occurrence of a variable is important. A PL formula that has no free occurrences of any variable is a PL sentence, and a PL formula that has at least one free occurrence of a variable is not a PL sentence. PL formulas that are not sentences are called **open formulas**. Every PL formula is either an open formula or a sentence. The formula $((\forall x)(\neg A \wedge B^3xag^1y) \vee \neg(\exists z)(z = y \rightarrow (\exists v)D^3evy))$ is an open formula, since all the occurrences of the variable y are free. However, if we quantify over these occurrences, we obtain a PL sentence; the formula $(\forall y)((\forall x)(\neg A \wedge B^3xag^1y) \vee \neg(\exists z)(z = y \rightarrow (\exists v)D^3evy))$ is a PL sentence, since it contains no free occurrences of any variable. The importance of a free occurrence of a variable is that it supplies a place over which a quantifier might be applied.

The notion of a bound occurrence of a variable is also significant. Syntactically, a bound occurrence of a variable is not available for quantification. For instance, the occurrences of the variable v in the formula $\neg(\exists z)(z = y \rightarrow (\exists v)D^3evy)$ are not available for quantification. Bound variables have also semantical significance, even though they have no semantical content. They serve as *placeholders* for the quantifiers: they indicate the places to which the quantifiers apply. For instance, the existential quantifier in the sentence $(\exists z)(\forall x)S^2zx$ applies to the first place of the predicate S^2zx and the universal quantifier applies to the second place of the predicate S^2zx. If we interchange the variables of the quantifiers, we obtain the sentence $(\exists x)(\forall z)S^2zx$. In the latter, the existential quantifier applies to the second place of the predicate S^2zx and the universal quantifier to the first place. These two sentences are syntactically and semantically different from each other. To see this, let S^2zx translates the PL-English predicate 'z hates x'. The first sentence says in English that there is someone who hates everyone. The second says that there is someone such that everyone hates him or her, that is, there is someone who is hated by everyone.[6]

---

6   Since the quantifier 'everyone' covers all people, the first sentence implies that the one who hates everyone also hates himself or herself, and the second sentence implies that the one who is hated by everyone is also hated by himself or herself.

The scopes of quantifiers have semantical significance because they determine the range of the applicability of a quantifier. Occurrences of the same variable in different scopes of quantifiers are independent of each other, in the sense that the interpretations of the quantified clauses need not be dependent on each other. Let us consider an example. The sentences $(\exists x)A^1x \land (\exists x)B^1x$ and $(\exists x)(A^1x \land B^1x)$ are syntactically and semantically very different. In the first sentence the second occurrence of x is inside the scope of the first $(\exists x)$ and the fourth occurrence of x is inside the scope of the second $(\exists x)$. These occurrences are independent of each other. We could have used different variables for these clauses. The formula $(\exists z)A^1z \land (\exists y)B^1y$ has the same semantical significance as the first formula above: any PL interpretation that makes one of them true (or false) makes the other true (or false) as well. However, in the sentence $(\exists x)(A^1x \land B^1x)$ the second and third occurrences of the variable x are within the scope of the same quantifier, $(\exists x)$. These occurrences are not independent of each other. If we use different variables, we might change the syntax and semantics of the sentence. For example, the formula $(\exists z)(A^1z \land B^1y)$ is not a sentence; it is an open formula because the occurrence of y is free.

In order to make our previous discussion concrete, let us interpret the sentences in the preceding paragraph. Suppose A^1 stands for the property of being a whale and B^1 stand for the property of being a fish. The sentence $(\exists x)A^1x \land (\exists x)B^1x$ says, on this interpretation, that there is a whale and there is a fish, which is true in our world. On the other hand, the sentence $(\exists x)(A^1x \land B^1x)$ says that there is something that is both a whale and a fish, which is false in our world.

In most cases we shall employ the **conventions** of dropping the outermost parentheses of formulas, of writing predicates without superscripts, and of writing '$s \neq t$' instead of '$\neg s = t$'. Occasionally we will write function symbols without superscripts. We can always tell the number of places a predicate has by counting the PL terms that immediately follow the predicate.

1.2   The Semantics of PL

1.2.1   The meanings of PL sentences are given through interpretations. A PL interpretation of some set of PL sentences tells us what the sentences are asserting about the individuals in a given collection. In other words, an interpretation of some PL sentences is a possible situation that is described (correctly or incorrectly) by those sentences. The reader will notice that our definition of a PL interpretation introduces a linguistic component of any such interpretation. This is a feature of the semantical approach we adopt in this book. We will discuss this approach with more elaboration later in this section.

Let Γ be any set of PL sentences. The **vocabulary of Γ**, which is designated as **Voc(Γ)**, consists of the extra-logical vocabulary of which the members of Γ are composed together with the logical vocabulary of PL (i.e., the five sentential connectives, the two quantifiers, the identity predicate, the variables, and the parentheses). A **PL interpretation I** for Γ consists of a **vocabulary Voc(I)**, which includes a **list of names LN**, and it consists of a **universe of discourse** UD and of **semantical assignments SA**. Voc(I), UD, and LN must satisfy the following conditions: (1) Voc(I) includes Voc(Γ) and it may include PL function symbols and PL predicates that are not in Voc(Γ); (2) UD is a nonempty, finite or infinite, collection of **individuals**; and (3) LN consists of names for *all* the individuals in UD and it may contain non-PL names, that is, it may contain names that are not listed in 1.1.1a or it may contain complex PL singular terms. If LN contains non-PL names, these names are treated syntactically as PL names in forming functional and singular terms and formulas. The semantical assignments (SA) made by I are as follows.

1.2.1a   To every name in LN, I assigns exactly one individual in UD; and every individual in UD is assigned by I to at least one name in LN.[7] The individual I assigns to the name **s** is called the **referent** of **s** on I and is denoted as 'I(**s**)'.

1.2.1b   To every n-place function symbol f^n that belongs to Voc(I), I assigns exactly one **n-place function** on UD. The function I assigns to the n-place function symbol f^n is denoted as 'I(f^n)'.

1.2.1c   To every singular term $f^n t_1 t_2 t_3 \ldots t_n$, where f^n is an n-place function symbol and t_1, t_2, t_3, ..., and t_n are singular terms, I assigns the individual $F(\alpha_1, \alpha_2, \alpha_3, \ldots, \alpha_n)$, where F is the function I assigns to the n-place function symbol f^n, α_1 is the referent I assigns to t_1, α_2 is the referent I assigns to t_2, ..., and α_n is the referent I assigns to t_n. Symbolically, $I(f^n t_1 t_2 t_3 \ldots t_n)$ = $I(f^n)(I(t_1), I(t_2), I(t_3), \ldots, I(t_n)) = F(\alpha_1, \alpha_2, \alpha_3, \ldots, \alpha_n)$.[8]

1.2.1d   I assigns to the identity predicate, '=', the binary relation of **token identity**, which holds between every individual in UD and itself and does not hold between different individuals in UD.

1.2.1e   To every 1-place predicate that belongs to Voc(I), I assigns exactly one **bivalent property** on UD. The property I assigns to the 1-place predicate P^1 is denoted as 'I(P^1)'.

1.2.1f   To every n-place predicate (n is greater than 1) that belongs to Voc(I), I assigns exactly one **bivalent n-place relation** on UD. The relation I assigns to the n-place predicate R^n is denoted as 'I(R^n)'.

I is said to be an interpretation for a PL sentence if it is an interpretation for a set containing that sentence. If I is a PL interpretation **for** a set Γ of PL sentences, it is also described as **relevant to** Γ. We call the individuals, functions, properties, and relations of a PL interpretation I **the constituents of I**. We also describe the vocabulary of I as **the vocabulary interpreted by I**.

---

7   The requirement that every individual in the Universe of Discourse (UD) must receive a name in the List of Names (LN) leads to odd consequences. For instance, if UD consists of all the beetles that ever existed and will exist on Earth, then this requirement implies that all these beetles must be named. Since this is not feasible, the requirement also implies that LN in this case cannot be specified. These odd consequences will be addressed in Subsection 1.2.6, where several objections to substitutional quantification are discussed.

8   The concept of a function and the notation used in 1.2.1b and 1.2.1c will be fully explained below.

Before discussing PL interpretations, we need to explain some of the notions mentioned in the definition above. We first define the notions of a bivalent sentence and bivalent language. A bivalent sentence is a declarative sentence that is either true or false but not both in every logical possibility, and a bivalent language is a language whose declarative sentences are all bivalent. The notion of bivalence can be extended to properties and relations. A bivalent property is such that every individual in UD either has or lacks this property, but it cannot both have this property and lack it. To see the relation between bivalent sentences and bivalent properties, consider the following example. A sentence that might be proposed as an example of a non-bivalent sentence is 'Pat gave up smoking'. The truth and falsity of this sentence seem to presuppose that Pat has smoked during her life. If this is correct, then it follows that the property "x gave up smoking" (call this property Gx) is not bivalent. If Pat smoked during her life and then stopped smoking, then she has this property, and hence the sentence Gp, where p stands for Pat, is true. If Pat continues to smoke, then she lacks this property, and hence the sentence Gp is false. However, if Pat never smoked in her life, it seems that she neither has nor lacks this property. In this case the sentence Gp is neither true nor false. The point is that if we allow non-bivalent properties and relations, then we have to allow non-bivalent sentences. Since we do not allow non-bivalent sentences, we should not allow non-bivalent properties and relations.

The set of all the individuals that have a property P is called **the extension of** P. The extension of a 1-place PL predicate on any relevant PL interpretation I is the extension of the property that this predicate designates on I. A sequence of n individuals in some specific order is called **n-tuple**. For instance, the 2-tuple $\langle 0, 1 \rangle$ is an ordered pair consisting of the first two natural numbers ordered by the relation "less than," and the 3-tuple $\langle a, b, c \rangle$ is an ordered triple consisting of the first three letters of the English alphabet in their usual order. In general, an n-tuple of individuals $a_1, a_2, a_3, \ldots, a_n$ (not necessarily all distinct) is denoted as $\langle a_1, a_2, a_3, \ldots, a_n \rangle$. The individuals that constitute an n-tuple are called **coordinates**. A bivalent n-place relation is such that every n-tuple of individuals in UD either bears this relation or it does not; it cannot bear and not bear this relation. The **extension of an n-place relation** is the set that consists of all the n-tuples of individuals that bear this relation to each other. For example, the ordered pairs $\langle 0, 1 \rangle$, $\langle 5, 7 \rangle$, and $\langle 13, 22 \rangle$ are all in the extension of the binary relation "less than," and the ordered pairs $\langle \text{Turkey, Iraq} \rangle$ and $\langle \text{China, India} \rangle$ are in the extension of the binary relation "to the north of"; of course, the extensions of these relations contain many more ordered pairs.

An n-place function on UD may be roughly defined as a *rule that assigns to every n-tuple of individuals in* UD *a one and only one individual also in* UD. For example, if UD is the set of all the natural numbers, $\{0, 1, 2, 3, \ldots\}$, and A is defined as the rule "the sum of x and y," then A is a 2-place function on UD. A satisfies the definition of function: to *every* ordered pair $\langle n, m \rangle$ of numbers in UD, A assigns one and only one number in UD that is the sum of n and m. Let k be the sum of n and m. We express this fact as "A(n, m) = k." We refer to the numbers n and m as the **arguments** of A and to the number k as the **value** of A at those arguments. Using the standard mathematical symbol for addition, we may express this function by writing "A(n, m) = n+m." Since PL interpretations assign functions to PL function symbols and since PL function symbols are meant as counterparts of functional descriptions, we will in most cases use PL-English functional descriptions to define the functions that are assigned to function symbols. For instance, rather than defining the function A above as the rule that assigns to every ordered pair of natural numbers the sum of these numbers, we will define A by the functional description 'The sum of x and y'.

Precisely speaking, functions are relations of a special type; and hence they too have extensions. The extension of the function A consists of ordered triples ⟨n, m, k⟩, where the first two coordinates of the triples are the arguments of A and the last coordinate is the value of A at n and m. For instance, the ordered triples ⟨5, 7, 12⟩, ⟨13, 13, 26⟩, and ⟨0, 21, 21⟩ are all in the extension of A. As another example, consider the 3-place function on the positive integers, {1, 2, 3, ...}, that is defined by the functional description "The greatest common divisor of x, y, z." This is a function because any three positive integers have a unique greatest common divisor. For example, the greatest common divisor of 21, 35, and 49 is 7, the greatest common divisor of 7, 13, and 15 is 1, the greatest common divisor of 6, 12, and 18 is 6, and the greatest common divisor of 2, 3, and 3 is 1. Using the notation introduced in the preceding paragraph, we may define this function as "F(n, m, k) = the greatest common divisor of n, m, and k, where n, m, and k are any positive integers." The numbers n, m, and k are the arguments of F and F(n, m, k) is the value of F at these arguments. The extension of F consists of 4-tuples ⟨n, m, k, j⟩ such that the first three coordinates are the arguments of the function F and the last coordinate is its value at those arguments. In general, the extension of an n-place function consists of (n+1)-tuples such that the first n coordinates of each tuple are the arguments of the function and the n+1st (i.e., the last) coordinate of the tuple is the value of the function at those arguments. The definition below summarizes this discussion.

Function: An n-place function F on some set D is an (n+1)-place relation such that for every n-tuple ⟨$α_1$, $α_2$, $α_3$, ..., $α_n$⟩ of coordinates in D, there is one and only one individual β in D where the (n+1)-tuple ⟨$α_1$, $α_2$, $α_3$, ..., $α_n$, β⟩ belongs to the extension of F. The individuals $α_1$, $α_2$, $α_3$, ..., and $α_n$ are called "the arguments of F" and β "the value of F at the arguments $α_1$, $α_2$, $α_3$, ..., and $α_n$." β is denoted as F($α_1$, $α_2$, $α_3$, ..., $α_n$).

As defined previously, a PL singular term is a PL name or a PL term that contains a function symbol but no variables. Given that function symbols are interpreted as functions on UD, singular terms behave like names in that they also refer to unique individuals in UD. To see this, consider the following example. Suppose that we have the singular term f^2ag^1b, and that a PL interpretation I whose UD is the set of all positive integers, {1, 2, 3, ...}, assigns the function "the sum of x and y" to f^2xy, the function "the square of z" to g^1z, and the numbers 11 and 7 to the names a and b, respectively. The interpretation of the singular term f^2ag^1b on I is the definite description "the sum of 11 and the square of 7" (i.e., $11 + 7^2$), which is the number 60. Thus, on I, the singular term f^2ag^1b refers uniquely to the number 60. This is true of all PL singular terms. We state this fact as a theorem.

Theorem 1.2.1: Every PL singular term has one and only one referent (in UD) on every PL interpretation that is relevant to it (i.e., that interprets it).

If **t** is a PL singular term and I is a PL interpretation that is relevant to **t**, we designate the referent of **t** on I as I(t). This theorem follows from 1.2.1a–1.2.1c, and the definition of a function. Although the proof is straightforward, it invokes a certain mathematical principle called the "Principle of Mathematical Induction," which we will discuss in Chapter Two.

1.2.2   A PL interpretation I can be represented as an ordered triple ⟨UD, V, SA⟩, where UD is the universe of discourse, V is the vocabulary of the interpretation, including the list of names LN, and SA are the semantical assignments. There is something peculiar about the way we defined a PL interpretation. We required an interpretation to bring its own names for the individuals in its universe of discourse. Observe that we did not require a PL interpretation to interpret function symbols and predicates that are not part of the vocabulary of the PL set for which it is an interpretation. We only said that this might be the case. But this is not so with names. A PL interpretation *must* contain enough names to name all the individuals in UD. In other words, the individuals of an interpretation come labeled.

In practice the list of names seems extraneous to the function of an interpretation. In reality these names are essential for the *formal* truth conditions of the quantifiers that are given in this book. We will explain later that those truth conditions attribute a certain semantics to the quantifiers that is technically called **substitutional quantification**. There is another semantical approach to the quantifiers that is called **objectual quantification**, which does not require the individuals of an interpretation to be labeled. Once we give an interpretation for a PL sentence, the sentence acquires a specific meaning in that interpretation. We describe this meaning by translating the PL sentence into the language of the interpretation. This gives the sentence a "natural reading" on that interpretation. Although our formal truth conditions of the quantifiers are substitutional, almost always our natural readings of the quantifiers are objectual. We will define the formal truth conditions of the quantifiers in Subsection 1.2.5, but we will give a preliminary explanation of substitutional and objectual quantification in Subsection 1.2.3.

The semantical assignments of names to individuals ensure three things: (1) every name has a referent, that is, there are no non-referring names; (2) no name has more than one referent, that is, there are no ambiguous names; and (3) every individual has *at least one* name, that is, there are no unnamed individuals. Every function symbol and every predicate is assigned, on a PL interpretation, exactly one function and exactly one property or a relation, respectively. This ensures that no function symbol and no predicate are ambiguous.

The best way to explain the notion of a PL interpretation is to give a few examples of it. Later in this subsection we will be given sets of PL sentences and we will be asked to give interpretations for them. Some problems place certain conditions on these interpretations. Unless the problem requires an interpretation to make certain sentences true or false, the truth and falsity of the interpreted sentences are not a condition on the proposed interpretations. A PL interpretation gives semantical contents (i.e., meanings) to the relevant PL sentences; it need not make them true.

Once we specify a UD, we make sure that we have enough names to label all the individuals in our UD. Our list of names must include the names that occur in the PL sentences that we want to interpret. Then we assign to each 1-place predicate and each n-place predicate that occurs in a given PL sentence a bivalent property or a bivalent n-place relation that the individuals in UD might have. We need to make clear which individuals, if any, have these properties and relations and which individuals do not. We can accomplish this in one of three ways: (1) state explicitly "facts" that indicate the individuals that have this property or relation and the individuals that do not; (2) give the extension of this property or relation; and (3) rely on common knowledge to determine the individuals that have this property or relation.

In interpreting the predicates that occur in the given PL sentences, we can assign properties and relations, or we can assign the extensions of properties and relations without specifying the properties and relations that have these extensions. In fact, PL semantics is extensional: a property or a relation is reduced to its extension. Thus, according to this semantics, specifying the extension of a property or a relation is the same thing as specifying a property or a relation. However, we will find in practice that specifying a "natural-sounding" property or a relation, instead of just giving an extension, improves the natural readings of the PL sentences.

Function symbols are interpreted by assigning functions to them. Since the extension of a 1-place function consists of ordered pairs and, in general, of an n-place function consists of (n+1)-tuples, almost always defining functions by means of their extensions is a cumbersome exercise. As stated previously, we will define functions by stating functional descriptions. We usually use PL-English for this purpose.

When all the tasks described above are accomplished, the construction of a PL interpretation is completed. We do not need to determine the truth values of the given PL sentences on this interpretation, unless we are asked to do so. However, in order to gain some experience, we will always *try* to determine the truth values of the PL sentences on the interpretations we construct. We say that we will *try*, because determining the truth values of PL sentences on certain interpretations might be very difficult or even impossible given the state of our current knowledge. Although we have not yet stated the truth conditions of the PL sentences, it is usually sufficient to relay on the natural readings of the interpreted PL sentences.

If the problem requires that a PL interpretation make certain sentences true and others false, we, after determining the truth values of the interpreted PL sentences, try to manipulate various aspects of the interpretation in order to meet these requirements. A PL sentence might come false on our interpretation, but the problem requires that the sentence be true; so we "tinker" with the semantical assignments of functions, properties, and relations, or with the semantical assignments of referents, or with UD itself until we manage to make the sentence true on the modified interpretation. We should note that *every* change in an aspect of an interpretation, no matter how minor it might be, produces a new interpretation. We now demonstrate how to implement the procedure described in the preceding paragraphs. As usual, any semantical assignment that I makes is denoted as I(#), where # is either a name, a function symbol, or a predicate.

1.2.2a We will construct a PL interpretation for the following set of PL sentences on which the first five sentences are true and the sixth sentence is false.

S1 $Hc \vee De$
S2 $(\exists x)Hx \wedge (\exists y)Dy$
S3 $(\exists x)Hx \rightarrow Pc$
S4 $(\exists x)Dx \rightarrow Pe$
S5 $(\exists x)(Hx \wedge Dx) \rightarrow c \neq e$
S6 $c \neq e$

I am going to describe the family of one of my brothers-in-law. I will not pay attention first to the requirement stated in the problem. After I construct the interpretation, I will manipulate certain aspects of the interpretation in order to meet that requirement. We use the colon, instead of the identity sign '=', to indicate identity in the metalanguage in order to reserve the sign '=' for the object language.

The PL Interpretation I

UD: {Jim, Peggy, Andrea, Tom}
LN: c, e, j, p

Note: We have to include the names c and e, which occur in the set above, and two more names in order to have enough names for all the individuals in the UD.

Semantical assignments

I(c): Andrea; I(e): Tom; I(j): Jim; and I(p): Peggy
I(Dx): the property "x is a parent": {Jim, Peggy}
I(Hx): the property "x is a child": {Andrea, Tom}
I(Px): the property "x was born in Iowa": {Jim, Peggy}
I($x = y$): the relation "x is identical with y": {⟨Jim, Jim⟩, ⟨Peggy, Peggy⟩, ⟨Andrea, Andrea⟩, ⟨Tom, Tom⟩}

Note: For the sake of simplicity, we will omit in future interpretations the phrases 'the property' and 'the relation'. We determined which individuals have these properties and which do not by giving the extensions of the properties. It is important to note that **the quantifiers range over the UD.** Thus if we say "every individual" and "there is an individual" we mean "every individual in UD" and "there is an individual in UD."

The truth values of the PL sentences

Natural reading of (S1) $Hc \vee De$ on I: *Andrea is a child or Tom is a Parent.*
S1 is true on I because the first disjunct is true.

Natural reading of (S2) $(\exists x)Hx \wedge (\exists y)Dy$ on I: *There is a child and there is a parent.*
S2 is true on I because the extensions assigned to H and D are not empty.

Natural reading of (S3) $(\exists x)Hx \rightarrow Pc$ on I: *If there is a child, then Andrea was born in Iowa.*
S3 is false on I since the antecedent is true and the consequent is false.

Natural reading of (S4) $(\exists x)Dx \rightarrow Pe$ on I: *If there is a parent, then Tom was born in Iowa.*
S4 is false on I because the antecedent is true and the consequent is false.

Natural reading of (S5) $(\exists x)(Hx \wedge Dx) \rightarrow c \neq e$ on I: *If there is a child who is also a parent, then Andrea is not Tom.*
S5 is true on I because the antecedent is false on I. As we will see later, a conditional with a false antecedent is always true.

Natural reading of (S6) $c \neq e$ on I: *Andrea is not Tom.*
S6 is true on I.

    We need to modify I in order to meet the requirement stated in the problem. To make S6 false, we should assign the same referent to c and e. We use I* to designate the modified interpretation.

First modification   I*(c): Andrea; I*(e): Andrea

But now we need a name for Tom. So we add a fifth name to LN, say, t.

Second modification   I*(t): Tom

Since $c \neq e$ is false on I*, the antecedent and consequent of $(\exists x)(Hx \wedge Dx) \rightarrow c \neq e$ are false on I*; hence S5 is still true on I*. It remains to make S3 and S4 true. It suffices to make Pe true on I*. In order to make Pe true, the referent of e must be in the extension of P. So we add Andrea to the extension of P.

Third modification   I*(P): "x was born in Iowa" = {Jim, Peggy, Andrea}

In fact, I* no longer describes the actual situation: Andrea was born in Minnesota and not in Iowa. This is immaterial. PL interpretations describe actual or possible situations. We may design our interpretations as we please, as long as we adhere strictly to the definition of a PL Interpretation in order to ensure consistency. The description of I and the modifications listed above supply a complete description of the new interpretation, I*. As we explained above, S1–S5 come true and S6 come false on I*. Therefore the requirement of the problem has been met.

1.2.2b We will give two interpretations for the PL sentence S, one of which makes S true and the other makes it false. We will begin with *any* interpretation for S; S is bound to be either true or false on this interpretation. We then modify the interpretation in order to switch the truth value of S.

S $(\forall x)(Qx \rightarrow (Kx \land Dx)) \leftrightarrow (\exists v)(Dv \land Kv)$

The PL Interpretation I

UD: {Samantha, Christopher, Lisa, Darius}
LN: s, c, l, d

Semantical Assignments

$I(s)$: Samantha; $I(c)$: Christopher; $I(l)$: Lisa; $I(d)$: Darius
$I(Qx)$: "x is an English major": {Samantha, Darius}
$I(Dx)$: "x plays tennis": {Samantha, Darius, Lisa}
$I(Kx)$: "x plays the piano": {Samantha, Darius, Christopher}

Natural reading of $(\forall x)(Qx \rightarrow (Kx \land Dx)) \leftrightarrow (\exists v)(Dv \land Kv)$ on I:
All English majors play the piano and play tennis if and only if there is someone who plays tennis and plays the piano.

The left-hand side of the biconditional is true on I, since every individual in the extension of Q—namely, Samantha and Darius—is also in the extensions of D and K. The right-hand side of the biconditional is also true on I, since there is an individual—e.g., Samantha—who is in the extensions of D and K. Thus S, which is the biconditional sentence, is true on I.

In order to make S false we need to make one of the sides of the biconditional true and the other false. We will keep the right-hand side true and try to switch the truth value of the left-hand side. If there is an English major who does not play tennis or does not play the piano, the left-hand side, $(\forall x)(Qx \rightarrow (Kx \land Dx))$, would be false because this side asserts that every English major (Q) plays tennis (D) and plays the piano (K). Thus it is sufficient to remove, say Samantha, from the extension of D. In this case, Samantha, who is an English major, does not play tennis. We let I* be the modified interpretation. Since the left-hand side of S is false and the right-hand side is true on I*, S is false on I*.

1.2.2c We want to construct an interpretation for the following set of PL sentences, such that it makes every sentence in the set true and has an infinite universe of discourse.

S1 $(\forall x)\, o \neq gx$
S2 $(\forall x)(\forall y)(gx = gy \rightarrow x = y)$
S3 $(\forall x)(x \neq o \rightarrow (\exists z)\, x = gz)$

The PL Interpretation I

UD: The set of all the natural numbers: {0, 1, 2, 3, 4, …}
LN: $o, a_1, a_2, a_3, \ldots, a_n, \ldots$

Semantical Assignments

$I(o)$: 0; $I(a_1)$: 1; $I(a_2)$: 2; $I(a_3)$: 3; …; in general, $I(a_n)$: n
$I(gx)$: "The successor of x" (i.e., $I(gx)$ is $x+1$)

Natural reading of (S1) $(\forall x)\, o \neq gx$ on I:
0 is not the successor of any natural number.
S1 is true on I.

Natural reading of (S2) $(\forall x)(\forall y)(gx = gy \rightarrow x = y)$ on I:
Any two numbers that have the same successor are identical.
S2 is true on I.

Natural reading of (S3) $(\forall x)(x \neq o \rightarrow (\exists z)\, x = gz)$ on I:
Every natural number that is not 0 is the successor of some natural number.
S3 is true on I.

Since S1–S3 are true on I and UD is an infinite set, the requirement of the problem is met.

1.2.3 Let us look more closely at what we termed "natural readings" of the PL sentences we interpreted in the previous examples. Consider the sentence $(\forall x)(Qx \rightarrow (Kx \wedge Dx))$, which is the left-hand side of the biconditional $(\forall x)(Qx \rightarrow (Kx \wedge Dx)) \leftrightarrow (\exists v)(Dv \wedge Kv)$ interpreted in 1.2.2b. We said that the natural reading of $(\forall x)(Qx \rightarrow (Kx \wedge Dx))$ on I is: all English majors play the piano and play tennis. According to this reading, the universal quantifier is interpreted as ranging over the members of UD. To see the point clearly, consider the PL-English reading of this sentence: for every x in UD, if x is an English major, then x plays the piano and plays tennis. Thus 'for every x, …' is interpreted as making an assertion about all the individuals in UD. This way of interpreting the quantifiers is referred to as **objectual quantification** ('objectual' from 'object').

However, the formal truth conditions we will give for the quantifiers in Subsection 1.2.5 specify an interpretation of the quantifiers on which the quantifiers range over the names in LN. This way of interpreting the quantifiers is called **substitutional quantification**. In order to state this sort of quantification precisely, we need to introduce the notion of substitutional instance of a quantified sentence.

Substitutional Instance. The sentence **X[t]** is a substitutional instance of the quantified sentence $(\Theta z)\mathbf{X}$, where Θ is either the universal quantifier symbol $\forall$ or the existential quantifier symbol $\exists$, if and only if **X[t]** is obtained from **X** by replacing all the occurrences of the variable z in the formula **X** with the singular term **t**. If **t** is a name listed in the LN of a PL interpretation I that is relevant to **X[t]**, we refer to **X[t]** as a **basic substitutional instance** of $(\Theta z)\mathbf{X}$ on I.

This definition makes sense since once we remove the quantifier (Θz) the occurrences of the variable **z** in **X** become free, and hence they become available for substitution. For example, $(Ql \rightarrow (Kl \wedge Dl))$ is a substitutional instance of the universally quantified sentence $(\forall x)(Qx \rightarrow (Kx \wedge Dx))$. This substitutional instance is obtained by removing the universal quantifier from $(\forall x)(Qx \rightarrow (Kx \wedge Dx))$, resulting in the formula $(Qx \rightarrow (Kx \wedge Dx))$, and then substituting the name l for x in all its occurrences in $(Qx \rightarrow (Kx \wedge Dx))$.

On the interpretation I described in 1.2.2b, $(\forall x)(Qx \rightarrow (Kx \wedge Dx))$ has four substitutional instances because there are only four singular terms that are interpreted by I. These are the four names, s, c, l, and d, which are listed in LN. There are no other singular terms because there are no function symbols interpreted by I. In other words, the substitutional instance of $(\forall x)(Qx \rightarrow (Kx \wedge Dx))$ on I are precisely its *basic* substitutional instances on I. These substitutional instances are:

$(Qs \rightarrow (Ks \wedge Ds))$; $(Qc \rightarrow (Kc \wedge Dc))$; $(Ql \rightarrow (Kl \wedge Dl))$; and $(Qd \rightarrow (Kd \wedge Dd))$

If the universal quantifier $(\forall x)$ is interpreted substitutionally, the sentence $(\forall x)(Qx \rightarrow (Kx \wedge Dx))$ asserts that all its basic substitutional instances are true. This entails that in order for $(\forall x)(Qx \rightarrow (Kx \wedge Dx))$ to be true on I, the substitutional instances stated above must all be true on I. Since they are true on I,[9] the sentence $(\forall x)(Qx \rightarrow (Kx \wedge Dx))$ is true on I.

The objectual reading of the sentence $(\exists v)(Dv \wedge Kv)$ on I is the natural reading we stated in 1.2.2b—namely, that there is someone who plays tennis and plays the piano. The quantifier 'There is v, such that ...' is interpreted objectually as asserting that there is an individual in UD, such that ... . The sentence is true on I because there is such an individual in UD (e.g., Samantha). Substitutional quantification interprets $(\exists v)(Dv \wedge Kv)$ as asserting that there is a basic substitutional instance that is true on I. Just as $(\forall x)(Qx \rightarrow (Kx \wedge Dx))$, this sentence has four substitutional instances on I (LN consists of four names). Since the basic substitutional instance $(Ds \wedge Ks)$ is true on I, the sentence $(\exists v)(Dv \wedge Kv)$ is also true on I.

In general, the **objectual interpretation** of a PL sentence of the form $(\forall z)X$ is that "for every individual in UD, this individual *satisfies* the formula **X**";[10] and the objectual interpretation of $(\exists z)X$ is that "there is an individual in UD, such that this individual satisfies the formula **X**." The **substitutional interpretation** of $(\forall z)X$ is that "every basic substitutional instance of $(\forall z)X$ is true"; and the substitutional interpretation of $(\exists z)X$ is that "there is at least one basic substitutional instance of $(\exists z)X$ that is true." When we discuss the truth conditions of the quantifiers, we will explain that, as long as every individual in UD

---

9   They can easily be shown to be true on I by stating their natural readings on I.

   $(Qs \rightarrow (Ks \wedge Ds))$: *If Samantha is an English major, then she plays the piano and plays tennis.*
   $(Qc \rightarrow (Kc \wedge Dc))$: *If Christopher is an English major, then he plays the piano and plays tennis.*
   $(Ql \rightarrow (Kl \wedge Dl))$: *If Lisa is an English major, then she plays the piano and plays tennis.*
   $(Qd \rightarrow (Kd \wedge Dd))$: *If Darius is an English major, then he plays the piano and plays tennis.*

   Since Qc and Ql are false on I (Christopher and Lisa are not English majors), the conditionals $(Qc \rightarrow (Kc \wedge Dc))$ and $(Ql \rightarrow (Kl \wedge Dl))$ are true on I (as we will see, a conditional with a false antecedent is true). In the conditionals $(Qs \rightarrow (Ks \wedge Ds))$ and $(Qd \rightarrow (Kd \wedge Dd))$, the antecedents and consequents are true (Samantha and Darius are English majors *and* they play the piano and play tennis), so the conditionals are also true on I.

10 The relation of *satisfaction*, which can hold between individuals in UD and PL formulas, is a technical notion. We will not discuss this notion in this book. We only speak of an interpretation's satisfying a set of PL sentences, in the sense that every member of the set is true on that interpretation. This sense is different from the technical notion of satisfaction.

has a name in LN, both interpretations of the quantifiers deliver the same truth values for the same quantified sentences. In Subsection 1.2.6, we will revisit objectual and substitutional quantification, discuss possible objections to substitutional quantification, and offer replies to these objections.

1.2.4   The **size of a PL interpretation** is the **cardinality** of its universe of discourse. Intuitively, the cardinality of a set is the number of its members. Thus the cardinality of a set consisting of n members, where n is a non-negative integer, is n. Such a set is finite. For example, the sets ∅ (the empty set), {a, b, c}, {Abraham, Sarah, Hagar, Ishmael, Isaac}, and {x: x is a planet in our solar system} (i.e., the set of all x, such that x is a planet in our solar system) are finite and their cardinalities are, respectively, 0, 3, 5, and 8. A **finite interpretation** is an interpretation whose universe of discourse is a finite set. A set whose cardinality is at least as great as the cardinality of the set of all natural numbers, {0, 1, 2, 3, 4, ...} (call it ℕ), is an infinite set. An **infinite interpretation** is an interpretation whose universe of discourse is an infinite set. The interpretations described in 1.2.1a and 1.2.1b are finite. Since there are four individuals in each of their UDs, their size is four. On the other hand, the interpretation of 1.2.2c is infinite: its UD is the set ℕ, which is an infinite set.

We will discuss infinite sets with some elaboration in Chapter Two. At this stage we only note that infinite interpretations come in many sizes. There are infinite cardinalities that are larger than other infinite cardinalities. The cardinality of ℕ is the *smallest infinite cardinality*. Set theorists denote this cardinality as $\aleph_0$ (aleph-null). Hence every infinite subset of ℕ, such as the set of non-negative even numbers, has the cardinality $\aleph_0$.

Is there a familiar set that has a cardinality greater than $\aleph_0$? Indeed, the set of all real numbers ℝ is such a set. The real numbers are the numbers that can be assigned to all the points on an infinite straight line. Intuitively, every real number measures the distance that a point on a line has from the zero-point. Distances on the right of the zero-point are considered positive and distances on the left of the zero-point are considered negative.[11] It is clear that ℕ is a subset of ℝ. It is a mathematical theorem that the cardinality of ℝ, which is usually denoted as C, is greater than the cardinality of ℕ, $\aleph_0$.

We will prove this fact in Chapter Two. There is a point of special significance to our approach to PL semantics, which concerns cardinalities larger than $\aleph_0$. A set whose cardinality is less than or equal to $\aleph_0$ is called **countable**. All sets whose cardinalities are larger than $\aleph_0$ are called **uncountable**. The distinction is intuitive enough, so we can explain it here. A countable set is a set that permits a finite or an infinite process of *counting* its members. Every finite set is clearly countable, since the process of counting its members is finite, meaning that it terminates after finitely many steps. For example, the set {Abraham, Sarah, Hagar Ishmael, Isaac} can be counted via a 5-step counting process. This counting process reveals the cardinality of the set. The fact that the process terminates at the number

---

11   To be technically correct, we should have used the term 'displacement' instead of 'distance'. Distance is the length of the line that a point traversed; it has no direction, and hence it cannot be negative. Displacement, roughly speaking, is distance with direction, and hence it can be positive or negative.

5 shows that the cardinality of the set is 5. The set of all natural numbers, ℕ, can also be counted via an infinite process. We count the members of the set {0, 1, 2, 3, 4, 5, ...} as follows: we assigning the number one to 0, the number two to 1, the number three to 2, the number four to 3, and so on. The process never terminates, but it has a nice feature: every natural number, no matter how big it is, will eventually be counted via this process. The facts that the process never terminates and that it will eventually reach any arbitrary natural number allow us to say that this set is **countably infinite**. All finite sets, that is, sets whose cardinalities are less than $\aleph_0$, and all sets whose cardinality is $\aleph_0$ admit such a process, and hence they are countable. Every set whose cardinality is greater than $\aleph_0$ does not admit such counting process, and hence it is uncountable. Recall that all sets whose cardinalities are at least as great as $\aleph_0$ are infinite. Since C, the cardinality of the set of real numbers ℝ, is greater than $\aleph_0$, ℝ is uncountable. This means that no matter how we might arrange and rearrange the order of the points on a line, there is no infinite counting process for this set. Thus the size of an interpretation whose UD is ℝ is not only infinite but also uncountable. We will revisit this issue in Subsection 1.2.6.

1.2.5   The **truth conditions** of a PL sentence **X** are the conditions that determine its truth value on any given PL interpretation for it. In other words, **X** is true on a PL interpretation I if and only if I satisfies its truth conditions, and **X** is false on I if and only if I is *relevant* to **X** and does not satisfy its truth conditions. Here is a complete list of those truth conditions. Let **X** and **Y** be any PL sentences and I be any PL interpretation for them (i.e., I is an interpretation for a set containing both **X** and **Y**).

1.2.5a   If **X** is of the form **r** = **s** where **r** and **s** are PL singular terms, then **X** is true on I if and only if the referents of **r** and **s** on I are identical, that is, they are the same individual (formally, I(**s**) = I(**r**)).

1.2.5b   If **X** is of the form **P¹s** where **P¹** is a 1-place PL predicate and **s** is a PL singular term, then **X** is true on I if and only if the referent of **s** on I has the property that I assigns to **P¹** (formally, I(**s**) ∈ I(**P¹**)).

1.2.5c   If **X** is of the form $\mathbf{R^n t_1 t_2 ... t_n}$ where **R^n** is an n-place PL predicate (n is greater than 1) and $\mathbf{t_1, t_2, ..., t_n}$ are PL singular terms, then **X** is true on I if and only if the referents of $\mathbf{t_1, t_2, ..., t_n}$, in the order indicated, are related to each other according to the relation that I assigns to **R^n** (formally, ⟨I($\mathbf{t_1}$), I($\mathbf{t_2}$), ..., I($\mathbf{t_n}$)⟩ ∈ I(**R^n**)).

1.2.5d   ¬**X** is true on I   if and only if   **X** is false on I.
         (**X**∧**Y**) is true on I   if and only if   **X** is true on I and **Y** is true on I.
         (**X**∨**Y**) is true on I   if and only if   **X** is true on I or **Y** is true on I or both.
         (**X**→**Y**) is true on I   if and only if   **X** is false on I or **Y** is true on I or both.
         (**X**↔**Y**) is true on I   if and only if   **X** and **Y** are both true on I or both false on I.

1.2.5e   If **X** is of the form $(\forall \mathbf{y})\mathbf{Z}$, then **X** is true on I if and only if *all* its basic substitutional instances are true on I, that is, for *every* name **s** in LN, the sentence **Z[s]** is true on I, where **Z[s]** is formed by replacing all the occurrences of **y** in **Z** by **s**.

1.2.5f   If **X** is of the form $(\exists \mathbf{y})\mathbf{Z}$, then **X** is true on I if and only if *at least one* of its basic substitutional instances is true on I, that is, for *some* name **s** in LN, the sentence **Z[s]** is true on I, where **Z[s]** is formed by replacing all the occurrences of **y** in **Z** by **s**.

If every PL sentence in a set Γ is true on I, we say that I **satisfies** Γ or that I is a **model** of Γ.

If $\mathbf{P}^1$ is a 1-place predicate and **s** is a singular term, I assigns a property to $\mathbf{P}^1$, which is denoted as $I(\mathbf{P}^1)$, and a referent to **s**, which is denoted as $I(\mathbf{s})$. The atomic sentence $\mathbf{P}^1\mathbf{s}$ is true on I if and only if the individual $I(\mathbf{s})$ has the property $I(\mathbf{P}^1)$. Since PL is an extensional system, the property $I(\mathbf{P}^1)$ is identified with its extension on I. Thus to say that $I(\mathbf{s})$ has the property $I(\mathbf{P}^1)$ is to say that $I(\mathbf{s})$ belongs to (is a member of) the extension $I(\mathbf{P}^1)$. The relation "belong to" ("is a member of") is designated by the symbol '$\in$'. So '$I(\mathbf{s}) \in I(\mathbf{P}^1)$' means that the individual $I(\mathbf{s})$ belongs to (is a member of) the extension $I(\mathbf{P}^1)$. Using this symbolism and the extensionality of PL, we can state the truth conditions of the atomic sentence $\mathbf{P}^1\mathbf{s}$ as follows: $\mathbf{P}^1\mathbf{s}$ is true on I if and only if $I(\mathbf{s}) \in I(\mathbf{P}^1)$.

The truth conditions of the atomic sentences of the form $\mathbf{R}^n\mathbf{t}_1\mathbf{t}_2\ldots\mathbf{t}_n$ are slightly more involved. $\mathbf{R}^n$ is an n-place predicate (n is greater than 1) and $\mathbf{t}_1, \mathbf{t}_2, \mathbf{t}_3, \ldots,$ and $\mathbf{t}_n$ are (not necessarily all distinct) PL singular terms. To see an example of n singular terms that are not necessarily all distinct of each other, consider the PL-English predicate 'x loves y and hates z', and the "three" singular terms: 'Raoul', 'Raoul', and 'Raoul's neighbor'; the first two terms are not distinct from each other; if we apply the PL-English predicate to these singular terms, we obtain the following English sentence: 'Raoul loves himself and hates his neighbor'. I assigns an n-place relation to $\mathbf{R}^n$; we denote this relation as $I(\mathbf{R}^n)$. $I(\mathbf{t}_1), I(\mathbf{t}_2), \ldots,$ and $I(\mathbf{t}_n)$ are the individuals (i.e., the referents) in UD that I assigns to the singular terms $\mathbf{t}_1, \mathbf{t}_2, \mathbf{t}_3, \ldots,$ and $\mathbf{t}_n$. The sentence $\mathbf{R}^n\mathbf{t}_1\mathbf{t}_2\ldots\mathbf{t}_n$ is true on I if and only if the relation $I(\mathbf{R}^n)$ holds between the individuals $I(\mathbf{t}_1), I(\mathbf{t}_2), \ldots,$ and $I(\mathbf{t}_n)$, in the order indicated. For instance, if $I(T^2)$ is the relation "taller than," $I(a)$ is Anna, and $I(b)$ is Bethany, then T^2ba is true on I if and only if Bethany is taller than Anna. Observe that the order is very important: T^2ba says that Bethany is taller than Anna; it does not say that Anna is taller than Bethany. Again, since PL is an extensional system, $I(T^2)$ is identified with its extension on I. $I(T^2)$ is a binary relation (i.e., a 2-place relation); therefore its extension consists of all ordered pairs $\langle \alpha, \beta \rangle$ such that α and β are individuals in UD and α bears the relation $I(T^2)$ to β, that is, α is taller than β. Using this notation and the extensionality of PL, we say that T^2ba is true on I if and only if the ordered pair $\langle I(b), I(a) \rangle \in I(T^2)$, that is, the ordered pair $\langle$Bethany, Anna$\rangle$ belongs to the extension of T^2 on I. In general, the atomic sentence $\mathbf{R}^n\mathbf{t}_1\mathbf{t}_2\ldots\mathbf{t}_n$ is true on I if and only if the n-tuple $\langle I(\mathbf{t}_1), I(\mathbf{t}_2), \ldots, I(\mathbf{t}_n) \rangle$ belongs to the extension $I(\mathbf{R}^n)$, that is, $\langle I(\mathbf{t}_1), I(\mathbf{t}_2), \ldots, I(\mathbf{t}_n) \rangle \in I(\mathbf{R}^n)$.

It is customary to describe the truth conditions of the compound sentences $\neg \mathbf{X}$, $(\mathbf{X} \wedge \mathbf{Y})$, $(\mathbf{X} \vee \mathbf{Y})$, $(\mathbf{X} \to \mathbf{Y})$, and $(\mathbf{X} \leftrightarrow \mathbf{Y})$ by means of truth tables. Each truth table shows the possible assignments of truth values to **X** and **Y** and the corresponding truth values of the compound sentence. In classical logic there are two truth values, "True," which we denote as "T," and "False," which we denote as "F." We have five sentential connectives, so we construct five tables.

X	¬X
T	F
F	T

Truth table for ¬

X	Y	X∧Y
T	T	T
T	F	F
F	T	F
F	F	F

Truth table for ∧

X	Y	X∨Y
T	T	T
T	F	T
F	T	T
F	F	F

Truth table for ∨

X	Y	X→Y
T	T	T
T	F	F
F	T	T
F	F	T

Truth table for →

X	Y	X↔Y
T	T	T
T	F	F
F	T	F
F	F	T

Truth table for ↔

The truth table for → requires some explanation. But in order to understand the choices we made in this truth table, we must first discuss the truth table for ↔. ↔ is intended to capture the English biconditional connective 'if and only if'. An English biconditional of the form '**X** if and only if **Y**' *usually* asserts that **X** and **Y** are either both true or both false—that is, they have identical truth values. If **X** and **Y** have different truth values, the biconditional '**X** if and only if **Y**' is false. Since we want ↔ and 'if and only if' to have similar semantical roles, the truth table for ↔ is, therefore, justified. → is meant to capture the English 'if-then'. The first, second, and fourth rows of the truth table seem to have some intuitive appeal. If I say to my daughter, "If you wash the car, I'll give you $20," you will not accuse me of making a false promise if she washes the car and I give her $20 or if she doesn't wash the car and I don't give her $20. This explains the assignment of T to **X→Y** in the first and fourth rows of the truth table for →. But if she washes the car and I don't give her $20, I am guilty of making a false promise. This explains the assignment of F to **X→Y** in the second row of the truth table for →. The third row seems arbitrary. It is, however, an outcome of three factors: (1) **X→Y** is either true or false, (2) → is truth-functional, and (3) → is weaker in commitment than ↔. We intuitively expect my promise 'I'll give you $20 if you wash the car' to be less committed than 'I'll give you $20 if and only if you wash the car'. The first factor means that we have only two options for any relevant PL interpretation I: **X→Y** is true on I or **X→Y** is false on I. The truth-functionality of → entails that all PL interpretations that assign F to **X** and T to **Y** must assign the *same* truth value to **X→Y**. However, if we assign F to **X→Y** when **X** is false and **Y** is true, the truth table for → would be identical with the truth table for ↔. → would correspond, in this case, to 'if and only if'. But we want → to correspond to the weaker 'if-then'; hence we have no choice but to assign T to **X→Y** when **X** is false and **Y** is true. In other words, we see the difference between 'I'll give you $20 if you wash the car' and 'I'll give you $20 if and only if you wash the car' as residing solely in the case when I give my daughter $20 in spite of her not washing the car. The less committed promise of 'if-then' is true in this case and the more committed promise of 'if and only if' is false in this case. In all other cases they have identical truth values.

I said above that the first, second, and fourth rows of the truth table of → *seem to have some intuitive appeal*. I was intentionally guarded in my language. I did not say, for example, that these rows are natural or intuitive. The reason for this caution is that the relation between the English 'if-then' and the PL conditional is notoriously problematic. Logicians and philosophers produced over the years a large assortment of counterexamples that meant to show that the PL conditional does not represent correctly the English 'if-then'. One of the main difficulties is that the English 'if-then' seems to affirm a state of affairs between two logically or naturally related statements. This is not a requirement for the PL conditional. The antecedent and consequent of a PL conditional may have no relation to each other at all. All that matters is their truth values. For instance, it is not obvious at all that the following English sentences are true: 'If Venus is a planet, then bears are mammals' and 'If Barack Obama is not a graduate of Harvard, then the Beatles were not a famous band'. However, the PL translations of these sentences are true according to the truth table for →.

As we explained in 1.2.3, the formal truth conditions, we adopt in this book, for the quantified sentences are based on a **substitutional interpretation** of the quantifiers. Truth conditions of the quantified sentences that are based on **objectual interpretation** of the quantifiers would be: $(\forall y)Z$ is true on I if and only if every individual in UD satisfies the formula Z, and $(\exists y)Z$ is true on I if and only if there is at least one individual in UD that satisfies the formula Z.[12] We stated in note 10 that the notion of *satisfaction* invoked in objectual quantification is rather a technical notion, which we would like to avoid. In fact, one of the main reasons for choosing substitutional over objectual quantification in this book is to avoid the technicalities of the relation of satisfaction, which can hold between individuals in UD and PL formulas.

In order for substitutional quantification to deliver the right truth values for the quantified sentences, all the individuals in UD must be named. To see the point clearly, consider the sentence $(\forall x)(Qx \to (Kx \wedge Dx))$ and the interpretation I described in 1.2.2b. Whether we interpret the quantifier $(\forall x)$ objectually or substitutionally, the sentence comes true on I. Precisely speaking, the objectual reading of $(\forall x)(Qx \to (Kx \wedge Dx))$ on I asserts that for every individual x in UD, x satisfies the formula $(Qx \to (Kx \wedge Dx))$. Given the semantical assignments of I and the nature of the relation of satisfaction, this assertion may be interpreted on I as saying that for every individual in UD, if this individual is an English major, then he or she plays the piano and plays tennis. Since this assertion is true on I, the sentence $(\forall x)(Qx \to (Kx \wedge Dx))$ is also true on I. On the other hand, the substitutional reading

---

12 These are simplified statements of the objectual truth conditions of quantified sentences. The precise definition of the relation of satisfaction describes a certain relation between *sequences* of individuals in UD and PL formulas. However, our simplified definitions capture the essence of this relation.

of $(\forall x)(Qx \to (Kx \wedge Dx))$ on I asserts that for every name **t** in LN, the substitutional instance $(Qt \to (Kt \wedge Dt))$ is true on I. As explained previously, there are four substitutional instances of $(\forall x)(Qx \to (Kx \wedge Dx))$ on I. Since all of these sentences are true on I, the sentence $(\forall x)(Qx \to (Kx \wedge Dx))$ is also true on I.

Both interpretations of the quantifiers, objectual and substitutional, delivered the same truth value for $(\forall x)(Qx \to (Kx \wedge Dx))$ on I. The general point is that if every individual in UD is named, substitutional and objectual quantifications deliver the same truth values for the same quantified sentences. However, if we allow for unnamed individuals, objectual quantification gets things right while substitutional quantification might not. Let us modify the interpretation I of 1.2.2b in order to allow for unnamed individuals. We will call this "interpretation" J. The reader must be warned, however, that J does not satisfy the definition of PL interpretation stated in 1.2.1; it violates the condition 1.2.1a. Nevertheless, we will consider J in order to see what might happen if unnamed individuals are permitted.

The "interpretation" J

UD: {Samantha, Christopher, Lisa, Darius}
LN: *s, c, l*

Semantical Assignments

I(*s*): Samantha; I(*c*): Christopher; I(*l*): Lisa
I(*Qx*): "*x* is an English major": {Samantha, Darius}
I(*Dx*): "*x* plays tennis": {Samantha, Darius, Lisa}
I(*Kx*): "*x* plays the piano": {Samantha, Christopher}

On J, no name in LN refers to Darius, and Darius is removed from the extension of *K*. So in the world of J, Darius is an unnamed individual, and he does not play the piano.

The objectual reading of $(\forall x)(Qx \to (Kx \wedge Dx))$ may be interpreted on J as saying that for every individual in UD, if this individual is an English major, then he or she plays the piano and plays tennis. This assertion, of course, is false on J, since Darius is an English major who does not play the piano. Therefore, according to the objectual interpretation of $(\forall x)(Qx \to (Kx \wedge Dx))$, this sentence is false on J. The substitutional reading of $(\forall x)(Qx \to (Kx \wedge Dx))$ makes the same assertion as stated previously: for every name **t** in LN, the substitutional instance $(Qt \to (Kt \wedge Dt))$ is true on J. However, we now only have the three names *s, c,* and *l*; so we have the following substitutional instances: $(Qs \to (Ks \wedge Ds))$, $(Qc \to (Kc \wedge Dc))$, and $(Ql \to (Kl \wedge Dl))$. As the reader can easily verify, all of these substitutional instances are true on J. Hence $(\forall x)(Qx \to (Kx \wedge Dx))$ is true on J as well. Therefore, substitutional quantification delivers the wrong verdict regarding the truth value of the quantified sentence $(\forall x)(Qx \to (Kx \wedge Dx))$ on J. The problem is that the individual who is the sole counterexample to the universal claim made by $(\forall x)(Qx \to (Kx \wedge Dx))$ is Darius—he is the only individual in UD who is in the extension of *Q* but not in the intersection of the extensions of *K* and *D*, that is, he is the only English major who does not play both tennis and the piano. However, since Darius has no name in LN, there is no substitutional instance of $(\forall x)(Qx \rightarrow (Kx \wedge Dx))$ that applies to Darius. In other words, the individual in UD that demonstrates the falsity of $(\forall x)(Qx \rightarrow (Kx \wedge Dx))$ is not "disclosed" at the level of language. Any individual that has no name in LN is "hidden" from substitutional quantifications, and hence is not considered in determining the truth values of quantified sentences.

We end this section by stating precisely the relationship between the basic substitutional instances of a quantified sentence $(\Theta z)X$ (where 'Θ' stands for '$\forall$' or '$\exists$') and its substitutional instances in general. Recall that a *basic substitutional instance* of $(\Theta z)X$ on an interpretation I is obtained by substituting a *name* in LN of the interpretation for the variable **z** in all its occurrences in **X**. On the other hand, a *substitutional instance* of $(\Theta z)X$ is obtained by substituting a *singular term* for **z** in all its occurrences in **X**. Since every name is a singular term, it immediately follows that every basic substitutional instance is a substitutional instance. The converse, however, is not true. If there are function symbols, then there might be singular terms that are not names listed in LN; and hence there might be many substitutional instances that are not basic on the given interpretation. For instance, the PL sentences $Pe \rightarrow (Rea \wedge Te)$ and $Pg^1e \rightarrow (Rg^1ea \wedge Tg^1e)$ are both substitutional instances of the sentence $(\forall x)(Px \rightarrow (Rxa \wedge Tx))$. Every PL interpretation for these sentences must include the names *a* and *e* in its LN, but it is not required to include g^1e among its names. Assume it does not. On such an interpretation, $Pe \rightarrow (Rea \wedge Te)$ is a basic substitutional instance of $(\forall x)(Px \rightarrow (Rxa \wedge Tx))$, while $Pg^1e \rightarrow (Rg^1ea \wedge Tg^1e)$ is only a substitutional instance of $(\forall x)(Px \rightarrow (Rxa \wedge Tx))$. So we have two trivial implications: (1) if all the substitutional instances of $(\Theta z)X$ are true (or false) on some PL interpretation I, then all its basic substitutional instances are true (or false) on I; and (2) if one of the basic substitutional instances of $(\Theta z)X$ is true (or false) on I, then one of its substitutional instances is true (or false) on I. But are the converse of these claims true? The answer is "Yes." We state these facts as a theorem.

Theorem 1.2.2. If all the basic substitutional instances of $(\Theta z)X$ are true (or false) on a PL interpretation I, then all its substitutional instances are true (or false) on I; and hence if one of the substitutional instances of $(\Theta z)X$ is true (or false) on I, then one of its basic substitutional instances is true (or false) on I.

It is not hard to see why this theorem is true. Informally speaking, if all the basic substitutional instances of $(\Theta z)X$ are true on I, then the formula **X** is true of all the individuals in UD (since all the individuals in UD have names in LN). But according to Theorem 1.2.1, all singular terms refer to unique individuals in UD. Thus if we substitute a singular term for **z** in **X**, we obtain a true sentence on I. On the other hand if there is a true substitutional instance of $(\Theta z)X$ on I, then there is an individual in UD of which the formula **X** is true (because every singular term has a unique referent on I). But this individual must have a name in LN. This implies that if we substitute this name for **z** in **X**, we obtain a true sentence on I. The same informal reasoning works for the case of falsehood. This is not a precise proof, since we never defined what we mean by a formula being true of an individual. However, the idea is intuitive enough. The informal reasoning we just gave conveys the central idea of the proof.

1.2.6   There are several possible **objections to substitutional quantification**. One of them is that the requirement that every individual in UD have a name in LN creates (1) a linguistic aspect of every PL interpretation and (2) a dependency of the given PL language on its PL interpretations. The first point seems artificial and unrealistic. It seems artificial to require actual and possible realities to contain names for their individuals. Why should the planets of a distant solar system be labeled in order for such a system to count as a possible situation? Furthermore, there seem to be many situations in which the requirement is practically impossible to fulfill. How do we name all the stars in the universe, all the grains of sand on the beaches of Florida, and all the bees in Africa? However, an advocate of substitutional quantification might respond to these objections in this way. The requirement of including names for all the individuals of a PL interpretation is a formal requirement that is neither meant to be authentic nor realistic. Also, we do not assume that there is a "practical" method of assigning names to all the individuals of an interpretation. One is free to consider an interpretation whose UD is the set of all the stars in the universe. Naming these stars is simply an *in-principle* requirement. There is nothing contradictory (and hence, logically impossible) about assigning names to all these stars in the universe, and there is no shortage of names: we can use subscripts to generate as many names as needed. PL only requires that *in principle* there is a way of naming all these individuals. In practice, when such an interpretation is described, it is enough to state that there is a name for every star in the universe. Even if UD is infinite, the use of subscripts ensures that there are plenty of names to label all the individuals in UD.

The second point, which is the dependency of the given PL language on its PL interpretations, is a more serious problem for substitutional quantification. Assume that Γ is a set of PL sentences. These sentences consist of basic PL vocabulary, which defines the PL language in which the members of Γ are constructed. A PL interpretation J for Γ brings its own set of names. Almost always many of these names do not occur in members of Γ. In fact, these names need not be part of the standard PL names listed in 1.1.1a. These names are, therefore, new items that are to be added to the basic vocabulary of the language of Γ. Adding these names would expand the language of Γ. This means that every PL interpretation for Γ might change the language of Γ in a particular way. The language of Γ would not be a fixed language but a "dynamic" language that changes with almost every PL interpretation for it. The situation becomes more troubling if we consider an interpretation whose UD is an uncountable set, say, the set $\mathbb{R}$ of all the real numbers. In this case, LN would also be uncountable. Therefore the PL language of Γ would have to be expanded beyond any countable limit.

Again, an advocate of substitutional quantification might be able to respond to these objections. We do not need to consider the language of Γ dynamic, changing with every PL interpretation for Γ. The vocabulary of the language of Γ consists of the extra-logical vocabulary that occurs in the members of Γ and of the logical vocabulary of PL (i.e., the eight logical symbols, the variables, and the parentheses). We previously designated this vocabulary as Voc(Γ). Thus the language of Γ is the PL language whose basic vocabulary is

Voc(Γ). This language does not change if an interpretation J for Γ brings with it additional names. The new names do not belong to the language of Γ, they belong to the language of J. Every PL interpretation has its own language, whose vocabulary is the PL logical vocabulary together with the extra-logical vocabulary that J interprets. We denoted this vocabulary as Voc(J). In fact, the definition of PL interpretation given in this book permits an interpretation for Γ to interpret not only new names but even new function symbols and predicates, which are not part of the language of Γ. So there are two languages: the language of the set Γ, whose basic vocabulary is Voc(Γ), and the language of the PL interpretation J for Γ, whose basic vocabulary is Voc(J). Γ has its own fixed vocabulary and every interpretation for Γ has its own vocabulary. There is a relation between those two vocabularies: the vocabulary of Γ is always a subset of the vocabulary of any of its PL interpretations. It should be clear from the preceding remarks that the language of Γ is a fixed PL language whose basic vocabulary is Voc(Γ); and this language remains the same no matter what interpretation is given of the extra-logical vocabulary in Voc(Γ).

The vocabulary of an interpretation of uncountable size is also uncountable, since its LN consists of uncountably many names in order to name all the individuals in its UD. Since the standard PL vocabulary we described in 1.1.1 is countable, most of the names in LN of an uncountable interpretation are not standard PL names. But this is perfectly all right. There is nothing in principle wrong about having languages whose vocabulary is uncountable. The same formation rules apply for terms and formulas. The only difference is that the sets of terms and of formulas of such a language are uncountable as well. As for finding enough subscripts to index the uncountable vocabulary of a PL interpretation whose size is uncountable, we can use UD itself as the source of these subscripts. For instance, if the set $\mathbb{R}$ is the UD, we might let LN consist of all the names that can be generated by indexing the letter 'c' with a real number r, that is, for every real number r, c_r is a name in LN.

All the proofs we will give in Chapter Three presuppose that the extra-logical vocabulary of PL is countable. The fact that there might be uncountable interpretations for this vocabulary is not relevant to the size of the PL extra-logical vocabulary, since these uncountable interpretations only require that their own vocabularies, that is, the vocabularies that these interpretations interpret, are uncountable. We explained that there are two sets of vocabulary here: the vocabulary of which collections of PL sentences are composed, and the vocabulary that is interpreted by some PL interpretation. The former is always countable (in this book), while the latter might be uncountable.

We mention one last objection to substitutional quantification. Substitutional quantification interprets the quantified sentences as making assertions about the language itself. $(\forall \mathbf{z})\mathbf{X}$ is interpreted as saying that for every name **s** in LN, the substitutional instance **X[s]** is true on I, and $(\exists \mathbf{z})\mathbf{X}$ as saying that there is at least one name **s** in LN, such that the substitutional instance **X[s]** is true on I. But this is counterintuitive. For example, when we assert $(\forall x)Px$, we ordinarily mean that every individual in UD has the property I(P). We make an assertion about the individuals of an interpretation and the property that the predicate P designates on I. We do not intend to make an assertion about the language in question—about its names and the substitutional instances of its formulas.

Two responses might be offered to this objection. First, there is no reason to take the truth conditions of a sentence as determining the assertion made by the sentence. The truth conditions of $(\forall z)X$ and $(\exists z)X$ involve the truth status of their basic substitutional instances. However, these truth conditions need not constitute the assertions made by these sentences. We may assume that $(\forall z)X$ and $(\exists z)X$ make assertions about the individuals in UD, but their truth conditions are given in terms of the truth status of their basic substitutional instances. Many philosophers would not find this response very persuasive, since it is typically assumed that there is a close relation between the truth conditions of a sentence and the assertion made by the sentence. An influential philosophical position, which was championed by the American Philosopher Donald Davidson (1917–2003), affirms that the meaning of a sentence is given by its truth conditions. If we equate the meaning of a sentence with the assertion it makes, we can say that the truth conditions of a sentence constitute the assertion it makes. According to this position, the assertion that $(\forall z)X$ makes is that every basic substitutional instance of it is true, and the assertion $(\exists z)X$ makes is that at least one of its basic substitutional instances is true. If this influential position is correct, then the first response is misguided.

The second response accepts this position—namely that the truth conditions of a sentence constitute its meaning; and hence it accepts that the assertions made by $(\forall z)X$ and $(\exists z)X$ are about the truth status of their basic substitutional instances. However, this response argues that objectual quantification fares no better. In order to appreciate the full force of this response, some knowledge of the technical relation of satisfaction is required. At any rate, we can describe the main idea. We only need to be aware that this relation is defined to hold between sequences of individuals in UD and PL formulas. So a defender of substitutional quantification might respond that the objectual readings of $(\forall z)X$ and $(\exists z)X$ do not escape the complaint of this objection; for the standard account of objectual quantification interprets $(\forall z)X$ as asserting that every sequence of individuals in UD that meets a certain condition *satisfies the formula* X, and $(\exists z)X$ as saying that there is a sequence of individuals in UD that meets a certain condition and which *satisfies the formula* X. These readings invoke the syntactical notion of a PL formula, and hence the assertions they make are, in part, about the language of PL.

A possible rebuttal to this response is that once the details of the relation of satisfaction are spelled out and once we invoke the interpretations, supplied by the PL interpretation I, of the basic vocabulary, an assertion that a quantified sentence makes is ultimately reduced to assertions solely about the individuals in UD and their properties and relations. However, a defender of substitutional quantification might say the same thing about a substitutional reading of a quantified sentence. She might say that, similarly, once the semantical contents of the basic substitutional instances are fully cashed out in terms of the interpretations supplied by I, the assertion that the quantified sentence makes is reduced to assertions about the individuals of I and the properties and relations they might have.

We will not attempt to settle the debate between advocates of substitutional quantification and advocates of objectual quantification. This debate is rather philosophically and logically involved. Our goal, for the sake of intellectual honesty, is to make clear that the semantical approach we adopted is controversial and to give a feel for some of the issues that are involved in the controversy. However, we should stress that even though we gave substitutional truth conditions for the quantifiers, in practice almost always we determine the truth values of quantified sentences by examining the assertions they make about the individuals in UD and their properties and relations. We also register a happy note—namely, that all the metatheorems we will prove in Chapters Three and Five are compatible with objectual as well as substitutional interpretation of the quantifiers.

1.3 Logical Concepts in PL

1.3.1 A PL **argument** is a nonempty collection of PL sentences: one of these sentences is the **conclusion** of the argument and the others are its **premises**. If Γ is the set of the premises of an argument whose conclusion is **X**, we may represent the argument schematically as Γ/X or $\frac{\Gamma}{X}$. An argument has exactly one conclusion, while its set of premises Γ might be empty or consist of finitely or infinitely many PL sentences.

1.3.2 The concepts of **validity** and **logical consequence** are equivalent; they are two sides of the same coin. An argument Γ/X is valid if and only if its conclusion, **X**, is a logical consequence of its set of premises, Γ. We introduce the following notation and terminology.

1.3.2a "$\Gamma \vDash X$" is read "**X** is a logical consequence of Γ," or "**X** logically follows from Γ," or "Γ logically implies **X**."

1.3.2b "$\Gamma \nvDash X$" is read "**X** is not a logical consequence of Γ," or "**X** does not logically follow from Γ," or "Γ does not logically imply **X**."

1.3.3a A PL argument Γ/X is **valid** (or $\Gamma \vDash X$) if and only if on every PL interpretation for Γ/X on which all the members of Γ are true **X** is true as well (or simply, if and only if **X** is true on every model of Γ that is relevant to **X**).

1.3.3b A PL argument Γ/X is **valid** (or $\Gamma \vDash X$) if and only if there is no PL interpretation on which all the members of Γ are true and **X** is false (or simply, if and only if there is no model of Γ on which **X** is false).

We will show that the argument below is valid.

S1 $(Raa \land Pc) \rightarrow (\exists x)Ax$
S2 $(\forall y)\neg Ay$
S3 Pc
___
S4 $(\exists x)\neg Rxx$

We will prove that every PL interpretation for S1–S4 that satisfies S1–S3 makes S4 true. We will use a style of proof that we may term **arbitrary-object proof**.[13] The strategy of this proof style is to take an object as an arbitrary representative of a certain type T, and to show that this arbitrary object meets a certain condition C. It is essential that in this proof no assumption about this arbitrary object be invoked other than that it is of type T. An arbitrary-object proof allows us to conclude that *all* objects of type T meet the condition C. In the following proof the arbitrary object of type T is a PL interpretation for S1–S4 on which the premises, S1–S3, are true. We invoke only the assumption that I is a PL interpretation for S1–S4 that makes S1–S3 true, the definition of PL interpretation given in 1.2.1, and the formal truth conditions listed in 1.2.5.

Proof that {S1, S2, S3}/S4 is valid

1. Assume that I is an arbitrary PL interpretation for the sentences S1–S4, such that S1–S3 are true on I, that is, the following sentences are true on I:
    1.1. $(Raa \land Pc) \rightarrow (\exists x)Ax$
    1.2. $(\forall y)\neg Ay$
    1.3. Pc
2. From 1.2 by the truth conditions of the universal quantifier: every basic substitutional instance of $(\forall y)\neg Ay$ is true on I. This implies that every sentence of the form $\neg A\mathbf{s}$, where $\mathbf{s}$ is any name in LN, is true on I.
3. From 2 by the truth conditions of the negation: for every name $\mathbf{s}$ in LN, $A\mathbf{s}$ is false on I.
4. From 3 by the truth conditions of the existential quantifier: $(\exists x)Ax$ is false on I.
5. From 1.1 and 4 by the truth conditions of the conditional: $Raa \land Pc$ is false on I.
6. From 5 by the truth conditions of the conjunction: either Raa is false on I or Pc is false on I.
7. From 1.3 and 6: Raa is false on I.
8. From 7 by the truth conditions of the negation: $\neg Raa$ is true on I.
9. From 8 by the truth conditions of the existential quantifier: $(\exists x)\neg Rxx$ is true on I.
10. From 1 through 9: if I is a PL interpretation for S1–S4 that makes S1–S3 true, then I makes S4 true as well.

___

13   The technical name of this proof style is 'universal generalization'. We will formalize this type of proof in Section 1.4.

11. From 10: given that I is arbitrary, it follows that *every* PL interpretation for S1–S4 that makes the premises, S1–S3, true also makes the conclusion, S4, true.
12. From 11 by Definition 1.3.3a: the PL argument {S1, S2, S3}/S4 is valid.

1.3.4 A PL argument Γ/X is **invalid** (or Γ⊭X) if and only if there is a PL interpretation on which the members of Γ are all true and X is false (or simply, if and only if there is a model of Γ on which X is false).

We will prove that the following PL argument is invalid by constructing a PL interpretation I that satisfies the premises, S1–S5, and on which the conclusion, S6, is false.

S1 $(\neg Ha \vee \neg Kb) \to a = gb$
S2 $a \neq gb$
S3 $(Ha \wedge Kb) \to (\exists x)(a = gx \vee x = gb)$
S4 $(\forall x)((x = b \vee x = c) \to Sagx)$
S5 $(\forall z)(\forall w)(z = gw \to Swb)$

S6 $\neg Saa \vee \neg Sbb$

The PL Interpretation I

UD: The set of all positive even numbers: {2, 4, 6, ...}
LN: $a, b, c, d, e_2, e_4, e_6, ..., e_n, ...$; where n is any positive even number

Semantical Assignments

I(a): 16; I(b): 2; I(c): 4; I(d): 8; I(e_n) = n
I(gz): "The square of z"
I(Hx): "x is even"
I(Kv): "v is prime"
I(Sxy): "x is divisible by y"

Natural reading of (S1) $(\neg Ha \vee \neg Kb) \to a = gb$ on I:
If 16 is not even or 2 is not prime, then 16 is the square of 2.
The sentence is true on I, because the antecedent and consequent are false on I.

Natural reading of (S2) $a \neq gb$ on I:
16 is not the square of 2.
The sentence is true on I.

PL-English reading of (S3) $(Ha \wedge Kb) \to (\exists x)(a = gx \vee x = gb)$ on I:
If 16 is even and 2 is prime, then there is a positive even number x, such that 16 is the square of x and x is the square of 2.
The sentence is true on I, since the antecedent and the consequent are true on I. The consequent is true on I because 16 is the square of 4 and 4 is the square of 2.

Natural reading of (S4) $(\forall x)((x = b \vee x = c) \rightarrow Sagx)$ on I:
16 is divisible by the square of 2 and by the square of 4.
The sentence is true on I.

PL-English reading of (S5) $(\forall z)(\forall w)(z = gw \rightarrow Swb)$ on I:
For any positive even numbers z and w, if z is the square of w, then w is divisible by 2.
The sentence is true on I, since all the numbers in UD are divisible by 2.

Natural reading of (S6) $\neg Saa \vee \neg Sbb$ on I:
16 is not divisible by 16 or 2 is not divisible by 2.
The sentence is false on I because both disjuncts are false on I.

Since there is a PL interpretation that makes the premises of the PL argument true and its conclusion false, by Definition 1.3.4, the argument is invalid.

1.3.5a   A PL sentence is **valid** if and only if it is true on every PL interpretation for that sentence.

1.3.5b   A PL sentence is **valid** if and only if there is no PL interpretation on which it is false.

Most introductory logic textbooks use the term **logically true** to designate sentences that fulfill this definition. In metalogic textbooks, however, it is common to describe these sentences as **valid**. This use of the term 'valid' is different from its use to describe arguments whose conclusions are logical consequences of the sets of premises (see 1.3.3 above). There is a relation between the two uses. A valid PL sentence is the conclusion of a valid PL argument whose set of premises is empty. We will explain later that any logical consequence of the empty set is true on all relevant PL interpretations. In this book, we will follow the usual practice of metalogic books of using the term 'valid' to describe sentences that are true on all relevant PL interpretations. In my introductory text, *An Introduction to Logical Theory*, those sentences are called "logically true," which is the standard practice of introductory logic textbooks. The reader should be careful to note, on the one hand, this divergence between *An Introduction to Logical Theory* and this book, and, on the other hand, the difference between calling an argument "valid" and calling a sentence "valid."

We will show that the PL sentence $(\forall z)Gzd \rightarrow Gdd$ is valid by showing that it satisfies Definition 1.3.5b. We use an **indirect-proof** strategy. In this strategy an assumption is made, which is called **the reductio assumption**,[14] and a contradiction is inferred from this assumption. This allows us to conclude that the reductio assumption is false. In the proof below we only invoke the reductio assumption, the definition of PL interpretation, and the truth conditions of PL sentences.

---

14   It is so called because the technical name of indirect proof is 'Reductio Ad Absurdum'.

Proof that $(\forall z)Gzd \rightarrow Gdd$ is a valid sentence

1. Reductio Assumption: there is a PL interpretation I on which the sentence $(\forall z)Gzd \rightarrow Gdd$ is false.
2. From 1 by the truth conditions of the conditional: $(\forall z)Gzd$ is true on I, and Gdd is false on I.
3. From 2 by the truth conditions of the universal quantifier: every basic substitutional instance of $(\forall z)Gzd$ is true on I, that is, for every name **s** in LN, Gsd is true on I.
4. From 3: Gdd is true on I.
5. From 2 and 4: Gdd is both true and false on I, which is a contradiction.
6. From 1 through 5: the Reductio Assumption is false. Hence, there is no PL interpretation on which $(\forall z)Gzd \rightarrow Gdd$ is false.
7. From 6 by Definition 1.3.5b: $(\forall z)Gzd \rightarrow Gdd$ is a valid sentence.

1.3.6a A PL sentence is **contradictory** if and only if it is false on every PL interpretation for that sentence.

1.3.6b A PL sentence is **contradictory** if and only if there is no PL interpretation on which it is true.

Similar to the case of valid sentences, in introductory logic textbooks, such as my *An Introduction to Logical Theory*, sentences that fulfill these definitions are called **logically false**. In books on metalogic, they are usually called "contradictory sentences" or simply "contradictions." We will reserve the word 'contradiction' for the conjunction or the presence of a sentence and its negation (i.e., **X** and ¬**X**), and we will follow the traditional usage of metalogic textbooks by calling sentences that are false on all relevant PL interpretations "contradictory."

We will demonstrate that the PL sentence $\neg(\exists v)Rvv \wedge (Red \wedge e = d)$ is contradictory. We use an indirect proof to establish that there is no PL interpretation on which $\neg(\exists v)Rvv \rightarrow (Red \wedge e = d)$ is true. We only invoke the reductio assumption, the definition of PL interpretation, and the formal truth conditions.

Proof that $\neg(\exists v)Rvv \wedge (Red \wedge e = d)$ is a contradictory sentence

1. Reductio Assumption: there is a PL interpretation I on which the sentence $\neg(\exists v)Rvv \wedge (Red \wedge e = d)$ is true.
2. From 1 by the truth conditions of the conjunction: the following sentences are true on I:
   2.1. $\neg(\exists v)Rvv$
   2.2. Red
   2.3. $e = d$
3. From 2.1 by the truth conditions of the negation: $(\exists v)Rvv$ is false on I.

4. From 3 by the truth conditions of the existential quantifier: every basic substitutional instances of $(\exists v)Rvv$ is false on I, that is, for every name **s** in LN, R**ss** is false on I.
5. From 4: *Ree* is false on I.
6. From 2.3 by the truth conditions of the identity predicate: I(e) is the same individual as I(d).
7. From 2.2 by the truth conditions of '*Red*': $\langle I(e), I(d) \rangle \in I(R)$.
8. From 6 and 7: $\langle I(e), I(e) \rangle \in I(R)$.
9. From 8 by the truth conditions of atomic sentences: *Ree* is true on I.
10. From 5 and 9: *Ree* is both true and false on I, which is a contradiction.
11. From 1 through 10: the Reductio Assumption is false; hence we conclude that there is no PL interpretation on which the sentence $\neg(\exists v)Rvv \wedge (Red \wedge e = d)$ is true.
12. From 11 by Definition 1.3.6b: $\neg(\exists v)Rvv \wedge (Red \wedge e = d)$ is a contradictory sentence.

1.3.7 A PL sentence is **contingent** if and only if it is true on some PL interpretations and false on other interpretations.

Almost every PL sentence we write without much thought is contingent. We establish the contingency of a sentence by presenting two PL interpretations one of which makes the sentence true and the other makes it false. We will show that $(\forall x)(Bx \vee Kx) \leftrightarrow (\exists v)(Bv \wedge Kv)$ is contingent.

The PL Interpretation I

UD: {John, Mary, Bill, Sarah}
LN: j, m, b, s

Semantical Assignments

I(j): John; I(m): Mary; I(b): Bill; I(s): Sarah
I(Bx): "x is male": {John, Bill}
I(Kx): "x is female": {Mary, Sarah}

Natural reading of $(\forall x)(Bx \vee Kx) \leftrightarrow (\exists v)(Bv \wedge Kv)$ on I:
Everyone is either male or female if and only if there is someone who is both male and female.
The sentence is false on I because the left-hand side of the biconditional is true and the right-hand side is false.

The PL interpretation I*

UD: {2, 3, 4, 5}
LN: a, b, c, d

Semantical Assignments

I*(a): 2; I*(b): 3; I*(c): 4; I*(d): 5

I*(Bx): "x is prime": {2, 3, 5}
I*(Kx): "x is even": {2, 4}

Natural reading of $(\forall x)(Bx \vee Kx) \leftrightarrow (\exists v)(Bv \wedge Kv)$ on I*:
Every number is either prime or even if and only if there is a number that is both prime and even.
This is true on I* since both sides of the biconditional are true.

Note: The sentence that every number is either prime or even is false on the set of all natural numbers, ℕ, since, e.g., 9 is neither prime nor even. But as we stated before, the quantifiers always range over the individuals in UD of the given interpretation. Based on the UD above, the sentence 'every number is either prime or even' is true.

According to Definition 1.3.7, since the sentence $(\forall x)(Bx \vee Kx) \leftrightarrow (\exists v)(Bv \wedge Kv)$ is false on I and true on I*, it is contingent.

1.3.8a Two PL sentences are **logically equivalent** if and only if on every PL Interpretation for these sentences they have identical truth values.

1.3.8b Two PL sentences are **logically equivalent** if and only if there is no PL Interpretation on which they have different truth values.

We will show that the PL sentences

S1 $\neg(\forall v)(Dv \to Ce)$, and
S2 $(\exists z)(Dz \wedge \neg Ce)$

are logically equivalent. We first prove that every PL interpretation for S1 and S2 that makes S1 true makes S2 true as well. We use the strategy of arbitrary-object proof, introduced in 1.3.3, to establish this claim. We only invoke the assumption that I is a PL interpretation for S1 and S2 on which $\neg(\forall v)(Dv \to Ce)$ is true, the definition of PL Interpretation, and the formal truth conditions.

Proof 1

1. Assume that I is an arbitrary PL interpretation for S1 and S2 on which $\neg(\forall v)(Dv \to Ce)$ is true.
2. From 1 by the truth conditions of the negation: $(\forall v)(Dv \to Ce)$ is false on I.
3. From 2 by the truth conditions of the universal quantifier: there is a basic substitutional instance of $(\forall v)(Dv \to Ce)$ that is false on I, that is, there is a name **s** in LN such that the substitutional instance $D\mathbf{s} \to Ce$ is false on I.
4. From 3 by the truth conditions of the conditional: $D\mathbf{s}$ is true on I and Ce is false on I.
5. From 4 by the truth conditions of the negation: $D\mathbf{s}$ and $\neg Ce$ are true on I.
6. From 5 by the truth conditions of the conjunction: $D\mathbf{s} \wedge \neg Ce$ is true on I.
7. From 6 by the truth conditions of the existential quantifier: $(\exists z)(Dz \wedge \neg Ce)$ is true on I.

8. From 1 through 7: if I is a PL interpretation for S1 and S2 on which S1 is true, S2 is true on I as well.
9. From 8: since I is arbitrary, every PL interpretation for S1 and S2 that makes S1 true makes S2 true as well.

We now prove that every PL interpretation for S1 and S2 that makes S1 false also makes S2 false.

Proof 2

1. Assume that I is an arbitrary PL interpretation for S1 and S2 on which $\neg(\forall v)(Dv \rightarrow Ce)$ is false.
2. From 1 by the truth conditions of the negation: $(\forall v)(Dv \rightarrow Ce)$ is true on I.
3. From 2 by the truth conditions of the universal quantifier: every basic substitutional instance of $(\forall v)(Dv \rightarrow Ce)$ is true on I, that is, for every name **s** in LN, the substitutional instance $Ds \rightarrow Ce$ is true on I.
4. From 3 by the truth conditions of the conditional: for every name **s** in LN, either Ds is false on I or Ce is true on I.
5. From 4 by the truth conditions of the negation: for every name **s** in LN, either Ds is false on I or $\neg Ce$ is false on I.
6. From 5 by the truth conditions of the conjunction: for every name **s** in LN, $Ds \wedge \neg Ce$ is false on I.
7. From 6 by the definition of basic substitutional instance: all basic substitutional instances of $(\exists z)(Dz \wedge \neg Ce)$ are false on I.
8. From 7 by the truth conditions of the existential quantifier: $(\exists z)(Dz \wedge \neg Ce)$ is false on I.
9. From 1 through 8: if I is a PL interpretation for S1 and S2 on which S1 is false, S2 is false on I as well.
10. From 9: since I is arbitrary, every PL interpretation for S1 and S2 that makes S1 false also makes S2 false.
11. From 10 and the conclusion of Proof 1 by Definition 1.3.8a: S1 and S2 are logically equivalent.

1.3.9 A set of PL sentences is **satisfiable** if and only if there is a PL interpretation on which every member of the set is true, that is, there is a PL interpretation that satisfies the set (or simply, if and only if the set has a model).

It is common in introductory logic books, including my *An Introduction to Logical Theory*, to call a satisfiable set **consistent**. In metalogic books, however, a set that has a model is described as satisfiable and the term 'consistent' is reserved for sets that are proof-theoretically consistent, that is, no contradiction is formally derivable from them. This is the terminology that we will use. We will define 'consistent set' in the next section when we study PL proof theory.

We will demonstrate that the set of PL sentences below is satisfiable. We will construct a PL interpretation that satisfies the set.

S1 $(\forall x)(\forall y)(Bxy \lor Byx)$
S2 $(\exists x)(\exists y)(Bxy \land x \neq y)$
S3 $(\forall x)(\forall y)((Bxy \land Byx) \rightarrow x = y)$
S4 $(\forall x)(\forall y)(\forall z)((Bxy \land Byz) \rightarrow Bxz)$

We will show that the PL interpretation below is a model of the set above.

The PL Interpretation I

UD: {Anna, Margaret, Patricia, Jennifer}
LN: a, m, p, j

Semantical Assignments

  I(a): Anna; I(m): Margaret; I(p): Patricia; I(j): Jennifer
  I(Bxy): "x likes y": {⟨Anna, Anna⟩, ⟨Margaret, Margaret⟩, ⟨Patricia, Patricia⟩,
    ⟨Jennifer, Jennifer⟩, ⟨Anna, Margaret⟩, ⟨Margaret, Patricia⟩,
    ⟨Patricia, Jennifer⟩, ⟨Anna, Patricia⟩, ⟨Anna, Jennifer⟩,
    ⟨Margaret, Jennifer⟩}
  I($x = y$): the relation of token identity

Natural reading of (S1) $(\forall x)(\forall y)(Bxy \lor Byx)$ on I:
For any two people, either the first likes the second or the second likes the first. (The "two" people could be one person.)
S1 is true on I.

Natural reading of (S2) $(\exists x)(\exists y)(Bxy \land x \neq y)$ on I:
There are two distinct people such that the first likes the second.
S2 is true on I.

Natural reading of (S3) $(\forall x)(\forall y)((Bxy \land Byx) \rightarrow x = y)$ on I:
For any two people, if the first likes the second and the second likes the first, then they are the same person.
S3 is true on I.

Natural reading of (S4) $(\forall x)(\forall y)(\forall z)((Bxy \land Byz) \rightarrow Bxz)$ on I:
For any three people, if the first likes the second and the second likes the third, then the first likes the third.
S4 is true on I.

1.3.10a A set of PL sentences is **unsatisfiable** if and only if on every PL interpretation for that set at least one member of the set is false.

1.3.10b A set of PL sentences is **unsatisfiable** if and only if there is no PL interpretation on which all the members of the set are true, that is, there is no PL interpretation that satisfies the set (or simply, if and only if the set has no model).

In introductory logic books, including my *An Introduction to Logical Theory*, unsatisfiable sets are called **inconsistent**. We will follow in this book the common practice of metalogic books, and reserve the term 'inconsistent' for sets from which a contradiction is derivable. This is a proof-theoretic notion, which we will introduce in the next section.

We will prove, using an indirect proof, that the following set of PL sentences is unsatisfiable. The reductio assumption is that there is a PL interpretation I that satisfies the set. We will show that this assumption leads to a contradiction; hence the assumption is false.

S1 Ec
S2 $(\forall w)(Ew \rightarrow Le)$
S3 $\neg(\exists v)Lv$

Proof that the set {S1, S2, S3} is unsatisfiable

1. Reductio Assumption: there is a PL interpretation I on which S1–S3 are true, that is, the following sentences are true on I:
    1.1. Ec
    1.2. $(\forall w)(Ew \rightarrow Le)$
    1.3. $\neg(\exists v)Lv$
2. From 1.2 by the truth conditions of the universal quantifier: every basic substitutional instance of $(\forall w)(Ew \rightarrow Le)$ is true on I, that is, for every name **s** in LN, the substitutional instance $Es \rightarrow Le$ is true on I.
3. From 2: $Ec \rightarrow Le$ is true on I.
4. From 1.1 and 3 by the truth conditions of the conditional: Le is true on I.
5. From 4 by the truth conditions of the existential quantifier: $(\exists v)Lv$ is true on I.
6. From 1.3 by the truth conditions of the negation: $(\exists v)Lv$ is false on I.
7. From 5 and 6: $(\exists v)Lv$ is both true and false on I, which is a contradiction.
8. From 1 through 7: the Reductio Assumption is false; hence there is no PL interpretation on which S1–S3 are all true.
9. From 8 by Definition 1.3.10b: the set {S1, S2, S3} is unsatisfiable.

1.3.11 A concept is **decidable** if and only if there is an **effective decision procedure** for determining whether something is subsumed under the concept or not; and a decision procedure is effective if and only if it can, in principle, be followed mechanically and after a finite number of deterministic steps leads to the correct answer, whether it is "Yes" or "No." Not all effective procedures are decision procedures. We call a procedure **effective** just in case it is mechanical (i.e., it involves no creative steps) and generates the desired result after finitely many deterministic steps. There are effective procedures that produce the answer "Yes" when and only when the correct answer is "Yes," but that might not produce any answer when the correct answer is "No," and others that produce the answer "No" when and only when the correct answer is "No," but that might not produce any answer when the correct answer is "Yes." The former procedure is called a **Yes-procedure** and the latter a **No-procedure**. Our definition of an effective *decision* procedure makes it clear that such a procedure must produce *both* answers. A concept that has only a Yes-procedure is described as **semidecidable**.

There is no effective *decision* procedure for answering *all* the questions of the form: "Is this PL sentence, or is this set of PL sentences, subsumed under this concept, where 'concept' here refers to one of the eight logical concepts defined in 1.3.3–1.3.10?" Even finite sets of PL sentences or sets that have effective decision procedures for determining membership in them typically fail to have an effective decision procedure for determining the applicability of the eight logical concepts to them or to their members.[15] This entails that these logical concepts are undecidable in PL. Some of these concepts are only semidecidable. In the last chapter of this book we will establish that some logical concepts of PL, such as validity and unsatisfiability, are not decidable but only semidecidable. This result establishes that other logical concepts of PL, such as invalidity and satisfiability, are not even semidecidable. The general conclusion is that none of the eight logical concepts defined above is decidable in PL. We will demonstrate this general conclusion in Chapter Five.

1.4 PL Proof Theory

1.4.1 A concept can be **universal** or **existential**. In general, a universal concept is a concept whose applicability to any object requires that *all* objects of a certain type meet certain conditions. On the other hand, an existential concept is a concept whose applicability to any object requires the *existence* of an object of a certain type. The denial of a universal concept is an existential concept, and the denial of an existential concept is a universal concept. For instance, the logical concepts of valid argument, valid sentence, contradictory sentence, and unsatisfiable set are universal concepts, while the logical concepts of invalid argument, contingent sentence, and satisfiable set are existential concepts. In the previous section, we discussed examples of invalid arguments, of contingent sentences, and of satisfiable sets of sentences. In all these examples, we established the applicability of these concepts by establishing the existence of PL interpretations that fulfill certain conditions. We typically established the existence of those interpretations by presenting examples of them: we "constructed" PL interpretations. On

---

15 A set for which there is an effective decision procedure for determining whether an object is a member of the set or not is said to be **decidable**.

the other hand, when we wanted to show that a universal concept is applicable to some set of PL sentences, we reasoned about all PL interpretations of certain types. For instance, to show that some set of PL sentences is unsatisfiable or that some PL argument is valid, we *demonstrated* that *all* the relevant PL interpretations fulfill certain conditions. Such a **demonstration** is typically a **proof** that proceeds from a certain set of premises and terminates with the desired conclusion. Between the initial and terminal stages, a proof comprises consecutive steps in which intermediate conclusions are inferred according to specific modes of reasoning. We call such proofs **demonstrative proofs** or simply **demonstrations**.[16] A **proof theory** is introduced precisely for the purpose of developing the necessary understanding and tools for constructing **formal representations** of demonstrative proofs. These formal representations are referred to as **derivations**. In this section, we will study PL derivations.

A demonstrative proof consists of **inferential steps**. Almost all inferential steps consist of three components: (1) the **antecedents**, (2) the **conclusion**, and (3) the **inferential license**. The term 'antecedent' here has a different meaning from the term 'antecedent' that we used in Section 1.1 to refer to the first clause of a conditional, i.e., to the **X** of the conditional **X→Y**. To avoid confusion, we will refer to the antecedents of an inferential step as *the inference's antecedents*; and in order to distinguish the conclusion of an inferential step from the (final) conclusion of the demonstration, we will refer to it as *the inference's conclusion*. The inferential license explains how this conclusion is inferred. It comprises references to the inference's antecedents and the name of the rule of inference that is used to infer this conclusion from those antecedents. Formal derivations, on the other hand, consist solely of symbolic sentences. In practice, we will write a formal derivation as a series of inferential steps. However, these inferential steps are for our consumption; they are not proper parts of the derivation. The derivation itself consists only of symbolic sentences. These sentences are either premises, or assumptions and conclusions that are licensed by specific formal rules of inference. All the additional information required to recognize an inferential step—information such as the number associated with a stage or its inferential license—is extraneous to the derivation: it is added for the benefit of the agent who is constructing or reading a derivation. This information tells the agent the place of a sentence in a derivation and the inferential license used to justify the introduction of this sentence at that stage of the derivation.

Assume that **L** is a symbolic logical system and **DS** (for "Deduction System") is a collection of formal rules of inference. Let Γ be a set of **L** sentences and **X** an **L** sentence.

Definition of Formal Derivation: An **L** derivation of **X** from Γ is a finite sequence of **L** sentences such that the last sentence of the sequence is **X** and every sentence in the sequence is either a member of Γ or is licensed by one of the rules in **DS**.

---

16 It is very common to use the word 'proof' in the sense of a demonstrative proof. In this book, we follow this common practice: almost always we refer to demonstrative proofs as simply "proofs."

If there is an **L** derivation of **X** from Γ, we say that **X** is **derivable** from Γ in **L** or that **X** is a **theorem** of Γ in **L**. Symbolically, we write 'Γ ⊢_L **X**'. In this chapter, **L** is PL. Since PL is our default system, we will write '⊢' instead of '⊢_PL'. The collection of inference rules that we will use in this Section is called the **Natural Deduction System** (NDS) and is described in Subsection 1.4.5.

A formal derivation is a sequence of sentences that typically begins with a list of premises and terminates with the desired conclusion. Following the standard practice in logic, the premises are listed at the start of the derivation, which we also call **the zero stage**. The sentences at the non-zero stages of a derivation are licensed by specific rules of inference. These rules are formal, in the sense that their applicability is determined by the **forms** of the inference's antecedents or the inference's conclusion. The form of a sentence may be defined as the *syntactical structure* of the sentence. This is not a precise definition. According to this definition, a sentence may have several forms. For instance, we can recognize at least four syntactical structures for the PL sentence $(Pa \rightarrow Gb) \vee \neg Ka$: (1) **X**∨**Y**, (2) **X**∨¬**Y**, (3) (**X**→**Y**)∨**Z**, and (4) (**X**→**Y**)∨¬**Z**. The first syntactical structure of this sentence is the coarsest and the other three are more refined. In applying formal inference rules, in most cases 'the form of a sentence or a formula' refers to its coarsest syntactical structure: the main connective or operator together with the immediate component(s).

1.4.2   We state two important theorems that link the proof theory and the semantics of PL. Here is the first theorem.

The Soundness Theorem for PL: For every set Γ of PL sentences and every sentence **X** of PL, if **X** is a theorem of Γ, then **X** is a logical consequence of Γ. Symbolically, if Γ ⊢ **X**, then Γ ⊨ **X**.

The inference rules of NDS are **truth-preserving**, that is, if the inference's antecedents are true on any PL interpretation I, then the inference's conclusion is also true on I. Logicians describe a formal inference rule that is truth-preserving as **sound**. The Soundness Theorem establishes not only that every NDS rule of inference is truth-preserving but also that every PL derivation is truth-preserving. Proving that every rule is truth-preserving is only part of proving that every derivation is truth-preserving. Every derivation involves a combination of inference rules applied in some order. To show that the proof theory of PL is sound is to show that no matter how these rules are combined the resulting derivations are truth-preserving. Later in this book, we will present a complete proof of the Soundness Theorem for PL.

Now we state the second theorem.

The Completeness Theorem for PL: For every set Γ of PL sentences and every sentence **X** of PL, if **X** is a logical consequence of Γ, then **X** is a theorem of Γ. Symbolically, if Γ ⊨ **X**, then Γ ⊢ **X**.

The Completeness Theorem establishes that the NDS rules of inference are **adequate**. The adequacy of these rules means that if a sentence of PL is a logical consequence of a set of premises, then it is possible to produce a demonstrative proof that establishes this fact and that can be represented by a formal derivation. A completeness theorem for any logical system has no significance without an accompanying soundness theorem. To see the point, suppose that we have only one rule of inference: for any sentence **X**, infer **X**. It is a silly rule, but it is complete; it allows us to derive any logical consequence since it allows us to derive any sentence whatsoever. Any proof theory that contains this rule is totally insignificant because it is unsound. It is clear that this rule is not truth-preserving. For instance, it licenses the inference of $Pa \land \neg Pa$, which is a contradiction. The difficult task is to construct a proof theory that is both sound and complete. The Soundness and Completeness Theorems for PL demonstrate that NDS accomplishes this task. They show that the rules of NDS are truth-preserving and adequate. We will devote a large portion of this book to a proof of the Completeness Theorem for PL that is due to the American logician Leon Henkin (1921–2006).[17]

There is a very significant implication of these theorems: the relation of logical consequence in PL is **formalizable**. In order to explain this fact, we first combine these two theorems below.

The Soundness and Completeness Theorems for PL: For every set Γ of PL sentences and every PL sentence **X**, $\Gamma \vDash X$ if and only if $\Gamma \vdash X$.

The 'if' direction is the Soundness Theorem, and the 'only-if' direction is the Completeness Theorem. The combined statement above shows that the notions of logical consequence and derivability are equivalent in PL: every logical consequence of Γ is derivable from it, and every sentence that is derivable from Γ is a logical consequence of it. In other words, the logical consequences of Γ are precisely its theorems. The notion of derivability is a formal notion. Whether **X** is derivable from Γ or not depends solely on the forms of the members of Γ and **X**. To say that a certain logical notion is formalizable is to say that it is equivalent to a formal notion. Since the notion of logical consequence in PL is equivalent to the notion of derivability and since the latter is formal, the notion of logical consequence in PL is formalizable.

This fact justifies our description of PL as a formal logic. Not every logical system is formal in this sense. Second-Order Predicate Logic (PL²), which we will describe later, is not a formal system since its notion of logical consequence is not formalizable. It is known that no "reasonable" collection of sound rules of inference for PL² can be complete. We will prove this fact in Chapter Five. This fact is sometimes described by saying that PL² is an "incomplete" logic or that it is an "essentially semantical" logic. The idea is that reasoning about logical consequence in PL² must make use of semantical notions, such as PL²

---

17   The Austrian logician Kurt Gödel (1906–78) proved the Completeness Theorem in 1929. Henkin gave a different proof of the same theorem in 1949. Henkin's proof is easier to describe, and its central ideas can be applied to other systems.

interpretation and truth; this reasoning cannot be reduced in general to PL² derivations or facts about PL² derivations.

1.4.3 The Soundness and Completeness Theorems for PL have many important **corollaries**. Some of these corollaries link certain logical concepts to the notion of derivability. In this subsection we list four corollaries. But first we introduce some terminology. A PL sentence that is derivable from the empty set is called a **logical theorem**. An **inconsistent** set of PL sentences is a set from which a sentence and its negation are derivable. A **consistent** set of PL sentences is a set that is not inconsistent. Observe that this use of the terms 'inconsistent' and 'consistent' is at odds with how they are used in many introductory logic books, including my *An Introduction to Logical Theory*. In these books 'inconsistent' means unsatisfiable and 'consistent' means satisfiable. But in metalogic books the terms 'inconsistent' and 'consistent' are used in their proof-theoretic sense. If a sentence X is derivable from a sentence Y, and Y is derivable from X, we say that X and Y are **interderivable**. Now we state the corollaries.

1.4.3a A set of PL sentences is **unsatisfiable** if and only if it is **inconsistent.**

1.4.3b Two PL sentences are **logically equivalent** if and only if they are **interderivable**.

1.4.3c A PL sentence is **valid** if and only if it is a **logical theorem**.

1.4.3d A PL sentence is **contradictory** if and only if a sentence and its negation are derivable from it.

We will prove that all of these statements follow from the Soundness and Completeness Theorems. Since all of these statements are biconditionals, we need to prove two parts for each one: an 'if' direction and an 'only-if' direction.

Proof of 1.4.3a: the 'if' direction

1 Let Γ be an inconsistent PL set, that is, Γ ⊢ X and Γ ⊢ ¬X, for some PL sentence X.

2 From 1 by the Soundness Theorem: Γ ⊨ X and Γ ⊨ ¬X.

3 Reductio Assumption: Γ is satisfiable.

4 From 3 by the definition of satisfiability: there is a PL interpretation I that is a model of Γ.

5 If I is not relevant to X, we expand I into I* such that I* interprets all the extra vocabulary in X without changing any of the semantical assignments made by I. Since I is a model of Γ, I* is also a model of Γ.

6 From 2 and 5 by the definition of logical consequence: X and ¬X are true on I*, which is impossible.

7 From 3 through 6: the Reductio Assumption is false; hence Γ is unsatisfiable.

Proof of 1.4.3a: the 'only-if' direction
1. Let Γ be an unsatisfiable set of PL sentences, and let **X** be any PL sentence.
2. From 1 by the definition of unsatisfiability: there is no PL interpretation that satisfies Γ.
3. In order for **X** not to be a logical consequence of Γ, **X** must be false on a PL interpretation that satisfies Γ.
4. From 2 and 3: Γ ⊨ **X** and Γ ⊨ ¬**X**.
5. From 4 by the Completeness Theorem: Γ ⊢ **X** and Γ ⊢ ¬**X**.
6. From 5 by the definition of inconsistency: Γ is inconsistent.

Proof of 1.4.3b: the 'if' direction
1. Suppose that **X** and **Y** are any interderivable PL sentences, that is, {**X**} ⊢ **Y** and {**Y**} ⊢ **X**.[18]
2. From 1 by the Soundness Theorem: {**X**} ⊨ **Y** and {**Y**} ⊨ **X**.
3. Let I be any PL interpretation for **X** and **Y** on which **X** is true.
4. From 2 ({**X**} ⊨ **Y**) and 3 by the definition of logical consequence: **Y** is true on I too.
5. Now let I be a PL interpretation for **X** and **Y** on which **X** is false.
6. From 2 ({**Y**} ⊨ **X**) and 5 by the definition of logical consequence: **Y** is false on I.
7. From 3 through 6: **X** and **Y** have identical truth values on every PL interpretation for them.
8. From 7 by the definition of logical equivalence: **X** and **Y** are logically equivalent.

Proof of 1.4.3b: the 'only-if' direction
1. Suppose that **X** and **Y** are logically equivalent PL sentences.
2. From 1 by the definition of logical equivalence: there is no PL interpretation on which **X** is true and **Y** is false, or **Y** is true and **X** is false.
3. From 2 by the definition of logical consequence: {**X**} ⊨ **Y** and {**Y**} ⊨ **X**.
4. From 3 by the Completeness Theorem: {**X**} ⊢ **Y** and {**Y**} ⊢ **X**.
5. From 4 by the definition of interderivable sentences: **X** and **Y** are interderivable.

Proof of 1.4.3c: the 'if' direction
1. Let **X** be any PL sentence that is a logical theorem, that is, ∅ ⊢ **X** ('∅' is the symbol for the empty set).
2. From 1 by the Soundness Theorem: ∅ ⊨ **X**.
3. Let I be any PL interpretation for **X**. Since there is no sentence in ∅ that is false on I (because there are no sentences in ∅), I satisfies ∅.

---

18. The notation {**X**} designates the set whose sole member is **X**.

4   From 2 and 3 by the definition of logical consequence: **X** is true on I.
5   From 3 and 4: since I is arbitrary, **X** is true on every PL interpretation for it.
6   From 5 by the definition of valid sentence: **X** is valid.

Proof of 1.4.3c: the 'only-if' direction
1   Let **X** be any valid PL sentence.
2   From 1 by the definition of valid sentence: **X** is true on every PL interpretation for it.
3   By the reasoning in step 3 of the preceding proof: every PL interpretation satisfies ∅.
4   From 1 and 2 by the definition of logical consequence: ∅ ⊨ **X** (since every PL interpretation for **X** that satisfies ∅ makes **X** true).
5   From 4 by the Completeness Theorem: ∅ ⊢ **X**.
6   From 5 by the definition of logical theorem: **X** is a logical theorem.

Proof of 1.4.3d: the 'if' direction
1   Let **Y** be a PL sentence such that {**Y**} ⊢ **X** and {**Y**} ⊢ ¬**X**.
2   From 1 by the Soundness Theorem: {**Y**} ⊨ **X** and {**Y**} ⊨ ¬**X**.
3   Reductio Assumption: there is a PL interpretation I on which **Y** is true.
4   If I is not relevant to **X**, we expand I into I* such that I* interprets all the extra vocabulary in **X** without changing any of the semantical assignments made by I. Since **Y** is true on I, it is true on I*.
5   From 2 and 4 by the definition of logical consequence: **X** and ¬**X** are true on I*, which is impossible.
6   From 3 through 5: the Reductio Assumption is false; and hence there is no PL interpretation on which **Y** is true.
7   From 6 by the definition of contradictory sentence: **Y** is contradictory.

Proof of 1.4.3d: the 'only-if' direction
1   Let **Y** be any contradictory PL sentence and **X** any PL sentence.
2   From 1 by the definition of contradictory sentence: there is no PL interpretation on which **Y** is true.
3   In order for **X** not to be a logical consequence of **Y**, it must be false on a PL interpretation on which **Y** is true.
4   From 2 and 3: {**Y**} ⊨ **X** and {**Y**} ⊨ ¬**X**.
5   From 4 by the Completeness Theorem: {**Y**} ⊢ **X** and {**Y**} ⊢ ¬**X**.

1.4.4   The **rules of inference** we adopt in this book are divided into three groups: normal rules, hypothetical rules, and replacement rules. In the next subsection we will describe a relatively large collection of rules. Many of these rules can be derived from other rules. We could have elected to give a "minimal" collection of rules, in the sense that these rules are adequate and none of them could be derived from the others. While such a strategy gives us fewer rules, in practice it is more difficult to construct derivations using very few rules. We would like to have rules that correspond to our natural modes of reasoning; and the fewer rules we have, the fewer modes of reasoning we can represent in our proof theory. On the other hand, if we have too many rules, it would become difficult to manage and remember all of these rules. The rules included in our deduction system are all traditional ones, which can be found in many logic books. In Subsection 1.4.7, however, we will describe a deduction system that has a relatively small collection of rules, from which all the rules presented in the next subsection can be derived. That system has philosophical significance.

Before delving into the rules in the following subsection, we will illustrate how to interpret these schematic descriptions with a few examples. We begin with two normal rules.

Constructive Dilemma (CD)

[n
  ⋮
  h $X \vee Y$
  ⋮
  i $X \to Z_1$
  ⋮
  j $Y \to Z_2$
  ⋮
  k $Z_1 \vee Z_2$ h, i, j, CD
  ⋮
n]

Disjunctive Syllogism (DS)

[n
  ⋮
  h $X \vee Y$
  ⋮
  i $\neg X$ (or $\neg Y$)
  ⋮
  k Y (or X)   h, i, DS
  ⋮
n]

The letters 'n', 'h', 'i', 'j', and 'k' are replaced with numerals in actual derivations. The lines h, i, and j are the inference's antecedents and the line k is the inference's conclusion. To the right of the inference's conclusion is the inferential license. Thus the license "h, i, j, CD" abbreviates "From lines h, i, and j by Constructive Dilemma," and "h, i, DS" says "From lines h and i by Disjunctive Syllogism." The metalinguistic variables 'X', 'Y', 'Z_1', and 'Z_2' stand for any PL sentences. The order of the lines h, i, and j is irrelevant. The fact that line k comes after the other lines is essential to the correctness of the rule. The parenthetical clause '(or ¬Y)' represents an alternative antecedent and the clause '(or X)' represents the inference's conclusion given the alternative antecedent. Thus the schematic description of DS really represents two rules. The brackets to the left represent the block within which the rule is applied. Normal rules must be applied within open blocks only. The meaning of this and the use of blocks will be explained later. We only note here that a left bracket followed by a numeral n represents the first line of the nth block, and a numeral n followed by a right bracket represents the last line of the nth block. Prior to '[n' the nth block has not been opened yet and after 'n]' the nth block is closed.

Below is an example of how CD and DS may be applied in an actual derivation.

[2
⋮
6     $Ps \rightarrow (\exists y)By$
7     $\neg(\forall x)\neg Dx$
8     $Ps \vee (Rab \wedge Qc)$
9     $(Rab \wedge Qc) \rightarrow (\forall x)\neg Dx$
10    $(\exists y)By \vee (\forall x)\neg Dx$ 6, 8, 9, CD
11    $(\exists y)By$ 7, 10, DS
⋮
2]

In this example the portion of the derivation displayed is part of the second block. Observe that the rules are applied fully within an open block: this block is opened at some point prior to the 6th line and is closed after the 11th line. CD is applied at line 10. It has three antecedents: two conditionals of the form X→Z_1 and Y→Z_2 on lines 6 and 9 and a disjunction of the form X∨Y on line 8. These are the forms of the antecedents required for an application of CD. The conclusion of CD is a sentence of the form $Z_1 \vee Z_2$, which is the sentence that occurs on line 10. On line 11, DS is applied. It has two antecedents: a disjunction on line 10 and the negation of one of the disjuncts on line 7. The conclusion of CD is the other disjunct, which is the sentence that occurs on line 11. It is clear from this example that any conclusion we infer may be used as an antecedent for a later inference *provided that they all occur in an open block.*

We now consider an example of the following replacement rules as they may be applied in an actual derivation.

De Morgan's Laws (DeM):[19] $\neg(X \wedge Y) \Leftrightarrow \neg X \vee \neg Y$
$\neg(X \vee Y) \Leftrightarrow \neg X \wedge \neg Y$

Material Conditional (MC): $X \rightarrow Y \Leftrightarrow \neg X \vee Y$

[1
⋮
6    $Ps \rightarrow (\forall x)\neg(Rab \wedge Dx)$
7    $Ps \rightarrow (\forall x)(\neg Rab \vee \neg Dx)$ 6, DeM
8    $Ps \rightarrow (\forall x)(Rab \rightarrow \neg Dx)$ 7, MC
9    $\neg Ps \vee (\forall x)(Rab \rightarrow \neg Dx)$ 8, MC
⋮
1]

The double arrow '⇔' in the descriptions of the replacement rules indicates that each side might be replaced by the other. Replacement rules are powerful inference rules because, unlike normal and hypothetical rules, they can be applied to formulas and sentences, and to whole lines and their parts. Normal and hypothetical rules can only be applied to whole lines (and hence they can only be applied to sentences). We observe that DeM was applied not to the whole conditional on line 6 but only to part of its consequent, which is an open formula. We replaced $\neg(Rab \wedge Dx)$ with $(\neg Rab \vee \neg Dx)$, and copied the rest of the sentence as is, obtaining the sentence on line 7. On line 8 we made an inference using MC by applying the rule to $(\neg Rab \vee \neg Dx)$, which is only part of line 7. The rule allows us to replace $(\neg Rab \vee \neg Dx)$ with $(Rab \rightarrow \neg Dx)$. Note that we applied the reverse direction of MC (i.e., $X \rightarrow Y \Leftarrow \neg X \vee Y$). Finally, we applied MC to the whole sentence on line 8, replacing the conditional $Ps \rightarrow (\forall x)(Rab \rightarrow \neg Dx)$ with the disjunction $\neg Ps \vee (\forall x)(\neg Rab \rightarrow Dx)$, which is the sentence on line 9. As the case with normal rules, replacement rules must be applied within open blocks.

The last example we discuss is of a hypothetical rule. Below are the rule of Conditional Proof and an application of it.

Conditional Proof (CP): 'CPA' stands for 'Conditional Proof Assumption'.

[n
⋮
[n+1   h   X   CPA
⋮
n+1]   k   Y
k+1 $X \rightarrow Y$ h–k, CP
⋮
n]

---

[19] These laws are named after their discoverer, the British mathematician and logician Augustus De Morgan (1806–71).

```
[1
    ⋮
 [2  i    Ps                          CPA
         ⋮
 2]  k    (∀x)¬(Rab∧Dx)
    k+1   Ps→(∀x)¬(Rab∧Dx)    i–k, CP
    ⋮
1]
```

Blocks are opened at two stages only: the zero stage and a stage at which the assumption of a hypothetical rule occurs. The block that is opened at the zero stage is the **0-block** and it encloses the **main derivation**. The 0-block closes at the conclusion of the derivation. A block that is opened due to the application of some hypothetical rule closes at a line permitted by that specific rule. The inference's conclusion of a hypothetical rule is stated immediately after the line at which its block is closed. The inference's antecedent of a hypothetical rule is the derivation enclosed in its block. Blocks are numbered according to the order in which they are opened. Thus if the last block opened carries the numeral n, the next block to be opened carries the numeral n+1. A block B* is a **subblock** of a block B if and only if B* is opened after B is opened and before B is closed. Hence all the blocks of a derivation, other than the 0-block itself, are subblocks of the 0-block. Blocks that are ordered in a series $B_1, B_2, B_3, ..., B_n$ such that B_{k+1} is a subblock of B_k, for all k = 1, 2, 3, ..., n–1, are called **nested blocks**. *Nested blocks are stacks*, that is, the last block to be opened is the first to be closed. Hence a block can be closed only if all its subblocks are closed. A very important fact to remember about subblocks is that once a subblock is exited, the hypothetical rule that initiated that block permits a *one-time* reference to the subderivation enclosed in that subblock; after that none of the lines of this subderivation can serve as an antecedent for any subsequent rule. In other words, once a subblock is exited and a reference is made to its lines as dictated by the hypothetical rule that initiated the subblock, the lines of the subblock become *inaccessible* to all subsequent stages of the derivation. The assumption of a hypothetical rule is "active" only within the block of its rule. After the block is exited the assumption becomes "inactive"; we say in this case that the assumption has been **discharged**; so an "active" assumption is an **undischarged** assumption. A discharged assumption no longer has a role in the derivation. Since all the subblocks of a block must be closed before the block can be closed, all the assumptions of the hypothetical rules invoked in some derivation must be discharged before inferring the conclusion of the main derivation.

The rule CP allows us to start a new block with any assumption we may want. There is no restriction on the form of the CP Assumption. A CP block may be closed at any line after its initial line. There is no restriction on the form of the last line of a CP block. However, the inference of CP is highly restricted: it is the conditional whose antecedent is the CP Assumption and whose consequent is the last line of the CP block. In the example above, a CP block was opened at line i with the CP Assumption '*Ps*'. The CP block is the

2nd block. It closes at line k with the line '$(\forall x)\neg(Rab \wedge Dx)$'. The CP inference is the conditional $Ps \rightarrow (\forall x)\neg(Rab \wedge Dx)$ on line k+1. The license "i–k, CP" abbreviates "From the derivation on lines i–k by Conditional Proof." Of course, the derivation on lines i–k is the derivation enclosed in the 2nd subblock.

1.4.5 The **Natural Deduction System (NDS)** we adopt in this book consists of seventeen **normal rules**, three **hypothetical rules**, and thirteen **replacement rules**.

1.4.5a   Normal Rules: *Every normal rule must be applied fully within an open block.* **X, Y, Z, Z_1, and Z_2** are PL sentences (or, where noted, PL formulas). In each of the rules of this category the order of the lines h, i, and j is irrelevant; the only order that matters is that line k comes after lines h, i, and j.

Reiteration (**Reit**):

```
[n
    ⋮
    h      X
    ⋮
    k      X      h, Reit
    ⋮
 n]
```

Conjunction (**Conj**):

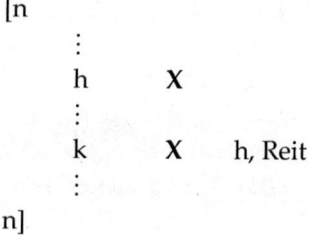

```
[n
    ⋮
    h      X
    ⋮
    i      Y
    ⋮
    k      X∧Y    h, i, Conj
    ⋮
 n]
```

Simplification (**Simp**): This is a two-part rule.

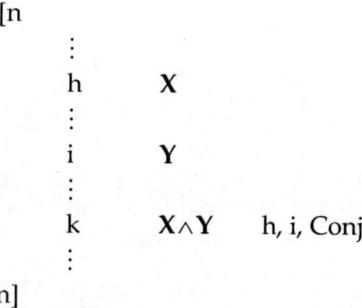

```
[n
    ⋮
    h      X∧Y
    ⋮
    k      X (or Y)   h, Simp
    ⋮
 n]
```

Addition (**Add**): This is a two-part rule.

    [n
        ⋮
        h   **X**
        ⋮
        k   **X**∨**Y** (or **Y**∨**X**)   h, Add
        ⋮
    n]

Disjunctive Syllogism (**DS**): This is a two-part rule.

    [n
        ⋮
        h   **X**∨**Y**
        ⋮
        i   ¬**X** (or ¬**Y**)
        ⋮
        k   **Y** (or **X**)   h, i, DS
        ⋮
    n]

Modus Ponens (**MP**):

    [n
        ⋮
        h   **X**→**Y**
        ⋮
        i   **X**
        ⋮
        k   **Y**   h, i, MP
        ⋮
    n]

Modus Tollens (**MT**)

[n
⋮
    h   X→Y
⋮
    i   ¬Y
⋮
    k   ¬X   h, i, MT
⋮
n]

Biconditional Modus Ponens (**BcMP**): This is a two-part rule.

[n
⋮
    h   X↔Y
⋮
    i   X (or Y)
⋮
    k   Y (or X)   h, i, BcMP
⋮
n]

Biconditional Modus Tollens (**BcMT**): This is a two-part rule.

[n
⋮
    h   X↔Y
⋮
    i   ¬X (or ¬Y)
⋮
    k   ¬Y (or ¬X)   h, i, BcMT
⋮
n]

FIRST-ORDER PREDICATE LOGIC (PL)

Hypothetical Syllogism (**HS**)

```
[n
  ⋮
  h      X→Y
  ⋮
  i      Y→Z
  ⋮
  k      X→Z       h, i, HS
  ⋮
n]
```

Constructive Dilemma (**CD**)

```
[n
  ⋮
  h      X∨Y
  ⋮
  i      X→Z₁
  ⋮
  j      Y→Z₂
  ⋮
  k      Z₁∨Z₂     h, i, j, CD
  ⋮
n]
```

where the subscripts are rendered as $X \to Z_1$, $Y \to Z_2$, and $Z_1 \vee Z_2$.

Explosion (**Expl**): **Y** is any PL sentence.

```
[n
  ⋮
  h      X∧¬X
  ⋮
  k      Y          h, Expl
  ⋮
n]
```

Universal Instantiation (**UI**): **s** is any PL *singular term*, **z** is a PL variable, and **X** is a PL formula that contains occurrences of **z** but no **z**-quantifiers. **X[s]** is the PL sentence formed by replacing *all* the occurrences of **z** in **X** by **s**.

[n
    ⋮
    h   (∀z)X
    ⋮
    k   X[s]   h, UI
    ⋮
n]

Universal Generalization (**UG**): **X** is a PL sentence, **s** is a PL *name* that occurs in **X**, and **z** is a PL variable that does not occur in **X**. **s** is **arbitrary** at line h, that is, **s** does not occur in any premise or undischarged assumption listed on line h or prior to it. **X[z]** is the PL formula formed by replacing *all* the occurrences of **s** in **X** by **z**.

[n
    ⋮
    h   X
    ⋮
    k   (∀z)X[z]   h, UG
    ⋮
n]

Existential Generalization (**EG**): **X** is a PL sentence, **s** is a PL *singular term* that occurs in **X**, and **z** is a PL variable that does not occur in **X**. **X[z, s]** is a PL formula formed by replacing *one or more* of the occurrences of **s** in **X** by **z**.

[n
    ⋮
    h   X
    ⋮
    k   (∃z)X[z, s] h, EG
    ⋮
n]

Identity (**Id**): **r** is any PL *singular term*. Id does not require an antecedent.

[n
    ⋮
    k   **r = r**   Id
    ⋮
n]

Substitution (**Sub**): **s** and **t** are PL *singular terms* and **X** is a PL sentence that contains occurrences of **s**. **X[t, s]** is a PL sentence formed by replacing *one or more* of the occurrences of **s** in **X** by **t**. This is a two-part rule.

```
[n
         ⋮
         h        s = t (or t = s)
         ⋮
         i        X
         ⋮
         k        X[t, s]       h, i, Sub
         ⋮
n]
```

1.4.5b Hypothetical Rules: Every hypothetical rule starts a new block and adds an assumption, and it terminates with exiting the block and discharging the assumption. **X** and **Y** are PL sentences (or, where noted, PL formulas). In these rules the order of the lines h and k is relevant and, therefore, must be observed.

Reductio Ad Absurdum (**RAA**): 'RA' stands for 'Reductio Assumption'; this is a two-part rule.

```
     [n
              ⋮
[n+1     h        X (or ¬X)         RA
              ⋮
         k–1      Y
  n+1]   k        ¬Y
         k+1      ¬X (or X)         h–k, RAA
              ⋮
     n]
```

Conditional Proof (**CP**): 'CPA' stands for 'Conditional Proof Assumption'.

```
     [n
              ⋮
[n+1     h        X           CPA
              ⋮
  n+1]   k        Y
         k+1      X→Y         h–k, CP
              ⋮
     n]
```

Existential Instantiation (**EI**): **s** is a PL *name*, **z** is a PL variable, and **X** is a PL formula that contains occurrences of **z** but no **z**-quantifiers. **s** satisfies three conditions: (1) it does not occur in any premise or undischarged assumption prior to line h, (2) it does not occur in (∃z)X, and (3) it does not occur in Y. X[s] is the PL sentence formed by replacing *all* the occurrences of **z** in **X** by **s**. 'EIA' stands for 'Existential Instantiation Assumption'.

```
[n
    ⋮
         h−1      (∃z)X
[n+1  h         X[s]       EIA, s
    ⋮
 n+1]  k         Y
         k+1     Y          h−1, h−k, EI
    ⋮
 n]
```

1.4.5c   **Replacement Rules**: *Every one of these rules must be applied fully within an open block.* Replacement rules may be applied to a complete sentence or to a component of a sentence. All replacements may be performed in the forward or reverse direction. **X**, **Y**, and **Z** are PL sentences or PL formulas.

Double Negation (**DN**):	¬¬X	⇔	X
Idempotence (**Idem**):	X∧X	⇔	X
	X∨X	⇔	X
Commutation (**Com**):	X∧Y	⇔	Y∧X
	X∨Y	⇔	Y∨X
Association (**Assoc**):	X∧(Y∧Z)	⇔	(X∧Y)∧Z
	X∨(Y∨Z)	⇔	(X∨Y)∨Z
Distribution (**Dist**):	X∧(Y∨Z)	⇔	(X∧Y)∨(X∧Z)
	X∨(Y∧Z)	⇔	(X∨Y)∧(X∨Z)
De Morgan's Laws (**DeM**):	¬(X∧Y)	⇔	¬X∨¬Y
	¬(X∨Y)	⇔	¬X∧¬Y
Material Conditional (**MC**):	X→Y	⇔	¬X∨Y
Negated Conditional (**NC**):	¬(X→Y)	⇔	X∧¬Y

Contraposition (**Cont**):	$X \rightarrow Y$	$\Leftrightarrow$	$\neg Y \rightarrow \neg X$
Exportation (**Expr**):	$X \rightarrow (Y \rightarrow Z)$	$\Leftrightarrow$	$(X \wedge Y) \rightarrow Z$
Biconditional (**Bc**):	$X \leftrightarrow Y$	$\Leftrightarrow$	$(X \rightarrow Y) \wedge (Y \rightarrow X)$
Negated Biconditional (**NBc**):	$\neg(X \leftrightarrow Y)$	$\Leftrightarrow$	$\neg X \leftrightarrow Y$
	$\neg(X \leftrightarrow Y)$	$\Leftrightarrow$	$X \leftrightarrow \neg Y$
Negated Quantifiers (**NQ**):	$\neg(\forall z)X$	$\Leftrightarrow$	$(\exists z)\neg X$
	$\neg(\exists z)X$	$\Leftrightarrow$	$(\forall z)\neg X$

Replacement rules are unique in their applicability to components of sentences and formulas. The rest of the NDS rules must only be applied to whole sentences that occupy complete lines. Applications that violate this condition might not be sound. Consider, for instance, the following three inferences using normal rules.

1. From $(Ae \wedge Be) \rightarrow Ce$ by Simplification (applied to $Ae \wedge Be$): $Ae \rightarrow Ce$.
2. From $Pa \wedge \neg(\forall x)Rx$ by Universal Instantiation (applied to $(\forall x)Rx$): $Pa \wedge \neg Ra$.
3. From $\neg(Pa \rightarrow Ra)$ by Existential Generalization (applied to Ra): $\neg(Pa \rightarrow (\exists x)Rx)$.

All of these inferences are unsound. If we assume that Ae is true and Be and Ce are false, $(Ae \wedge Be) \rightarrow Ce$ is true and $Ae \rightarrow Ce$ is false. Hence the first inference is not truth-preserving. For the second inference, let a stand for John, P for being tall, and R for being taller than 6 feet. Suppose that John is actually taller than 6 feet. The first sentence says that John is tall and not everyone is taller than 6 feet, which is true, and the second says that John is tall but he is not taller than 6 feet, which is false. So the inference is not truth-preserving. For the last inference, let a stand for number 7, P for being odd, and R for being divisible by 2. The first sentence asserts that it is not the case that if 7 is odd then it is divisible by 2, which is true, and the second asserts that it is not the case that if 7 is odd then there is a number that is divisible by 2, which is clearly false. Again, this inference is not truth-preserving. These examples are sufficient to show that if a normal (or a hypothetical) rule is applied to a component of a sentence, the inference might not be sound.

1.4.6   All of the normal, hypothetical, and replacement rules can be justified on semantical grounds. More precisely, it can be shown that all of them are sound on the basis of the truth conditions of the connectives and quantifiers that appear in these rules. For instance, the soundness of the rules Conjunction and Simplification is an immediate consequence of the truth conditions of the conjunction: if **X** and **Y** are true on a PL interpretation I, then so is $X \wedge Y$; and if $X \wedge Y$ is true on I, then so are **X** and **Y**. In the next chapter, we will present an economical version of PL and NDS, which is equivalent to the full version given in this chapter. In Chapter Three we will prove the Soundness and Completeness Theorems for PL using this economical version.

In this subsection we would like to take a closer look at the rule Explosion. The justification for Explosion is less straightforward and less intuitive than the rest of the normal rules. Each one of these rules of inference is truth-preserving, which means that any PL interpretation on which the inference's antecedents are true makes the inference's conclusion true as well. But the inference's antecedent of Explosion is a sentence of the form $X \land \neg X$, which is a contradiction. So it is unclear what it means to say that if the antecedent is true, then the conclusion is true as well. I will give two justifications for this rule: the first justification is the standard one, but the second is more intuitive even though it is a form of "logic fiction." Let us examine carefully the definition of "truth-preserving." To simplify matters, we will use the notation "$J(\Sigma) = T$" to mean that every member of the set Σ is true on J (i.e., J is a model of Σ). We assume that Σ consists of the antecedents of some inference. We have the following definition:

An inference $\Sigma \vdash Y$ is **truth-preserving** if and only if for every PL interpretation J for Σ and Y, if $J(\Sigma) = T$, then $J(Y) = T$.

In the case of Explosion, Σ consists of only one member, which is a sentence of the form $X \land \neg X$. All PL interpretations make this sentence false, that is, for every PL interpretation J for X, $J(X \land \neg X) = F$. If the English conditional in the statement above (i.e., if $J(\Sigma) = T$, then $J(Y) = T$) is given the same truth conditions as the connective '$\rightarrow$', then this conditional is true for any given PL interpretation. It follows that the following universal conditional is true: for every PL interpretation J for X and Y, if $J(X \land \neg X) = T$, then $J(Y) = T$, no matter what Y is. The truth of the conditional is guaranteed by the fact that its antecedent is contradictory. I stated previously that according to the truth table for the connective '$\rightarrow$', a PL conditional is true if its antecedent is false. Since the true conditional "if $J(X \land \neg X) = T$, then $J(Y) = T$" is the right-hand side of the definition above, hence the inference $\{X \land \neg X\} \vdash Y$ is a truth-preserving inference according to that definition. An important observation to make is that the correctness of this reasoning depends on attributing truth conditions to the English conditional above, which is the right-hand side of the definition, similar to the truth conditions of the connective '$\rightarrow$'. This, in fact, is a standard feature of the metatheory of classical logic. The connectives of the metatheory, such as 'if-then' and 'or', have the same truth conditions as their formal counterparts, such as '$\rightarrow$' and '$\lor$'.

The second justification presupposes an impossible state of affairs: there is a PL interpretation that makes a sentence of the form $X \land \neg X$ true. This, of course, is logic fiction, for the semantics of PL does not allow for the existence of such PL interpretation. But we will indulge ourselves in assuming that such a PL interpretation exists. We call it O. If Explosion is a sound rule, then *every* sentence Y that is interpretable by O must be true on O. This is because Explosion licenses the inference of any sentence from a contradiction. We will show that this is the case by invoking a famous argument that is due to the American philosopher and logician C.I. Lewis (1883–1964).

Lewis's Argument for Explosion
1. Suppose that a sentence of the form $X \land \neg X$ is true on some PL interpretation O.
2. From 1: since the inference rule Simplification is sound, X is true on O.
3. From 2: since the inference rule Addition is sound, $X \lor Y$ is true on O, for any sentence Y that is interpretable by O.
4. From 1: since the inference rule Simplification is sound: $\neg X$ is true on O.
5. From 3 and 4: since the inference rule Disjunctive Syllogism is sound, Y is true on O, which is the desired conclusion.

This argument is standardly presented without the logic fiction. It is employed to show that the inference rule Explosion is derivable from the rules of Simplification, Addition, and Disjunctive Syllogism. The argument shows that if these three rules are justified, then Explosion is equally justified. We modified the setting of the argument to show that if these three rules are truth-preserving, then Explosion is also truth-preserving without assuming that the English conditional has the same truth conditions as the conditional connective '$\to$'. Lewis's original argument is stronger: it aims to show that if we have reason to believe that Simplification, Addition, and Disjunctive Syllogism are correct rules of inference, then we have no option but to accept the correctness of Explosion. Counterarguments against Lewis's argument have traditionally focused on showing that Disjunctive Syllogism is unjustifiable. We will not be able to consider any of these counterarguments here, for they rely on logical resources that are beyond the scope of this book.

1.4.7 The Natural Deduction System (NDS) we described in 1.4.5 is not the **standard deduction system** that is given in many logic textbooks. The standard deduction system for PL consists of seventeen rules only. Our system contains thirty-three rules (some of which are two-part rules). All these rules are traditional rules of inference. There is a philosophical reason for including almost all the traditional rules of formal classical logic in our NDS. Different logical systems might question the justification of some of these rules. Not all of these rules are included in the standard deduction system, which we will describe in this subsection. To be sure, many logical systems focus their disagreement with classical logic on some of the rules of the standard deduction system. But these rules do not exhaust all the debates. For instance, there are non-classical logical systems that challenge the correctness of Disjunctive Syllogism. As we saw, Disjunctive Syllogism is employed in an argument that is meant to justify Explosion. There are many non-classical systems that reject Explosion. Thus these systems must address the correctness of Disjunctive Syllogism. Disjunctive Syllogism, however, is not part of the standard deduction system. A possible reaction to this point is that although Disjunctive Syllogism is not part of the standard rules of inference, it can be derived from them. Hence anyone who rejects Disjunctive Syllogism must identify the standard rules that give rise to Disjunctive Syllogism and explain what is wrong with them. This is a reasonable response, as all the rules we included in our NDS are derivable from the standard rules, which we will describe in this subsection. However, there is still value to NDS that is not fully exhibited by the standard rules. Let us continue with our example of DS. One might be willing to reject some of the standard rules that give rise to DS but still accept DS on independent grounds. Hence someone who rejects Lewis's argument for Explosion would have to address the correctness of DS independently of the standard rules of inference.

The standard deduction system was discovered by the German mathematician and logician Gerhard Gentzen (1909–45). We will refer to it as the **Gentzen Deduction System** (GDS). The reader, however, must be warned that many logic textbooks use the title "natural deduction system" to describe this system instead of the one we introduced as NDS in the preceding subsection. This system consists of seventeen *primitive* rules: Reiteration and two rules for each connective and quantifier and two rules for the identity predicate. The pair of rules that is associated with each of these eight logical symbols consists of a rule that *introduces* the symbol into its conclusion and a rule that *eliminates* the symbol from its conclusion. Accordingly, these rules are called "introduction and elimination rules." They are abbreviated as #I and #E, respectively, where # is replaced by a sentential connective, a quantifier, or the identity predicate. We have already studied several of these rules in our NDS.

We list the rules of GDS. (1) Reiteration. (2–3) **Conjunction Introduction** ($\wedge$I) is Conj, and **Conjunction Elimination** ($\wedge$E) is Simp. (4–5) **Conditional Introduction** ($\rightarrow$I) is CP, and **Conditional Elimination** ($\rightarrow$E) is MP. (6–7) **Universal Introduction** ($\forall$I) is UG, and **Universal Elimination** ($\forall$E) is UI. (8–9) **Existential Introduction** ($\exists$I) is EG, and **Existential Elimination** ($\exists$E) is EI. (10–11) **Identity Introduction** (=I) is Id, and **Identity Elimination** (=E) is Sub. (12–13) **Negation Introduction** ($\neg$I) is the first part of RAA (where the conclusion is $\neg$X), and **Negation Elimination** ($\neg$E) is the second part of RAA (where the conclusion is **X**). (14–15) **Disjunction Introduction** ($\vee$I) is Add, and **Disjunction Elimination** ($\vee$E) is the following hypothetical rule (j is an integer greater than 1):

```
[n              
        ⋮
        h       X∨Y
[n+1    h+1     X           ∨E Assumption
        ⋮
n+1]    m       Z
[n+j    m+1     Y           ∨E Assumption
        ⋮
n+j]    k       Z
        k+1     Z           h, (h+1)–m, (m+1)–k, ∨E
        ⋮
n]
```

(16–17) **Biconditional Elimination** (↔E) is BcMP, and **Biconditional Introduction** (↔I) is the following hypothetical rule (j is an integer greater than 1):

```
[n      ⋮
[n+1    h      X           ↔I Assumption
                ⋮
 n+1]   m      Y
[n+j    m+1    Y           ↔I Assumption
                ⋮
 n+j]   k      X
        k+1    X↔Y         h–m, (m+1)–k, ↔I
                ⋮
 n]
```

These are the seventeen rules of GDS. All the NDS rules introduced in Subsection 1.4.5 can be derived from these seventeen rules. The reverse is also true. All the GDS rules are either NDS rules or can be derived from NDS rules. Thus from a logical point of view the two systems, NDS and GDS, are equivalent. The first philosophically significant characteristic of GDS is that it focuses the debate between systems of non-classical logic and classical logic on some of these introduction and elimination rules. The second philosophically significant characteristic of GDS is that the GDS rules may be considered as "defining" the sentential connectives, quantifiers, and the identity predicate. We defined all the logical symbols of PL semantically by giving their truth conditions. We then invoke these truth conditions to justify the NDS rules. The Soundness Theorem for PL offers complete semantical justifications for all the NDS and GDS rules. There is, however, a different approach to the "nature" of these logical symbols. Rather than defining them semantically by giving their truth conditions, this approach defines the logical symbols of PL by giving their proof-theoretic roles, that is, their introduction and elimination rules. For example, the conditional may be defined as the binary sentential connective that can be introduced into any conclusion by Conditional Proof (i.e., Conditional Introduction) and that can be eliminated from any conclusion by Modus Ponens (i.e., Conditional Elimination). This formal approach to the logical symbols of PL reverses the direction of justification of the semantical approach. The main point is that the meaning of these logical symbols can be "fixed" by their introduction and elimination rules.[20]

---

20 We say "fixed" with caution, since it has been pointed out by many logicians and philosophers that the introduction and elimination rules are not strong enough to fix the semantics of these logical symbols uniquely. However, a defender of the formal approach most likely would not be moved by this objection, since she might insist that whatever semantical significance these introduction and elimination rules attribute to the logical symbols is the entirety of their semantical content. If this content is not fixed uniquely, then this fact is part of the semantical significance of these logical symbols.

1.5 Exercises

Notes: All answers must be adequately justified. The solutions to the starred exercises are appended to this chapter.

1.5.1* Construct two PL interpretations of the set below such that they are substantively different from each other, each one of them is a model of the set, neither assigns the empty extension to any predicate, and one is finite and the other is infinite.

$Hm \wedge Sb$
$(\forall x)(Gx \rightarrow Hx)$
$(\forall x)(Rx \rightarrow Gx)$
$\neg(\exists z)(Sz \wedge Rz)$

1.5.2 Give three PL interpretations of the following set of PL sentences such that the first satisfies the set and is of size 3, the second also satisfies the set but it is of an infinite size, and the third is of size 4 and satisfies only the first four sentences. Is the last sentence a logical consequence of the other sentences?

$(\forall x)((x = a \vee x = b) \rightarrow Lx)$
$a \neq b$
Pe
$Oea \wedge Oeb$
$(\forall z)((Pz \wedge (Oza \wedge Ozb)) \rightarrow z = e)$

1.5.3 Demonstrate that '$s = t$' is not a logical consequence of the set Γ whose members are the following PL sentences.

$(\forall x)(Ax \rightarrow \neg Gx)$
$(\exists x)Ax \leftrightarrow Gs$
$(\exists w)\neg Gw \rightarrow Kt$
$(\exists x)(Gx \wedge Kx)$

1.5.4* Let Γ be the set consisting of the following three PL sentences.

$(\forall x)\neg Sox$
$(\forall x)(\exists u)((u \neq x \wedge Sux) \wedge (\forall z)(Szx \rightarrow z = u))$
$(\forall x)(x \neq o \rightarrow (\exists u)Sxu)$

1.5.4a Construct a model of Γ that is of size of 3.
1.5.4b Construct a model of Γ that is of size $\aleph_0$.
1.5.4c Is it possible for Γ to have models of size 1 or 2? Explain.

1.5.5   Show that the following sets of PL sentences are consistent.

1.5.5a* $(\forall y)(Py \rightarrow (\exists x)(\exists z)Bxyz)$
$(\forall x)(\forall y)(\forall z)((Bxyz \land Px) \rightarrow (Py \land Pz))$
$\neg Pn$
$(\exists x)(\exists z)Bxnz$

1.5.5b $(\neg Ha \lor \neg Kb) \rightarrow a = gb$
$a \neq gb$
$(\exists x)(a = gx \land x = gb)$
$(\forall x)(\forall y)((x = gy \land (x = a \land y = b)) \rightarrow (Sax \land Syb))$
$(\forall u)(Sau \rightarrow Hu)$
$(\forall z)(Sbz \rightarrow Kz)$

1.5.5c $(\forall x)Gox$
$(\forall x)(x \neq o \rightarrow ((\exists y)Gxy \land (\exists v)Lxv))$
$(\forall x)(\forall y)(\forall z)((Gxy \land Lzy) \rightarrow x \neq z)$
$(\forall x)(\forall y)((\exists z)(Gxz \land Lyz) \rightarrow Gxy)$

1.5.6   Establish that each of the sets below is satisfiable. Explain why all the models of these sets are infinite.

1.5.6a* $(\forall x)(\forall y)(\forall z)((Kxy \land Kyz) \rightarrow Kxz)$
$(\forall x)(\exists y)Kxy$
$(\forall x)(\forall y)(Kxy \rightarrow \neg Kyx)$[21]

1.5.6b $(\forall x)(\forall y)(\forall z)((Gxz \land Gyz) \rightarrow x = y)$
$(\forall x)(\exists y)Gxy$
$(\forall x)\neg Gxe$

1.5.7   Construct appropriate PL interpretations to establish that the PL sentences in each of the pairs below are not interderivable.

1.5.7a* $(\exists v)(Hv \land Fv)$ and $(\exists v)Hv \land (\exists v)Fv$
1.5.7b $(\forall u)(Nu \rightarrow Ca)$ and $(\forall u)Nu \rightarrow Ca$
1.5.7c $(\forall w)(Dw \lor Sw)$ and $(\forall w)Dw \lor (\forall w)Sw$

1.5.8   Show that the PL sentences below are contingent.

1.5.8a* $(Em \land Pn) \rightarrow (\exists y)(Ey \land Py)$
1.5.8b $(\forall x)(\exists y)Bxy \land \neg(\exists y)(\forall x)Bxy$

---

[21] This sentence could be replaced with a simpler one: $(\forall x)\neg Kxx$. The resulting set is also satisfiable and all its models are infinite.

1.5.9   Give PL derivations to establish the following.

1.5.9a $\{\neg Bs \land (At \leftrightarrow Bs), Dr \rightarrow At\} \vdash \neg Dr$
1.5.9b* $\{(\neg At \lor Bs) \rightarrow Dr, At \rightarrow Eg, \neg Eg\} \vdash Dr$
1.5.9c $\{\neg At \rightarrow (Dr \lor Hd), At \rightarrow Eg, \neg Eg, \neg Hd\} \vdash Dr$
1.5.9d* $\{\neg At \land (Dr \lor Hd), At \leftrightarrow Eg, Hd \rightarrow Eg\} \vdash Dr \land \neg At$
1.5.9e $\{At \land (Dr \rightarrow Hd), At \leftrightarrow \neg Eg, Hd \rightarrow Eg\} \vdash \neg Dr \land At$
1.5.9f* $\{At \lor (Dr \leftrightarrow \neg Hd), Sm \land Kc, Sm \rightarrow (Kc \rightarrow \neg At), (Sm \lor Rn) \rightarrow \neg \neg Hd\} \vdash$
          $\neg Dr \land (Kc \rightarrow \neg At)$
1.5.9g $\{At \lor Dr, Sm \land Kc, Sm \rightarrow (At \rightarrow (Fq \land Rn)), Kc \rightarrow (Dr \rightarrow \neg Lo)\} \vdash (Fq \land Rn) \lor \neg Lo$
1.5.9h $\{\neg At \lor Bs, Bs \rightarrow \neg(Cp \rightarrow Cp)\} \vdash \neg At$
1.5.9i* $\{(Rn \land \neg Lo) \rightarrow \neg Nj, Nj \leftrightarrow Rn, (Me \land \neg Lo) \rightarrow Nj, Me \lor Nj\} \vdash Lo$
1.5.9j $\{\neg Ae, (\forall v)(Bv \rightarrow Av), (\forall x)(Bx \lor Dx)\} \vdash (\exists z)Dz$
1.5.9k* $\{Rst, (\forall u)(\forall w)((Ruw \land Gwu) \rightarrow u = w), (\forall x)(\forall z)(Rxz \leftrightarrow Gzx)\} \vdash s = t$
1.5.9l $\{Rst, (\forall u)(\forall w)((Ruw \land Gwu) \rightarrow u = w), (\forall x)(\forall z)(Rxz \leftrightarrow Gzx)\} \vdash (\exists y)Gyy$
1.5.9m $\{(\forall x)\neg Kxx, (\forall x)(\forall w)(Gwx \rightarrow Kwx), (\forall z)(Gzz \lor Dz)\} \vdash Db$
1.5.9n $\{(\forall x)\neg Kxx, (\forall x)(\forall w)(Gwx \rightarrow Kwx), (\forall z)(Gzz \lor Dz)\} \vdash (\forall x)Dx$
1.5.9o* $\{(\forall x)(\forall y)(x = y \rightarrow \neg Sxy), Pe, (\forall w)((Pw \land Lw) \rightarrow Swe)\} \vdash \neg Le$
1.5.9p* $\{(\forall x)(x = q \lor Nxq), (\forall x)(\forall z)(Nxz \rightarrow x = z)\} \vdash (\forall x)(\forall y)x = y$
1.5.9q $\{(\forall z)(Jz \lor \neg Hz), (\forall z)(Jz \rightarrow z \neq d)\} \vdash \neg Hd$
1.5.9r* $\{(\exists u)Au \rightarrow (\exists u)Bu, \neg(\exists z)(Az \land Bz), Ac\} \vdash (\exists x)(\exists y)x \neq y$

1.5.10   Give PL derivations to demonstrate the truth of the following statements.

1.5.10a* The PL sentence '$(\forall u)(Rub \land \neg Qu) \land (\neg Qa \rightarrow \neg Rab)$' is contradictory.
1.5.10b   The PL sentence '$(\forall z)(Lz \rightarrow (\exists x)(Mx \land Ezx)) \land ((\forall x)(\forall y)\neg Exy \land (\exists x)Lx)$' is contradictory.
1.5.10c   The PL sentence '$(\forall x)(\forall y)((Dxgy \rightarrow Agx) \land Drgs) \rightarrow (\exists z)(z = gr \land Az)$' is valid.
1.5.10d* The PL sentence '$\neg(\exists v)Mvv \rightarrow (\forall x)(\forall y)(\neg Mxy \lor x \neq y)$' is valid.
1.5.10e   The PL sentences '$(\forall z)(Ezz \lor (\exists v)(Ezv \land z \neq v))$' and '$(\forall z)(\exists v)Ezv$' are logically equivalent.

1.5.11   Give PL derivations to show that the following sets of PL sentences are unsatisfiable.

1.5.11a $\{(\forall x)(Ux \rightarrow (\forall z)Cxz), (\forall x)(\forall y)(Cxy \rightarrow (x \neq y \lor \neg Uy)), (\exists w)Uw\}$
1.5.11b $\{(\forall x)(\forall y)(Kxy \rightarrow \neg Kyx), (\exists w)(\forall y)Kwy\}$
1.5.11c* $(\forall u)(Du \lor Eu), (\exists v)(\exists x)Qvx, (\forall x)(\forall y)(Qxy \rightarrow (Ax \lor By)),$
          $(\forall z)(Az \rightarrow (\neg Dz \land \neg Ez)), (\forall z)(Bz \rightarrow (\neg Dz \land \neg Ez))\}$

1.5.12   Prove that the arguments below are valid.

1.5.12a*

S1 $\neg(\forall x)(Rx \to x = j)$
S2 $(\forall x)((Rx \land Tx) \leftrightarrow x = j)$

S3 $(\exists x)(Rx \land \neg Tx)$

1.5.12b

S1 Lm
S2 Sb
S3 $(\forall x)(Lx \to Gx)$
S4 $(\forall y)(Gy \to Ry)$
S5 $(\forall z)(Sz \to \neg Rz)$

S6 $m \neq b$

1.5.12c

S1 $(\forall z)(Rz \to Az)$

S2 $(\forall x)(\forall z)((Rx \land Hzx) \to (\exists y)(Ay \land Hzy))$

1.5.12d*

S1 $(\forall y)((Ciy \land \neg Ay) \to Riy)$
S2 $\neg Ad$
S3 $(\forall y)(Riy \to Ey)$
S4 $Lib \to Cib$
S5 $Cib \to Cid$
S6 $Ed \to Fb$

S7 $Lib \to Fb$

1.5.12e

S1 $(\forall x)(Rx \leftrightarrow (\forall y)(Sxy \leftrightarrow ((My \land Ly) \land \neg Syy)))$
S2 $(\forall z)(Rz \to Mz)$
S3 $(\forall v)(Rv \to Lv)$

S4 $\neg(\exists x)Rx$

1.5.12f

S1 $(\forall x)(\forall y)(\neg Mxy \to \neg Rxy)$
S2 $(\forall z)(Mzj \to (\forall v)(Cv \to Fzv))$
S3 $(\exists w)(Cw \land (Ptw \lor Ltw))$
S4 $(\exists w)(Cw \land Ptw)) \to (\exists v)(Cv \land \neg Ftv)$
S5 $(\exists w)(Cw \land Ltw)) \to (\exists v)(Cv \land \neg Ftv)$

S6 $\neg Rtj$

1.5.12g*

S1 $(\forall x)(Exx \lor (\exists y)(Exy \land x \neq y))$
S2 $(\forall z)(Sz \leftrightarrow Ezz)$
S3 $(\forall z)(Mz \leftrightarrow (\exists v)(Ezv \land z \neq v))$
S4 $(\forall x)(Mx \to (\exists y)(Sy \land Exy))$

S5 $(\exists w)Sw$

1.5.12h

S1 $(\forall x)(\exists y)Exy$
S2 $(\forall z)(Sz \leftrightarrow Ezz)$
S3 $(\forall z)(Mz \leftrightarrow (\exists v)(Ezv \land z \neq v))$
S4 $(\forall x)(Mx \to (\exists y)(Sy \land Exy))$

S5 $(\exists w)Sw$

1.5.12i

S1 $(\forall x)(\forall y)((Hx \land Cyx) \to Ry)$
S2 $(\forall x)(Rx \to (\forall y)(Ixy \to Ry))$
S3 $(\forall x)(Bx \to (\exists y)(My \land Cyx))$
S4 $(\forall x)(Mx \to (\exists y)(Ey \land Ixy))$
S5 $(\forall x)(Bx \to Hx)$
S6 Bq

S7 $(\exists x)(Ex \land Rx)$

1.5.12j*

S1 Tn
S2 $e = gn \land Rpe$
S3 Dp
S4 $(\forall x)(\forall y)((Mxy \land Dx) \to Ly)$
S5 $(\forall x)(\forall y)((x = gy \land Lx) \to \neg Ty)$

S6 $\neg(\forall x)(\forall y)(Rxy \leftrightarrow Mxy)$

1.5.12k

S1 $(\forall x)\neg Mex$
S2 $(\forall x)(\forall y)(\forall z)((Mzx \land Mzy) \to x = y)$
S3 $(\forall w)(Tw \to (\exists z)(Mwz \land Tz))$
S4 $(\neg Te \land (\forall y)(\forall z)((Myz \land \neg Tz) \to \neg Ty)) \to (\forall x)\neg Tx$

S5 $(\forall x)\neg Tx$

Solutions to the Starred Exercises

SOLUTION TO 1.5.1

The PL Interpretation I

UD: {Amber, Deborah, Rubin, Jolene, Augustine, Colin}
LN: b, m, d, r, j, a, c

Semantical Assignments

$I(b)$: Amber; $I(m)$: Amber; $I(d)$: Deborah; $I(r)$: Rubin; $I(j)$: Jolene; $I(a)$: Augustine; $I(c)$: Colin
$I(Hx)$: "x is a physics major": {Amber, Rubin, Jolene, Augustine}
$I(Sy)$: "y plays basketball": {Amber, Colin, Deborah}
$I(Gy)$: "y is interested in astronomy": {Rubin, Jolene, Augustine}
$I(Rz)$: "z owns a telescope": {Jolene}

Natural reading of $(Hm \land Sb)$ on I: *Amber is a physics major and plays basketball.*
True on I.

Natural reading of $(\forall x)(Gx \to Hx)$ on I: *All those who are interested in astronomy are physics majors.*
True on I.

Natural reading of $(\forall x)(Rx \to Gx)$ on I: *Everyone who owns a telescope is interested in astronomy.*
True on I.

Natural reading of $\neg(\exists z)(Sz \land Rz)$ on I: *No one who plays basketball owns a telescope.*
True on I.

Since UD consists of six individuals, I is finite. Also, I is a model of the set.

The PL Interpretation J

UD: The set of all even natural numbers: {0, 2, 4, 6, ...}
LN: $m, b, c_0, c_1, c_2, c_3, \ldots c_k, \ldots$

Semantical Assignments

$J(m)$: 18; $J(b)$: 12; $J(c_0)$: 0; $J(c_1)$: 2; $J(c_2)$: 4; $J(c_3)$: 6; ...; $J(c_k)$: 2k; ...
$J(Hx)$: "x is a multiple of 6": {6, 12, 18, 24, 30, ...}
$J(Sy)$: "y is a divisor of 60": {2, 4, 6, 10, 12, 20, 30, 60}
$J(Gy)$: "y is a multiple of 12": {12, 24, 36, 48, 60, ...}
$J(Rz)$: "z is a multiple of 24": {24, 48, 72, 96, 120, ...}

Natural reading of $(Hm \land Sb)$ on J: *18 is a multiple of 6 and 12 is a divisor of 60.*
True on J.

Natural reading of $(\forall x)(Gx \rightarrow Hx)$ on J: *All multiples of 12 are multiples of 6.*
True on J.

Natural reading of $(\forall x)(Rx \rightarrow Gx)$ on J: *All multiples of 24 are multiples of 12.*
True on J.

Natural reading of $\neg(\exists z)(Sz \land Rz)$ on J: *No divisor of 60 is a multiple of 24.*
True on J.

Since UD consists of infinitely many numbers, J is infinite; and J is a model of the set.

SOLUTIONS TO 1.5.4

1.5.4a   Here is a PL interpretation I of size 3 that satisfies the set Γ.

UD: {Sarah, John, Robert}
LN: o, j, r

Semantical Assignments

$I(o)$: Sarah; $I(j)$: John; $I(r)$: Robert
$I(Sxy)$: "x is a half-brother of y"

Facts of I

1. John and Robert are half-brothers (they share the same father but have different mothers).
2. John is a half-brother of Sarah (they share the some mother but have different fathers).
3. Robert is not a brother of Sarah (they have different parents).

The natural reading of $(\forall x)\neg Sox$ on I: *Sara is not a half-brother of anyone.*
True on I.

The natural reading of $(\forall x)(\exists u)((u \neq x \land Sux) \land (\forall z)(Szx \rightarrow z = u))$ on I:

Every person has exactly one half-brother, who is different from him or her.
True on I.

The natural reading of $(\forall x)(x \neq o \rightarrow (\exists u) Sxu)$ on I: *Everyone who is not Sarah is a half-brother of someone.*
True on I.

1.5.4b The following PL interpretation J satisfies the set Γ and is of size $\aleph_0$.

UD: The set of natural numbers: {0, 1, 2, 3, ...}
LN: $o, a_1, a_2, a_3, \ldots, a_k, \ldots$

Semantical Assignments

$J(o)$: 0; $J(a_k)$: k, for every k = 1, 2, 3, ...
$J(Sxy)$: "x is the successor of y" (i.e., $x = y+1$)

The natural reading of $(\forall x)\neg Sox$ on J: *0 is not the successor of any natural number.*
True on J.

The natural reading of $(\forall x)(\exists u)((u \neq x \land Sux) \land (\forall z)(Szx \rightarrow z = u))$ on J:
Every natural number has exactly one successor, which is different from it.
True on J.

The natural reading of $(\forall x)(x \neq o \rightarrow (\exists u) Sxu)$ on J:
Every natural number that is not 0 is the successor of some natural number.
True on J.

1.5.4c Let K be any PL interpretation of **size 1** that is relevant to the set Γ. This means that there is only one individual α in UD. Since every name has a referent and o is a name in LN, α must be named o. On K, the sentence $(\forall x)(\exists u)((u \neq x \land Sux) \land (\forall z)(Szx \rightarrow z = u))$ is false since the following basic substitutional instance is false: $(\exists u)((u \neq o \land Suo) \land (\forall z)(Szo \rightarrow z = u))$. The latter is false on K because it implies that there is an individual *other than* α that bears the relation K(S) to α, and of course there is no such individual. Therefore, no interpretation of size 1 can satisfy the sentence $(\forall x)(\exists u)((u \neq x \land Sux) \land (\forall z)(Szx \rightarrow z = u))$, and hence no interpretation of size 1 can satisfy the set Γ.

Now let H be any PL interpretation of **size 2** that is relevant to the set Γ. The UD of H consists of only two individuals α and β. Since o is a name in LN, one of these individuals must be named o; say α is named o. β must also have a name; let us call it e (the actual name is irrelevant as long as it is not o). Suppose that H satisfies the sentence $(\forall x)(\exists u)((u \neq x \land Sux) \land (\forall z)(Szx \rightarrow z = u))$. Thus all the basic substitutional instances of this sentence are true on H. There are at least two basic substitutional instances: $(\exists u)((u \neq o \land Suo) \land (\forall z)(Szo \rightarrow z = u))$ and $(\exists u)((u \neq e \land Sue) \land (\forall z)(Sze \rightarrow z = u))$. The truth of these instances entails that there is an individual other than α that bears the relation H(S) to α and that there is an individual other than β that bears the relation H(S) to β. Given that we only have the two individuals α and β, it follows that β bears the relation H(S) to α and that α bears the relation H(S) to β. In other words, the ordered pairs ⟨β, α⟩ and ⟨α, β⟩ belong to the extension H(S). By the truth conditions of the atomic sentences, the sentences *Seo* and *Soe* must be true on H. Now assume that H also satisfies the sentence $(\forall x)\neg Sox$. This implies that the substitutional instance ¬*Soe* is true on H. Therefore both *Soe* and ¬*Soe* are true on H, which is impossible. It follows that if H makes the sentence $(\forall x)(\exists u)((u \neq x \land Sux) \land (\forall z)(Szx \rightarrow z = u))$ true, it cannot make $(\forall x)\neg Sox$ true as well. In other words, H cannot satisfy the set Γ. This establishes that no PL interpretation of size 2 can satisfy Γ.

SOLUTION TO 1.5.5

1.5.5a   The PL Interpretation I

UD: {–1, 0, 1, 2, 3, 4, 5, ...}
LN: $a, n, c_1, c_2, c_3, ..., c_k, ...$

Semantical Assignments

$I(a)$: –1; $I(n)$: 0; $I(c_k)$: k, for every k = 1, 2, 3, 4, 5, ...
$I(Pz)$: "z is positive"
$I(Bxyz)$: "y is between x and z" (i.e., $x < y < z$)

Natural Reading of $(\forall y)(Py \rightarrow (\exists x)(\exists z)Bxyz)$ on I: *Every positive number is between two numbers.*
True on I.

PL-English reading of $(\forall x)(\forall y)(\forall z)((Bxyz \land Px) \rightarrow (Py \land Pz))$ on I:
For all numbers x, y, and z, if y is between x and z and x is positive, then y and z are positive as well.
True on I.

Natural reading of ¬*Pn* on I: *Zero is not positive.*
True on I.

Natural reading of $(\exists x)(\exists z)Bxnz$ on I: *Zero is between two numbers.*
True on I.

Since all the sentences are true on I, the set is satisfiable. By Corollary 1.4.3a, the set is consistent.

SOLUTION TO 1.5.6

1.5.6a   We will prove that this set is satisfiable by describing a PL interpretation J on which every member of the set is true.

UD: The set that consists of infinitely many planets that orbit a black hole. These planets revolve around the black hole in circular orbits such that no two planets share the same orbit, and for any two planets one of them must be inside the orbit of the other. The picture below depicts the relations between these heavenly bodies.[22] (We are assuming, of course, that no planet is inside its own orbit.)

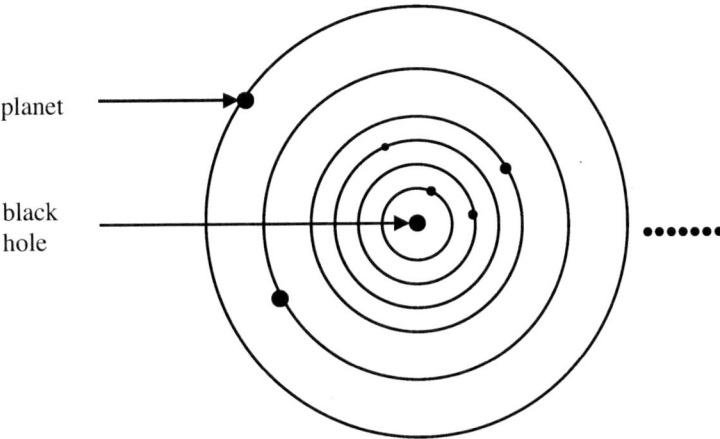

LN: $p_1, p_2, p_3, \ldots, p_k, \ldots$

Semantical Assignments

$J(p_1)$: The first planet from the black hole; $J(p_2)$: The second planet from the black hole; $J(p_3)$: The third planet from the black hole; ..., $J(p_k)$: The kth planet from the black hole
$J(Kxy)$: "x is inside the orbit of y"

Natural reading of $(\forall x)(\forall y)(\forall z)((Kxy \land Kyz) \rightarrow Kxz)$ on J:
For any three planets, if the first is inside the orbit of the second and the second is inside the orbit of the third, then the first is inside the orbit of the third.
True on J.

Natural reading of $(\forall x)(\exists y)Kxy$ on J: *Every planet is inside the orbit of some planet.*
True on J.

---

22 It might be objected that this PL interpretation is not physically possible since no black hole has a gravitational field that is so powerful as to keep infinitely many planets revolving in orbits around it. Besides, it is very likely that the universe contains only finitely many planets. However, physical possibility is not the same as logical possibility. There is no reason to suppose that logical possibilities must be constrained by the laws of physics or by physical facts. The only constraint on a logical possibility is that it must be consistent. The PL interpretation described above might not be physically possible but it appears to be perfectly consistent.

Natural reading of $(\forall x)(\forall y)(Kxy \rightarrow \neg Kyx)$ on J:
If a planet is inside the orbit of another planet, then the latter is not inside the orbit of the former.
True on J.

Let Φ be the set of the three PL sentences above. A very interesting fact about Φ is that every model of it is infinite. The set Φ consists of three sentences that involve a 2-place predicate K. Any PL interpretation that is relevant to Φ must assign a binary (i.e., 2-place) relation to K. Let M be any arbitrary PL interpretation that is a model of Φ. Thus M(K) is a binary relation on the UD of M. To simplify our notation, let us call this relation R. Since PL is an extensional system, we can think of M(K) as an extension whose members are ordered pairs. Intuitively speaking, an ordered pair $\langle \alpha, \beta \rangle$, where α and β are members of UD, belongs to the extension M(K) if and only if α bears the relation R to β. Rather than saying that $\langle \alpha, \beta \rangle$ belongs to the extension M(K), we will simply write 'α R β'.

At this stage, we do not know whether the extension M(K) is empty or not. What we know is that M is a model of Φ and that R is the binary relation that M assigns to the 2-place predicate K. Our goal is to show that the extension of R on M (i.e., M(K)) is an infinite set. This immediately entails that the UD of M is infinite; and hence M is an infinite PL interpretation.

Since M is a model of Φ, the sentences in Φ are all true on M. The truth of these sentences entails that R has certain properties. To be precise, R has three properties, each of which is entailed by one of the sentences in Φ. We discuss each one in turn.

Transitivity: For all α, β, and δ in UD, if α R β and β R δ, then α R δ. (R is said to be *transitive*.)

Extendibility: For every α in UD, there is β in UD, such that α R β. (R is said to be *extendible*.)

Asymmetry: For all α and β in UD, if α R β, then not-(β R α) (i.e., "it is not the case that β R α").[23] (R is said to be *asymmetric*.)

Transitivity follows from the truth of the sentence $(\forall x)(\forall y)(\forall z)((Kxy \wedge Kyz) \rightarrow Kxz)$ on M, extendibility from the truth of $(\forall x)(\exists y)Kxy$ on M, and asymmetry from the truth of $(\forall x)(\forall y)(Kxy \rightarrow \neg Kyx)$ on M.

We will now prove that R has an infinite extension. We begin by making a helpful observation. We can describe an asymmetric relation graphically by saying that such a relation does not allow "loops." To see this, we diagram α R

---

23 We included the clause 'in UD' for emphasis. It is understood that all quantifiers range over the UD.

β as two dots—one is α and the other is β—that are connected by an arrow, representing R and its direction from α to β. If we want, instead, to represent β R α, we draw the arrow from β to α. If we have α R β and β R α, then we obtain two arrows, one from α to β and the other from β to α, forming a loop. To say that R is asymmetric is to say that there are never such loops. Of course, there is also no loop between an individual and itself if R is asymmetric, since if α R α, then not-(α R α), which shows that it is impossible for α to bear the relation R to itself. Here are diagrams of a relation Z that exhibits two loops: δ Z δ; and α Z β and β Z α.

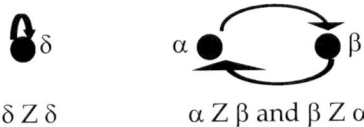

δ Z δ   α Z β and β Z α

As indicated above, an asymmetric relation cannot have any such loops. So it is clear that Z is not asymmetric.

As is the case with any PL interpretation, the UD of M cannot be empty. So there is at least one individual in UD; we call it ε_1. Since R is extendible, there is an individual in UD such that ε_1 bears the relation R to this individual. This individual cannot be ε_1 itself; otherwise we end up with the loop ε_1 R ε_1. Thus this individual must be something other than ε_1; call it ε_2. So far we have ε_1 R ε_2. Again, since R is extendible, ε_2 must relate to an individual in UD. It cannot relate to itself or to ε_1; for either case generates a loop. Hence it must relate to a third individual ε_3. We now have the following situation: ε_1 R ε_2 and ε_2 R ε_3. To simplify matters, we will write this sequence as 'ε_1 R ε_2 R ε_3'. Given that R is transitive, ε_1 bears the relation R to ε_3, i.e., ε_1 R ε_3. The diagram below describes the current situation.

Since R is extendible, ε_3 must relate to some individual. It cannot relate to itself, ε_2, or ε_1, because any such relation would create a loop. The only option we have is that ε_3 relates to a fourth individual ε_4. Thus we obtain that ε_1 R ε_2 R ε_3 R ε_4. By the transitivity of R, we have the following additional relations: ε_1 R ε_3, ε_1 R ε_4, ε_2 R ε_4. The diagram below describes all of the relations we have so far.

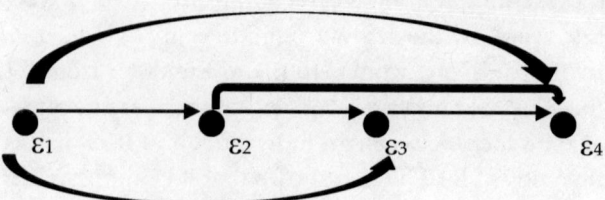

The same reasoning applies to ε₄. The property of extendibility forces ε₄ to bear R to some individual. This individual cannot be ε₄ or any of the previous individuals, since each one of these individuals relates to ε₄, and hence if ε₄ bears R to any one of them, a loop would be formed, which is ruled out by the asymmetry of R.

This pattern of reasoning applies to any finite sequence ε₁ R ε₂ R ε₃ R ε₄ R … R εₙ. The transitivity of R ensures that every εₖ bears the relation R to εₙ, where k is less than n–1. The extendibility of R requires εₙ to bear the relation R to some individual. But this individual cannot be ε₁, ε₂, ε₃, ε₄, …, or εₙ, since otherwise a loop would form. In other words, there are infinitely many individuals in UD that form an unending sequence "connected" by the binary relation R: ε₁ R ε₂ R ε₃ R ε₄ R … R εₙ R …. . For every εⱼ R εₖ, the ordered pair ⟨εⱼ, εₖ⟩ belongs to the extension of R on M. Thus the extension of R on M is infinite, which implies that UD is an infinite set. We conclude that M is infinite. But M was an arbitrary model of the set. Hence every model of this set is infinite.

Solution to 1.5.7

1.5.7a The following PL interpretation I makes the first sentence false and the second true. This proves that these sentences are not logically equivalent. Hence by Corollary 1.4.3b they are not interderivable.

The PL Interpretation I

UD: {Cow, Pig, Whale, Manatee}
LN: c, p, h, m

Semantical Assignments

$I(c)$: Cow; $I(p)$: Pig; $I(h)$: Whale; $I(m)$: Manatee
$I(Hx)$: "x is a farm animal": {Cow, Pig}
$I(Fv)$: "v is a marine mammal": {Whale, Manatee}

Natural reading of $(\exists v)(Hv \land Fv)$ on I: *There is a farm animal that is a marine mammal.*
False on I.

Natural reading of $(\exists v)Hv \wedge (\exists v)Fv$ on I: *There is a farm animal and there is a marine mammal.*
True on I.

SOLUTION TO 1.5.8

1.5.8a Consider the following two PL interpretations.

The PL Interpretation K

UD: {Minnesota, New Mexico, Arizona, Iowa}
LN: *m, n, a, i*

Semantical Assignments

K(*m*): Minnesota; K(*n*): New Mexico; K(*a*): Arizona; K(*i*): Iowa
K(*Ez*): "z is a Midwestern state": {Minnesota, Iowa}
K(*Py*): "y is a Southwestern state": {New Mexico, Arizona}

Natural reading of $(Em \wedge Pn) \rightarrow (\exists y)(Exy \wedge Py)$ on K:
If Minnesota is a Midwestern state and New Mexico is a Southwestern state, then there is a Midwestern state that is also Southwestern.
The sentence is false on K since its antecedent is true and its consequent is false.

PL Interpretation H

UD: {Minnesota, Wisconsin, Iowa, Arizona}
LN: *m, n, i, a*

Semantical Assignments

H(*m*): Minnesota; H(*n*): Wisconsin; H(*i*): Iowa; H(*a*): Arizona
H(*Ez*): "z is an agricultural state": {Minnesota, Wisconsin, Iowa}
H(*Py*): "y borders one of the Great Lakes": {Minnesota, Wisconsin}

Natural reading of $(Em \wedge Pn) \rightarrow (\exists y)(Ey \wedge Py)$ on H:
If Minnesota is an agricultural state and Wisconsin borders one of the Great Lakes, then there is an agricultural state that borders one of the Great Lakes.
This sentence is true on H since its antecedent and consequent are both true.

Given that $(Em \wedge Pn) \rightarrow (\exists y)(Ey \wedge Py)$ is false on one interpretation and true on another, it is contingent.

Solutions to 1.5.9

1.5.9b

[0	0.1	$(\neg At \vee Bs) \to Dr$	P
	0.2	$At \to Eg$	P
	0.3	$\neg Eg$	P
	1	$\neg At$	0.2, 0.3, MT
	2	$\neg At \vee Bs$	1, Add
0]	3	Dr	0.1, 2, MP

1.5.9d

[0	0.1	$\neg At \wedge (Dr \vee Hd)$	P
	0.2	$At \leftrightarrow Eg$	P
	0.3	$Hd \to Eg$	P
	1	$\neg At$	0.1, Simp
	2	$\neg Eg$	0.2, 1, BcMT
	3	$\neg Hd$	0.3, 2, MT
	4	$Dr \vee Hd$	0.1, Simp
	5	Dr	3, 4, DS
0]	6	$Dr \wedge \neg At$	1, 5, Conj

1.5.9f

[0	0.1	$At \vee (Dr \leftrightarrow \neg Hd)$	P
	0.2	$Sm \wedge Kc$	P
	0.3	$Sm \to (Kc \to \neg At)$	P
	0.4	$(Sm \vee Rn) \to \neg\neg Hd$	P
	1	Sm	0.2, Simp
	2	Kc	0.2, Simp
	3	$Kc \to \neg At$	0.3, 1, MP
	4	$\neg At$	2, 3, MP
	5	$Sm \vee Rn$	1, Add
	6	$\neg\neg Hd$	0.4, 5, MP
	7	$Dr \leftrightarrow \neg Hd$	0.1, 4, DS
	8	$\neg Dr$	6, 7, BcMT
0]	9	$\neg Dr \wedge (Kc \to \neg At)$	3, 8, Conj

1.5.9i

```
[0    0.1   (Rn∧¬Lo)→¬Nj          P
      0.2   Nj↔Rn                  P
      0.3   (Me∧¬Lo)→Nj            P
      0.4   Me∨Nj                  P
[1    1     ¬Lo                    RA
[2    2     Rn                     RA
      3     Rn∧¬Lo                 1, 2, Conj
      4     ¬Nj                    0.1, 3, MP
      5     ¬Rn                    0.2, 4, BcMT
2]    6     Rn∧¬Rn                 2, 5, Conj
      7     ¬Rn                    2–6, RAA
      8     ¬Nj                    0.2, 7, BcMT
      9     Me                     0.4, 8, DS
      10    Me∧¬Lo                 1, 9, Conj
      11    Nj                     0.3, 10, MP
1]    12    ¬Nj                    8, Reit
0]    13    Lo                     1–12, RAA
```

1.5.9k

```
[0    0.1   Rst                              P
      0.2   (∀u)(∀w)((Ruw∧Gwu)→u = w)        P
      0.3   (∀x)(∀z)(Rxz↔Gzx)                P
      1     (∀w)((Rsw∧Gws)→s = w)            0.2, UI
      2     (Rst∧Gts)→s = t                  1, UI
      3     (∀z)(Rsz↔Gzs)                    0.3, UI
      4     Rst↔Gts                          3, UI
      5     Gts                              0.1, 4, BcMP
      6     Rst∧Gts                          0.1, 5, Conj
0]    7     s = t                            2, 6, MP
```

1.5.9o

[0   0.1 $(\forall x)(\forall y)(x = y \rightarrow \neg Sxy)$ P
      0.2 Pe P
      0.3 $(\forall w)((Pw \wedge Lw) \rightarrow Swe)$ P
      1   $(\forall y)(e = y \rightarrow \neg Sey)$ 0.1, UI
      2   $e = e \rightarrow \neg See$ 1, UI
      3   $e = e$ Id
      4   $\neg See$ 2, 3, MP
      5   $(Pe \wedge Le) \rightarrow See$ 0.3, UI
      6   $\neg(Pe \wedge Le)$ 4, 5, MT
      7   $\neg Pe \vee \neg Le$ 6, DeM
      8   $\neg\neg Pe$ 0.2, DN
0]   9 $\neg Le$ 7, 8, DS

1.5.9p

[0   0.1 $(\forall x)(x = q \vee Nxq)$ P
      0.2 $(\forall x)(\forall z)(Nxz \rightarrow x = z)$ P
[1   1 $a \neq q$ RA
      2   $a = q \vee Naq$ 01, UI
      3   $(\forall z)(Naz \rightarrow a = z)$ 0.2, UI
      4   $Naq \rightarrow a = q$ 3, UI
      5   $\neg Naq$ 1, 4, MT
      6   $a = q$ 2, 5, DS
1]   7 $a \neq q$ 1, Reit
      8   $a = q$ 1–7, RAA
      9   $(\forall z)z = q$ 8, UG
      10   $b = q$ 9, UI
      11   $a = b$ 8, 10, Sub
      12   $(\forall y)a = y$ 11, UG
0]   13 $(\forall x)(\forall y)x = y$ 12, UG

1.5.9r

[0	0.1	$(\exists u)Au \to (\exists u)Bu$	P
	0.2	$\neg(\exists z)(Az \wedge Bz)$	P
	0.3	Ac	P
	1	$(\exists u)Au$	0.3, EG
	2	$(\exists u)Bu$	0.1, 1, MP
[1	3	Be	EIA, e
	4	$(\forall z)\neg(Az \wedge Bz)$	0.2, NQ
	5	$\neg(Ac \wedge Bc)$	4, UI
	6	$\neg Ac \vee \neg Bc$	5, DeM
	7	$\neg\neg Ac$	0.3, DN
	8	$\neg Bc$	6, 7, DS
[2	9	$c = e$	RA
	10	$\neg Be$	8, 9, Sub
2]	11	Be	3, Reit
	12	$c \neq e$	9–11, RAA
	13	$(\exists y)c \neq y$	12, EG
1]	14	$(\exists x)(\exists y)x \neq y$	13, EG
0]	15	$(\exists x)(\exists y)x \neq y$	2, 3–14, EI

SOLUTIONS TO 1.5.10

1.5.10a

[0	0.1	$(\forall u)(Rub \wedge \neg Qu) \wedge (\neg Qa \to \neg Rab)$	P
	1	$(\forall u)(Rub \wedge \neg Qu)$	0.1, Simp
	2	$\neg Qa \to \neg Rab$	0.1, Simp
	3	$Rab \wedge \neg Qb$	1, UI
	4	$Rab \to Qa$	2, Cont
0]	5	$\neg(Rab \to Qa)$	3, NC

Since a sentence and its negation are derivable from the sentence '$(\forall u)(Rub \wedge \neg Qu) \wedge (\neg Qa \to \neg Rab)$', it is contradictory by Corollary 1.4.3d.

1.5.10d

[0	0	∅	
[1	1	¬(∃v)Mvv	CPA
	2	(∀v)¬Mvv	1, NQ
	3	¬Maa	2, UI
[2	4	Mab	CPA
[3	5	a = b	RA
	6	Maa	4, 5, Sub
3]	7	¬Maa	3, Reit
2]	8	a ≠ b	5–7, RAA
	9	Mab → a ≠ b	4–8, CP
	10	¬Mab ∨ a ≠ b	9, NC
	11	(∀y)(¬May ∨ a ≠ y)	10, UG
1]	12	(∀x)(∀y)(¬Mxy ∨ x ≠ y)	11, UG
0]	13	¬(∃v)Mvv → (∀x)(∀y)(¬Mxy ∨ x ≠ y)	1–12, CP

Note on the zero stage of a derivation without premises: To derive **X** from the empty set is to derive it without premises. In such derivations, we indicate this fact by writing the symbol for the empty set, '∅', at the zero stage. We still start the main block with the line 0 but the actual first line of the derivation is line 1. Derivations without premises can only start by invoking the NDS rules Identity, Conditional Proof, or Reductio Ad Absurdum. All other rules require antecedents, and hence they are not suitable to initiate a derivation without premises. Since the sentence '¬(∃v)Mvv → (∀x)(∀y)(¬Mxy ∨ x ≠ y)' is derivable from the empty set, it is a logical theorem; and hence it is valid by Corollary 1.4.3c.

Solution to 1.5.11

1.5.11c

[0	0.1	$(\forall u)(Du \vee Eu)$	P
	0.2	$(\exists v)(\exists x)Qvx$	P
	0.3	$(\forall x)(\forall y)(Qxy \to (Ax \vee By))$	P
	0.4	$(\forall z)(Az \to (\neg Dz \wedge \neg Ez))$	P
	0.5	$(\forall z)(Bz \to (\neg Dz \wedge \neg Ez))$	P
	1	$(\exists v)(\exists x)Qvx$	0.2, Reit
[1	2	$(\exists x)Qsx$	EI, s
[2	3	Qst	EI, t
	4	$(\forall y)(Qsy \to (As \vee By))$	0.3, UI
	5	$Qst \to (As \vee Bt)$	4, UI
	6	$As \vee Bt$	3, 5, MP
	7	$As \to (\neg Ds \wedge \neg Es)$	0.4, UI
	8	$As \to \neg(Ds \vee Es)$	7, DeM
	9	$Ds \vee Es$	0.1, UI
	10	$\neg\neg(Ds \vee Es)$	9, DN
	11	$\neg As$	8, 10, MT
	12	Bt	6, 11, DS
	13	$Bt \to (\neg Dt \wedge \neg Et)$	0.5, UI
	14	$\neg Dt \wedge \neg Et$	12, 13, MP
	15	$\neg(Dt \vee Et)$	14, DeM
2]	16	$(\exists u)\neg(Du \vee Eu)$	15, EG
1]	17	$(\exists u)\neg(Du \vee Eu)$	2, 3–16, EI
	18	$(\exists u)\neg(Du \vee Eu)$	1, 2–17, EI
	19	$\neg(\forall u)(Du \vee Eu)$	18, NQ
0]	20	$(\forall u)(Du \vee Eu)$	0.1, Reit

Since a sentence and its negation are derivable from the set, it is inconsistent, and hence it is unsatisfiable by Corollary 1.4.3a.

Solutions to 1.5.12

All of these problems have the same justification: since the conclusion of the argument is derivable from its premises, the argument, by the Soundness Theorem, is valid.

1.5.12a

[0	0.1	$\neg(\forall x)(Rx \to x = j)$	P
	0.2	$(\forall x)((Rx \wedge Tx) \leftrightarrow x = j)$	P
	1	$(\exists x)\neg(Rx \to x = j)$	0.1, NQ
	2	$(\exists x)(Rx \wedge x \neq j)$	1, NC
[1	3	$Ra \wedge a \neq j$	EI, a
	4	Ra	3, Simp
	5	$a \neq j$	3, Simp
	6	$(Ra \wedge Ta) \leftrightarrow a = j$	0.2, UI
	7	$\neg(Ra \wedge Ta)$	5, 6, MT
	8	$\neg Ra \vee \neg Ta$	7, DeM
	9	$\neg\neg Ra$	4, DN
	10	$\neg Ta$	8, 9, DS
	11	$Ra \wedge \neg Ta$	4, 10, Conj
1]	12	$(\exists x)(Rx \wedge \neg Tx)$	11, EG
0]	13	$(\exists x)(Rx \wedge \neg Tx)$	2, 3–12, EI

1.5.12d

[0	0.1	$(\forall y)((Ciy \wedge \neg Ay) \to Riy)$	P
	0.2	$\neg Ad$	P
	0.3	$(\forall y)(Riy \to Ey)$	P
	0.4	$Lib \to Cib$	P
	0.5	$Cib \to Cid$	P
	0.6	$Ed \to Fb$	P
[1	1	Lib	CPA
	2	Cib	0.4, 1, MP
	3	Cid	0.5, 2, MP
	4	$(Cid \wedge \neg Ad) \to Rid$	0.1, UI
	5	$Cid \wedge \neg Ad$	0.2, 3, Conj
	6	Rid	4, 5, MP
	7	$Rid \to Ed$	0.3, UI
	8	Ed	6, 7, MP
1]	9	Fb	0.6, 8, MP
0]	10	$Lib \to Fb$	1–9, CP

1.5.12g

[0	0.1	$(\forall x)(Exx \lor (\exists y)(Exy \land x \neq y))$	P
	0.2	$(\forall z)(Sz \leftrightarrow Ezz)$	P
	0.3	$(\forall z)(Mz \leftrightarrow (\exists v)(Ezv \land z \neq v))$	P
	0.4	$(\forall x)(Mx \rightarrow (\exists y)(Sy \land Exy))$	P
	1	$Eii \lor (\exists y)(Eiy \land i \neq y))$	0.1, UI
[1	2	Eii	CPA
	3	$Si \leftrightarrow Eii$	0.2, UI
	4	Si	2, 3, BcMP
1]	5	$(\exists w)Sw$	4, EG
	6	$Eii \rightarrow (\exists x)Sx$	2–5, CP
[2	7	$(\exists y)(Eiy \land I \neq y)$	CPA
	8	$Mi \leftrightarrow (\exists v)(Eiv \land I \neq v)$	0.3, UI
	9	Mi	7, 8, BcMP
	10	$Mi \rightarrow (\exists y)(Sy \land Eiy)$	0.4, UI
	11	$(\exists y)(Sy \land Eiy)$	9, 10, MP
[3	12	$Sb \land Eib$	EIA, b
	13	Sb	12, Simp
3]	14	$(\exists w)Sw$	13, EG
2]	15	$(\exists w)Sw$	11, 12–14, EI
	16	$(\exists y)(Eiy \land i \neq y) \rightarrow (\exists w)Sw$	7–15, CP
	17	$(\exists w)Sw \lor (\exists w)Sw$	1, 6, 16, CD
0]	18	$(\exists w)Sw$	17, Idem

1.5.12j

[0	0.1	Tn	P
	0.2	$e = gn \land Rpe$	P
	0.3	Dp	P
	0.4	$(\forall x)(\forall y)((Mxy \land Dx) \to Ly)$	P
	0.5	$(\forall x)(\forall y)((x = gy \land Lx) \to \neg Ty)$	P
[1	1	$(\forall x)(\forall y)(Rxy \leftrightarrow Mxy)$	A
	2	$(\forall y)(Mpy \leftrightarrow Rpy)$	1, UI
	3	$Mpe \leftrightarrow Rpe$	2, UI
	4	Rpe	0.2, Simp
	5	Mpe	3, 4, BcMP
	6	$(\forall y)((Mpy \land Dp) \to Ly)$	0.4, UI
	7	$(Mpe \land Dp) \to Le$	6, UI
	8	$Mpe \land Dp$	0.3, 5, Conj
	9	Le	7, 8, MP
	10	$(\forall y)((e = gy \land Le) \to \neg Ty)$	0.5, UI
	11	$(e = gn \land Le) \to \neg Tn$	10, UI
	12	$e = gn$	0.2, Simp
	13	$e = gn \land Le$	9, 12, Conj
	14	$\neg Tn$	11, 13, MP
1]	15	Tn	0.1, Reit
0]	16	$\neg(\forall x)(\forall y)(Rxy \leftrightarrow Mxy)$	1–15, RAA

Chapter Two

Resources of the Metatheory

2.1 Linguistic and Logical Resources

2.1.1 We will spend the rest of this book studying the **metatheory** of First-Order Predicate Logic (PL). It is called a "metatheory" because it is a theory whose subject of study is a theory. We can think of a symbolic logical system as a collection of three theories: a syntactical theory that defines the symbolic language of the logical system, a semantical theory that assigns meanings and truth conditions to the sentences of the symbolic language and that defines the relation of logical consequence, and a proof theory that defines the relation of provability and that specifies the rules of inference allowed in that logical system. Some logical systems, such as Second-Order Predicate Logic, do not have sound and complete proof theories. Many logicians are inclined to say that such systems do not have proof theories, that they only have syntax and semantics. Others might be inclined to attribute a proof theory to every logical system. They might say that every logical system specifies a collection or collections of sound rules of inference, except that for some logical systems none of these collections is complete. It is not important to take sides on this issue. All that is required is to recognize that every symbolic logical system is a theory or a collection of theories. A theory in which a symbolic logical system is studied is the metatheory of that system. The theorems of the metatheory are called **metatheorems**. The metatheory of a logical system is also referred to as **metalogic**. In this book, when we use the term 'metatheory' and its concomitant terms, such as 'metalogic' and 'metatheorem', we always intend the metatheory of PL. We will frequently drop the prefix 'meta' when there is no cause for misunderstanding.

Some logicians offer an austere description of what a logical system is. For them, a logical system consists of a formal language and rules of inference. The problem with this conception of a logical system is that it does not comfortably accommodate logical systems that have no sound and complete set of inference rules. In a celebrated theorem the Swedish logician Per Lindström (1936–2009) characterizes a logical system in terms of its formal language and the notion of true-on-an-interpretation (precisely speaking, in terms of the relation of satisfaction between an interpretation and a formula). So this is a conception of a logical system that goes beyond the traditionally austere depiction.

In general, the metatheory of PL studies the properties and relations of the proof-theoretic, semantical, and logical concepts of PL. For example, while the fact that a certain sequence of PL sentences is a PL derivation of a sentence **X** from a set Γ belongs, on our conception, to PL proper, the assertion that this fact entails that this **X** is a logical consequence of Γ belongs to the metatheory of PL, since it describes a relation between two logical relations: provability and logical consequence. Hence the Soundness and the Completeness Theorems and all of their corollaries are metatheorems that are part of the metatheory of PL. All of these statements affirm certain relations between proof-theoretic and semantical concepts.

In the logical literature, the name 'proof theory' is standardly used to designate the branch of the metatheory in which the properties of formal derivations and their relations to semantical concepts are studied. The branch of the metatheory in which the semantics of a logical system is studied is typically referred to as **model theory**, since its main subject is the study of interpretations and models of collections of sentences of that logical system. It is important to emphasize that our usage of the term 'proof theory' in this book is ambiguous. We use it to refer to the proof-theoretic concepts and objects that belong to PL and to the metalogical study of these concepts and objects. Thus in Chapter One, we included under the heading 'Proof Theory' PL rules of inference and formal derivations as well as metatheorems, such as the Soundness and the Completeness Theorems. We rely on the context to disambiguate any specific use of the term 'proof theory'.

2.1.2   The language of the metatheory is **PL-English**, which is **English** augmented with **variables** and other **symbols**. For instance, we will have metalinguistic variables ranging over PL linguistic categories, such as terms, predicates, and sentences, and there will also be variables ranging over sets of sentences, PL interpretations, and PL derivations. Also we will use symbols designating some arithmetical and set-theoretical operations and functions. We will not at this point enumerate all the different types of variables and all the symbols we will use. We will introduce these variables and symbols and what they range over or designate as we need them.

The linguistic categories of PL have counterparts in the language of the metatheory of PL. For example, there are sentential connectives and quantifiers that correspond to the sentential connectives and quantifiers of PL, and there are singular terms, monadic and relational predicates, and function symbols as well. A salient feature of the metatheory of PL (and of many other systems of classical logic) is that the **truth conditions** of all these connectives and quantifiers are identical with the truth conditions of their formal counterparts. Hence, for instance, the 'if-then' connective at the meta-level is a material, truth-functional conditional: 'If A then B' is true if and only if either A is false or B is true. The same is true for negations, conjunctions, disjunctions, biconditionals, and the quantifiers. The biconditional 'if and only if' is usually abbreviated as 'iff'. We will follow this usage in the remainder of this book. The standard reading of the quantifiers at the meta-level is objectual and not substitutional. Parts of the vocabulary of the language of the metatheory are arithmetical and set-theoretic. This will become much clearer when we study the arithmetical and set-theoretic resources of the metatheory.

2.1.3   The metatheory has a good deal of **logical resources**. The most important part of metalogic is its metatheorems. These theorems require proofs, and proofs require logical resources, such as rules of inference. So what are the rules of inference that are available at the meta-level? The answer would seem at first surprising: all the rules of inference of the **Natural Deduction System** (NDS) described in Chapter One are part of the logical resources of the metatheory. Of course, these rules are modified to fit the language of the metatheory. In addition to these inference rules, the metatheory invokes rules that are particular to arithmetic and set theory. Traditionally these rules are not thought of as rules of inference but as arithmetical and set-theoretic principles, such as the Principle of Mathematical Induction and the Principle of Extensionality. Thus we will cover these principles when we discuss arithmetical and set-theoretic resources of the metatheory.

The presence of counterparts of the NDS rules of inference in the metatheory poses a philosophical problem. One of the important theorems that we will prove in this book is the Soundness Theorem. This theorem in a certain sense presents justifications for these rules: it shows that all PL derivations that are constructed by means of these rules are sound (i.e., truth-preserving). The problem is this: How is it possible to invoke the same rules in the metatheory to justify these rules when they are invoked in PL derivations? Isn't this circular reasoning? It is in fact **circular reasoning** but it is benign. The explanation is a little subtle. What the Soundness Theorem shows is that any repeated applications of these rules in any possible combination always produce sound derivations. This is a general statement about all possible PL derivations that can be constructed on the basis of these rules. We will establish this theorem by invoking a *specific finite sequence* of some of these rules. In other words, we will defend a very *general* claim about all possible combinations of these rules on the basis of a *particular* combination of some of these rules. All that is required of us is to make sure that every time we invoke one of these rules in the proof of the Soundness Theorem, we employ a correct application of it. Such an application can be justified on intuitive grounds given the truth-functional semantics of the logical connectives and quantifiers at the meta-level. The use of intuitive justification in this case is perfectly legitimate since it examines one specific application of a single rule given the semantics of its logical vocabulary. This is the best we can do. All justifications must start somewhere. Demonstrative justification invokes rules of inference. Hence to give demonstrative justifications of rules of inference, there is no alternative but to invoke rules of inference. The best one can do is to make sure that the specific employment of the rules of inference in the justification is secure. The proof of the Soundness Theorem is exactly of this sort. We employ finitely many applications of specific rules of inference in order to justify the soundness of infinitely many derivations that are constructed by infinite possible combinations of the NDS rules of inference.

2.2   Arithmetical Resources

2.2.1   The metatheory is endowed with informal arithmetical resources. We will assume the existence of a mathematical structure called the **structure of the natural numbers**. We denote this structure as N. It consists of an infinite set ℕ of all the familiar natural numbers, 0, 1, 2, 3, …, and the standard arithmetical properties, such as being even and being prime, operations, such as addition and multiplication, and relations, such as greater-than and less-than, which are defined on the members of ℕ. At this stage we will not attempt to design a formal arithmetical language and identify arithmetical axioms. We will leave this task until we introduce Peano Arithmetic later in this book. In the metatheory of PL, we will avail ourselves of all the arithmetical properties, relations, facts, and principles that we need to prove metatheorems about PL.

Since we have all studied arithmetic and some algebra, we are all quite familiar with this set and many of its properties. However, we will be a little more specific here and enumerate some of the facts about N, which we will invoke routinely in this book. First we have the relations "less than or equal to," "strictly less than," "greater than or equal to," and "strictly greater than." We use the standard symbols for these relations, respectively: $\leq$, $<$, $\geq$, and $>$. ℕ is naturally ordered by $<$ in the usual way; hence $0 < 1 < 2 < 3 < \ldots$. These relations have familiar properties, such as transitivity (see the solution to Exercise 1.5.6a). Here we will only mention the properties of $<$. It is transitive, asymmetric, and extendible. It is also irreflexive, that is, for every natural number n, not-(n < n); connex, that is, for any natural numbers n and m, either n < m, m < n, or n = m; and it has a minimal element (namely 0) and no maximal element. In fact, $<$ has a further feature that is stronger than the feature that it has a minimal element in ℕ. $<$ is **well-founded** on ℕ, that is, every nonempty subset of ℕ has a minimal element with respect to $<$. In simpler terms, every nonempty collection of natural numbers has a smallest member.

We introduced the concept of function in Chapter One. We will employ it here to describe the various operations defined on ℕ. First, there is the successor function S. S is a 1-place function that takes any natural number as an argument and returns its immediate successor as a value. So, S(0) = 1, S(1) = 2, S(2) = 3, and so on. To simplify our notation, we will write Sn instead of S(n). Many philosophers think that the set of the natural numbers is built from 0 and S, since every natural number can be obtained from 0 by repeated applications of the successor function; for instance, 3 is SSS0 and 5 is SSSSS0.

Addition and multiplication are 2-place functions: they take two numbers as arguments and return unique numbers as their values. The function of addition returns the sum of the two arguments and the function of multiplication returns the product of the two arguments. We will use the letter 'A' for the addition function and 'M' for the multiplication function. In standard notation, for all natural numbers n and m, A(k, n) = k+n and M(k, n) = k×n. Frequently, we will simply write k+n and k×n instead of A(k, n) and M(k, n). Addition and multiplication can be defined in terms of the successor function and 0. Below are the standard definitions of these functions.

2.2.1a For every natural number n, A(n, 0) = n (that is, n+0 = n).
2.2.1b For all natural numbers k and n, A(n, Sk) = S(A(n, k)) (that is, n+(k+1) = (n+k)+1).
2.2.1c For every natural number n, M(n, 0) = 0 (that is, n×0 = 0).
2.2.1d For all natural numbers k and n, M(n, Sk) = A(M(n, k), n) (that is, n×(k+1) = (n×k)+n).

Subtraction can also be defined, but it is not a function, since the subtraction of any two natural numbers is not always a natural number. For instance, there is no natural number that results from subtracting 5 from 2. Logicians call such functions **partial functions**. They are defined for certain members of the set but not for others. We can give the following definition of subtraction, which we will denote as 'D' (for 'difference').

2.2.1e For all natural numbers j, k, and m, D(j, k) = m iff A(k, m) = j (that is, j−k = m iff k+m = j); and D(j, k) is undefined otherwise.

Even the ordering relations mentioned above can be defined on the basis of these functions. For instance, we can define ≤ and < as follows.

2.2.1f For all natural numbers k and n, k ≤ n iff there is a natural number j such that A(k, j) = n (that is, k+j = n).

2.2.1g For all natural numbers k and n, k < n iff k ≤ n and k ≠ n.

2.2.2 The most important arithmetical resource of metalogic is the Principle of Mathematical Induction. This principle can be given two equivalent formulations. We will refer to the first formulation as the **Principle of Mathematical Induction** (PMI) and to the second as the **Principle of Complete Induction** (PCI). This principle allows us to infer that a certain statement X is true of all the natural numbers that are greater than or equal to a given natural number n_0, if two conditions are satisfied. The first condition is common to both formulations: X is true of n_0. This is called the **Base Step**. The second condition is a conditional. In the first formulation the conditional reads: for every natural number k ≥ n_0, if X is true of k, then X is true of Sk. In the second formulation the conditional reads: for every natural number k > n_0, if X is true of all m, where n_0 ≤ m < k, then X is true of k. Given these two conditions, the principle allows us to infer that X is true of all natural numbers n that are greater than or equal to n_0. The second condition, which is a conditional, is referred to as the **Inductive Step**; its antecedent is called the **Induction Hypothesis**.

The idea behind PMI is intuitive and simple. If one wants to show that X holds for all natural numbers ≥ n_0, it is sufficient to show that X holds for n_0 and that X is "hereditary," that is, it is always passed from any number ≥ n_0 to its successor. It is clear that if X holds for n_0 and it is hereditary, then it is passed to n_0+1; but now it can be passed from n_0+1 to n_0+2, from n_0+2 to n_0+3, from n_0+3 to n_0+4, and so on. No matter how large a number might be, X will eventually be passed to that number from its predecessor. The net result is that X holds for all the natural numbers that are greater than or equal to n_0. The same reasoning applies to PCI, except in this case 'hereditary' means that it is passed from a number's predecessors (instead of a single predecessor) to the number itself. Thus it is not sufficient that X holds for, say, 3 in order to be passed to 4; rather it must hold for 0, 1, 2, and 3 in order for 4 to inherit X. The same intuitive idea is applicable here. The Base Step shows that X holds for n_0; since n_0 is the only predecessor of n_0+1, X holds for n_0+1; given that X holds for the predecessors of n_0+2 (namely n_0 and n_0+1), it is passed to n_0+2; from n_0, n_0+1, and n_0+2 it is passed to n_0+3, and so on. Again the net result is that X holds for all natural numbers that are greater than or equal to n_0.

We need to clarify what we mean by saying that X holds for a natural number in order to give more precise formulations of these principles. We think of X as a metalinguistic formula that contains one free variable. In a manner of speaking X has a blank that can be filled by a numeral. We should really write X as X( ) in order to indicate that it is not a complete declarative sentence, but it would become one if its blank is filled with a numeral. Thus to say that X is true of 0 is to say that the declarative sentence X(0), which is obtained by replacing the blank in X( ) with the numeral '0', is true. Notice that we do not write X('0'). This is too cumbersome. Instead, we write X(0). In general, to say that X holds for any natural number n is to say that the declarative sentence X(n) is true for every natural number n. Now we are ready to state PMI and PCI. Let n_0 be any natural number.

Principle of Mathematical Induction (PMI): If $X(n_0)$, and for every natural number $k \geq n_0$, $X(Sk)$ when $X(k)$, then for every natural number $n \geq n_0$, $X(n)$. In simpler terms, if X is true of n_0 and if it is true of the successor of k whenever it is true of k, where $k \geq n_0$, then X is true of every natural number $n \geq n_0$.

Principle of Complete Induction (PCI): If $X(n_0)$, and for every natural number $k > n_0$, $X(k)$ when $X(m)$ for each m such that $n_0 \leq m < k$, then for every natural number $n \geq n_0$, $X(n)$. In other words, if X is true of n_0 and if it is true of a natural number $k > n_0$ whenever it is true of all the numbers that are greater than or equal to n_0 and less than k, then X is true of every natural number $n \geq n_0$.

These principles have strikingly wide applications, even in contexts in which numbers do not seem to be an obvious part. Numbers can be incorporated in many contexts that do not appear to be arithmetical at first. Let us prove a metatheorem about PL. We want to show that no matter how complex a PL sentence **X** might be, if **X** is quantifier-free and one of its *sentential components* is of the form **Y→Z**, then the PL sentence that is obtained by replacing **Y→Z** in **X** with ¬**Y**∨**Z** is logically equivalent to **X**. This metatheorem does not seem to involve any numbers in it. Nevertheless, a numerical variable can be introduced, and hence PMI or PCI might be applicable. Since we don't know how **X** looks, we will write **X[Y→Z]** to indicate that **X[Y→Z]** is some PL sentence of which **Y→Z** is a sentential component. Let us define *the complexity of* **X[Y→Z]** to be the number of connectives that appear in **X[Y→Z]** other than the → of **Y→Z**. **X[Y→Z]** might contain many connectives; for example, it might be $(\neg(At \vee (Cg \rightarrow \neg Dr)) \rightarrow (\neg Ge \wedge Ks))$. Our theorem would establish that this sentence is logically equivalent to $(\neg(At \vee (\neg Cg \vee \neg Dr)) \rightarrow (\neg Ge \wedge Ks))$, which is obtained from the previous one by replacing $(Cg \rightarrow \neg Dr)$ with $(\neg Cg \vee \neg Dr)$. We want to prove that for every natural number n, if n is the complexity of **X[Y→Z]**, then **X[Y→Z]** is logically equivalent to **X[¬Y∨Z]**, where **X[Y→Z]** is quantifier-free and the latter is obtained from the former by replacing the sentential component **Y→Z** with **¬Y∨Z**. We will invoke PCI.

The Base Step: Let the complexity of **X[Y→Z]** be 0. Hence there are no connectives in this sentence other than the → of **Y→Z**. It follows that this sentence is simply **Y→Z**. We know from the truth conditions of the sentential connectives that this is logically equivalent to **¬Y∨Z**. So the Base Step is established.

The Inductive Step: Suppose that for every m < k, where k is some non-zero natural number, any quantifier-free sentence **X[Y→Z]** whose complexity is m is logically equivalent to **X[¬Y∨Z]**. This supposition is the **Induction Hypothesis**. We want to show that the theorem holds for a quantifier-free sentence **W[Y→Z]** whose complexity is k. Since k is not zero, **W[Y→Z]** contains at least one additional connective other than the → of **Y→Z**. Hence **W[Y→Z]** could be a negation, a conjunction, a disjunction, a conditional, or a biconditional. Let us consider each case in turn.

(a) Suppose that **W[Y→Z]** is a negation. Its form therefore is ¬**V**. **Y→Z** must be a sentential component of **V**. We can write **V** as **V[Y→Z]**. But now the latter has a complexity less than k; so the Induction Hypothesis applies to it. It follows that **V[Y→Z]** is logically equivalent to **V[¬Y∨Z]**. This implies that ¬**V[Y→Z]** is logically equivalent to ¬**V[¬Y∨Z]**. Since **W[Y→Z]** is ¬**V[Y→Z]** and **W[¬Y∨Z]** is ¬**V[¬Y∨Z]**, we obtain that **W[Y→Z]** is logically equivalent to **W[¬Y∨Z]**.

(b) Now we suppose that **W[Y→Z]** is a conjunction. Hence it has the form **V**∧**U**. Since **Y→Z** is a sentential component of **V**∧**U**, it must be either a sentential component of **V**, **U**, or both. Without loss of generality, assume it is a sentential component of **V**. Thus we can write **V** as **V[Y→Z]**. Given that the complexity of **V**∧**U** is k, the complexity of **V[Y→Z]** is less than k. The Induction Hypothesis therefore applies to **V[Y→Z]**. It follows that **V[Y→Z]** is logically equivalent to **V[¬Y∨Z]**. This entails that **V[Y→Z]**∧**U**, which is **W[Y→Z]**, is logically equivalent to **V[¬Y∨Z]**∧**U**, which is **W[¬Y∨Z]**. If **Y→Z** is a sentential component of both **V** and **U**, then we can denote them as **V[Y→Z]** and **U[Y→Z]**. Since both of **V** and **U** have complexities less than k, the Induction Hypotheses applies to each of them. It follows that **V[Y→Z]** and **U[Y→Z]** are logically equivalent to **V[¬Y∨Z]** and **U[¬Y∨Z]**, respectively. Hence **V[Y→Z]**∧**U[Y→Z]**, which is **W[Y→Z]**, is logically equivalent to **V[¬Y∨Z]**∧**U[¬Y∨Z]**, which is **W[¬Y∨Z]**.

(c) The same reasoning applies to the cases of disjunction, conditional, and biconditional.

(d) (a)–(c) establish the Inductive Step. By PCI, every quantifier-free PL sentence (of any complexity) that contains a sentential component of the form **Y→Z** is logically equivalent to the PL sentence that is obtained from the original sentence by replacing **Y→Z** with **¬Y∨Z**.

The important thing to note about the metatheorem we proved is that it is not an arithmetical statement. It is a statement about the syntax and semantics of PL. However, we were able to introduce a numerical variable into it, and hence make it amenable to mathematical induction. This is a very common procedure, and we will apply it often in this book. We conclude this subsection with an application of PMI. We prove that the following arithmetical formula holds for every natural number n ≥ 1.

E(n)   1 + 2 + 3 + ... + n = n(n+1)/2

The Base Step: Let n = 1. The right-hand side of E(1) = 1(1+1)/2 = 1 = the left-hand side of E(1). This establishes the Base Step.

The Inductive Step: Let k ≥ 1, and suppose that E(k) is true, i.e., 1 + 2 + 3 + ... + k = k(k+1)/2. This is the Induction Hypothesis. We want to show that E(Sk) is true (recall that Sk is the successor of k, which is k+1). In other words, we want to prove

E(k+1)   1 + 2 + 3 + ... + k + (k+1) = (k+1)(k+2)/2.

Given the Induction Hypothesis, we have: the left-hand side of E(k+1) = 1 + 2 + 3 + ... + k + (k+1) = k(k+1)/2 + (k+1) = [k(k+1) + 2(k+1)]/2 = (k+1)(k+2)/2 = the right-hand side of E(k+1). This establishes the Inductive Step.

From the Base Step and the Inductive Step by PMI, E(n) is true for each n ≥ 1.

2.3   Set-Theoretic Resources

2.3.1   The metatheory makes extensive use of **set-theoretic concepts, objects, and principles**. In this section we will cover all the set-theoretic resources needed for future topics. **Set theory** is the modern foundation of almost all of mathematics and it is a highly developed discipline in its own right. In the literature set theory is developed as a theory of PL sentences that are derived from an infinite set of axioms, called **Zermelo-Fraenkel Axioms**, supplemented by the **Axiom of Choice**. The set theory that is based on these axioms is referred to as **ZFC Set Theory**. In this book we will not develop ZFC as an axiomatic PL theory. Rather, we will avail the metatheory of set-theoretic concepts, objects, and principles. We will give descriptions of these concepts, objects, and principles that are sufficiently precise to prove theorems and make definitions.

We take a very liberal attitude about sets. Sets are collections of objects—any objects that are posited by any discourse or a theory, including the metatheory of PL. These objects might be numbers, PL formulas, PL interpretations, formal derivations, geometric figures, people, stars, biological species, ideas, or what have you. Sets of objects can themselves be objects that belong to other sets, and sets of sets can also be objects that belong to yet other sets. The fundamental concept of set theory is the concept of membership. **Membership** is a relation that holds between a set and its members. If e is a member of a set B, we write 'e ∈ B'; we also say that e belongs to B, that e is an element of B, that e is in B, and that B contains

e. The negation of 'e ∈ S' is traditionally written as 'e ∉ S'. A set may be described by enclosing its members between braces or by stating a property that is exclusively shared by the members of the set. Thus we may write {1, 2, 3, 4} or {n: n is a positive integer that is greater than 0 and less than 5}. The last notation is read "The set of all n such that n is a positive integer that is greater than 0 and less than 5." There is a set that has no elements. It is called the **empty set** and is denoted as '∅'. The empty set is unique—that is, there is only one empty set. If x is an object, there is a unique set whose only member is x. We call this set the **singleton of** x and we denoted it as "{x}." In standard set theory, no object is identical with its singleton. Thus x and {x} are two different objects. If $x_1, x_2, x_3, \ldots,$ and x_n are n distinct objects, there is a unique set whose members are precisely those objects. This is the set $\{x_1, x_2, x_3, \ldots, x_n\}$. If n is 2, the set $\{x_1, x_2\}$ is called the **unordered pair of** x_1 **and** x_2.

Order and repetition are irrelevant for the identity of a set. Thus, {1, 1, 2, 3, 2, 5} = {1, 2, 3, 5} = {2, 3, 5, 1} = {5, 3, 2, 1}, and so on. It is possible to construct set-theoretic objects in which order and repetition matter. We call these objects **n-tuples**. An n-tuple whose components are $x_1, x_2, x_3, \ldots,$ and x_n is denoted as '$\langle x_1, x_2, x_3, \ldots, x_n \rangle$'. The components of an n-tuple are called its **coordinates**. Hence the coordinates of the 4-tuple $\langle 1, 2, 3, 5 \rangle$ are the numbers 1 2, 3, and 5 *in this order*. Since order and repetition matters for n-tuples, the following tuples are all different from each other: $\langle 1, 2, 3 \rangle$, $\langle 1, 3, 2 \rangle$, $\langle 2, 1, 3 \rangle$, and $\langle 1, 1, 3, 2 \rangle$. By the same logic, the ordered pair $\langle 1, 1 \rangle$ is neither identical with the singleton {1} nor with the ordered triple $\langle 1, 1, 1 \rangle$. 2-tuples are traditionally called **ordered pairs** and 3-tuples are called **ordered triples**. We state the following general principle.

The Principle of Ordered Tuples: For all n-tuples $\langle x_1, x_2, x_3, \ldots, x_n \rangle$ and $\langle y_1, y_2, y_3, \ldots, y_n \rangle$, they are identical iff $x_k = y_k$, for every k = 1, 2, 3, …, n. In other words, two n-tuples are identical iff their *corresponding* coordinates are identical.

If all the members of a set A are also members of a set B, we say that A is a **subset** of B, and we write 'A ⊆ B'. We also say that B *includes* A. Every set is a subset of itself and the empty set is a subset of every set. If a set A is a subset of a set B but it is not identical with it (i.e., if B contains an element that is not in A), we say that A is a **proper subset** of B, and we write 'A ⊂ B'. We also say that B *strictly includes* A. The negations of 'A ⊆ B' and of 'A ⊂ B' are standardly written as 'A ⊈ B' (A is not a subset of B) and 'A ⊄ B' (A is not a proper subset of B), respectively. Sets are determined solely by their members. Hence two sets that have the same members are identical. This is an important principle of set theory because it stipulates the identity condition for sets. We state it precisely below.

The Principle of Extensionality: For any two sets A and B, A = B just in case for every object x, x ∈ A iff x ∈ B.

It immediately follows that A = B iff A ⊆ B and B ⊆ A. The set that consists of the subsets of a set A is called the **powerset of A**. It is denoted as "$\mathcal{P}A$." Thus $\mathcal{P}A = \{B: B \subseteq A\}$ (i.e., the set of all B such that B is a subset of A). For instance, the powerset of {1, 2, 3} is {∅, {1, 2, 3}, {1}, {2}, {3}, {1, 2}, {1, 3}, {2, 3}}. It is clear that the powerset of {1, 2, 3} has 2^3 members. In general if a set has n members, where n is a natural number, its powerset has 2^n members. In particular, $\mathcal{P}\emptyset$ has only one member ($2^0 = 1$), which is $\emptyset$.

The **union** of two sets is the set that consists of the members of the two sets. Precisely speaking, the union of A and B = {x: x ∈ A or x ∈ B}. The union of A and B is denoted as "A∪B." For instance, {a, b, 2}∪{a, 2, 3} = {a, b, 2, 3}. The **intersection** of two sets is the set that consists of the elements that are common to both sets. The intersection of A and B is denoted as "A∩B" and it is the set {x: x ∈ A and x ∈ B}. For example, {a, b, 2}∩{a, 2, 3} = {a, 2}. The **complement** of B in A is the set whose members are precisely those elements that are contained in A but not in B. We denote this set as A–B. Formally, A–B = {x: x ∈ A and x ∉ B}. For example, {John, Mary, Sarah, William, Robert} – {Sarah, Mary, Elizabeth, Anna} = {John, William, Robert}. We can think of the complement operation as "subtraction": we subtracted Mary and Sarah from the set {John, Mary, Sarah, William, Robert}. The reason A–B is called "the complement of B in A," is because if we "add" the members of A–B to the set B, we obtain A∪B. In symbols, (A–B)∪B = A∪B. For instance, in the example above if we add John, William, and Robert to the set {Sarah, Mary, Elizabeth, Anna}, we obtain the set {Sarah, Mary, Elizabeth, Anna, John, William, Robert}, which is the union of the two original sets.

If two sets share no common elements, we say that they are **disjoint**. Using the notation of intersection, A and B are disjoint iff A∩B = $\emptyset$. A set whose members are also sets is usually called a **family of sets**. A family $\mathcal{F}$ of sets is said to be **pairwise disjoint** iff for all A and B in $\mathcal{F}$, A and B are disjoint, that is, no two members of $\mathcal{F}$ share common elements. In this case we also say that the members of $\mathcal{F}$ are **mutually exclusive**. The union of all the members of a family $\mathcal{F}$ of sets is denoted as "∪$\mathcal{F}$"; and the intersection of all the members of $\mathcal{F}$ is denoted as "∩$\mathcal{F}$." For example, if $\mathcal{F}$ is the set whose members are {2, 3, 5, 7, c, d}, {2, 7, c, f}, and {0, 2, 7, c}, then ∪$\mathcal{F}$ = {2, 3, 5, 7, c, d}∪{2, 7, c, f}∪{0, 2, 7, c} = {0, 2, 3, 5, 7, c, d, f} and ∩$\mathcal{F}$ = {2, 3, 5, 7, c, d}∩{2, 7, c, f}∩{0, 2, 7, c} = {2, 7, c}. If A is a nonempty set and $\mathcal{F}$ is a family whose members are subsets of A, we say that $\mathcal{F}$ is **exhaustive** (of A) or that the members of $\mathcal{F}$ are **collectively exhaustive** (of A) iff ∪$\mathcal{F}$ = A. If $\mathcal{F}$ is a family of mutually exclusive and collectively exhaustive nonempty subsets of a set A, $\mathcal{F}$ is said to be a **partition** of A. For instance, the following infinite family $\mathcal{F}$ of subsets of $\mathbb{N}$ is a partition of $\mathbb{N}$: $\mathcal{F}$ = {{m: $0 \leq m < 10$}, {m: $10 \leq m < 19$}, {m: $20 \leq m < 29$}, …, {m: $n \times 10 \leq m < (n \times 10 + 9)$}, …}.

If A and B are nonempty sets, their **Cartesian product** is the set that consists of all the ordered pairs whose first coordinates are members of A and whose second coordinates are members of B. The Cartesian product of A and B is denoted as "A×B." Thus, A×B = {⟨x, y⟩: x ∈ A and y ∈ B}. For example, if A is {1, 2, 3} and B is {e, d}, A×B = {⟨1, e⟩, ⟨1, d⟩, ⟨2, a⟩, ⟨2, d⟩, ⟨3, a⟩, ⟨3, d⟩}. If $A_1, A_2, A_3, …,$ and A_n are nonempty sets, their Cartesian product is the set $A_1 \times A_2 \times A_3 \times … \times A_n$ that consists of all the n-tuples whose k^{th} coordinates are members of A_k. Symbolically, $A_1 \times A_2 \times A_3 \times … \times A_n$ = {⟨$x_1, x_2, x_3, …, x_n$⟩: $x_k \in A_k$}. If $A_1 = A_2 = A_3 = … = A_n = Z$, the Cartesian product $A_1 \times A_2 \times A_3 \times … \times A_n = Z \times Z \times Z \times … \times Z$ (n times); it is denoted as "Z^n." Thus Z^2 is simply Z×Z.

2.3.2   Is there a **universal set**? In other words, is there a set that contains everything? If there is such a set, it would have to contain itself. It is not immediately obvious that there is something wrong with a set containing itself. In fact, there are versions of set theory that allow for "circular membership," that is, $A_1 \in A_2 \in A_3 \in \ldots \in A_n \in A_1$. If circular membership is allowed, there is no good reason to rule out $A \in A$. The standard ZFC does not allow circular membership. Hence, according to ZFC, there can be no universal set because no set may contain itself.

As it turns out, there is a much stronger reason for there being no universal set. The reason has to do with a very important discovery in the history of set theory that is called **Russell's Paradox** after its discoverer, the British philosopher, logician, and mathematician Bertrand Russell (1872–1970). Before the discovery of this paradox, it was thought that for any "concept" there is a unique set that is the extension of that concept (the extension of a concept is the collection that consists of all the objects that are subsumed under the concept). The notion of "concept" is philosophically loaded and philosophers do not agree on a single definition of this notion. But many philosophers and logicians understand "concept" as the thing that is designated by a well-defined predicate. Thus if C is a well-defined predicate, then C designates a concept. We can now define the extension of a predicate C as the extension of the concept that C designates. Since it was assumed that the extension of any concept is a set, we obtain the following principle: If C is a well-defined predicate, then there is a set A such that A is the extension of C. The paradox arises for the, presumably, well-defined predicate 'X is a set that does not belong to itself'. According to the former principle, there is a set R that is the extension of this predicate. In other words, for every X, X belongs to the set R iff X is a set that does not belong to itself. Since R is a set, either it belongs to itself or it does not. Assume that R belongs to itself. Hence R belongs to the extension of the predicate 'X is a set that does not belong to itself', since R is the extension of this predicate. But this means that the predicate 'X is a set that does not belong to itself' is true of R, which entails that R does not belong to itself. This contradicts the original assumption. Now assume that R does not belong to itself. It follows that the predicate 'X is a set that does not belong to itself' is true of R, which implies that R belongs to the extension of this predicate. Since R *is* the extension of this predicate, R must belong to itself after all, which contradicts the second assumption.

This paradox shows that the previous principle is false: it is not true that every well-defined predicate has an extension that is a set. The principle was replaced with another, more restrictive principle that is called the **comprehension scheme**. It asserts that for every well-defined predicate and every set X, there is a set A that consists of all the members of X of which the predicate is true. This scheme blocks Russell's Paradox, because it does not allow us to collect all the sets that don't belong to themselves into a single set. Rather we can collect such sets only if we are already given another set that contains those sets. Since no such set is given, the extension of the predicate 'X is a set that does not belong to itself' fails to form a set, according to the comprehension scheme. However, if there exists a universal set U, then the old, contradictory principle would follow from the comprehension scheme, and Russell's Paradox would reemerge. To see the point let C be any well-defined predicate. The comprehension scheme entails the existence of a set A whose members are all the objects that belong to U and of which C is true. But everything belongs to U (U is a universal set). It follows that A is a set that consists of all the objects of which C is true. In other words, A is a set that is the extension of C. Therefore, for every well-defined predicate C, there exists a set that is the extension of C. Since this principle was shown to be contradictory, we ought to deny that U exists. Simply stated, if we want the comprehension scheme to be a consistent principle, we must deny the existence of a universal set, that is, a set that contains everything.[1]

2.3.3   We are familiar with the notions of **relation** and **function** from Chapter One, and we discussed some of the properties of relations in the solution to Exercise 1.5.6a. We will revisit these notions here offering precise set-theoretic descriptions of them. Relations and functions are defined in set theory as certain types of sets of n-tuples. For example, a 2-place relation R on a set A is a set of ordered pairs of members of A, and a 3-place relation on A is a set of ordered triples of members of A. A function may be thought of as a relation between a set of "inputs" and a set of "outputs" that satisfies two conditions: An **existence condition** and a **uniqueness condition**. The existence condition says that for *every* input *there exists* an output; and the uniqueness condition says an input can have only *one* output.

A relation R between two sets A and B is a subset of the Cartesian product A×B, that is, it is a set that consists of ordered pairs whose first coordinates are members of A and whose second coordinates are members of B. If A and B are the same set, we say that R is a **relation on** A. In this case, $R \subseteq A^2$. In general, an n-place relation on A is a subset of A^n. If the ordered pair ⟨x, y⟩ belongs to a relation R, we usually write 'x R y'. If ⟨x, y⟩ does not belong to R, we write 'not-(x R y)'. A 2-place relation on a set A is commonly referred to as a **binary relation** on A. A binary relation may be of various types. We define these types below. Let R be any binary relation on some set A.

R is **reflexive** iff for every x in A, x R x.

R is **irreflexive** iff for every x in A, not-(x R x).

R is **symmetric** iff for all x and y in A, if x R y, then y R x.

R is **asymmetric** iff for all x and y in A, if x R y, then not-(y R x).

R is **antisymmetric** iff for all x and y in A, if x R y and y R x, then x = y.

R is **transitive** iff for all x, y, and z in A, if x R y and y R z, then x R z.

R is **extendible** iff for all x in A, there is y in A such that x R y.

R is **total** iff for all x and y in A, either x R y or y R x (we also say that R satisfies **dichotomy**).

---

[1]   I am grateful to an anonymous reviewer for suggesting that I include a discussion of Russell's Paradox.

R is **connex** iff for all x and y in A, either x R y, y R x, or x = y (we also say that R satisfies **trichotomy**).

R is **injective** iff for all x, y, and z in A, if x R z and y R z, then x = y.

R **has a minimal element in D**, where D is a subset of A, iff there is x in D, such that for every y in D, not-(y R x). (x is called "an R-minimal element in D.")

R **has a maximal element in D**, where D is a subset of A, iff there is x in D, such that for every y in D, not-(x R y). (x is called "an R-maximal element in D.")

A **function** F from a set A *into* a set B is a relation between A and B that satisfies the following condition: for *every* member x of A, there *exists* a *unique* member y of B, such that $\langle x, y \rangle \in F$. The "there exists" part of the preceding condition is the existence condition and the "unique" part is the uniqueness condition mentioned above. If $\langle x, y \rangle \in F$, we say that x is the **argument** of F and that y is the **value** of F at x. Since y is unique, the fact that $\langle x, y \rangle \in F$ is standardly described by writing "F(x) = y," that is, the value of F at the argument x is y. The existence condition ensures that every member of A is an argument of F, and the uniqueness condition ensures that every argument has one and only one value. In other words, F does not leave a member of A without a value assigned to it, nor does it assign more than one value to a member of A. Sometimes we allow functions to violate the existence condition. If F is a relation between A and B that satisfies the uniqueness condition but not the existence condition, then there is at least one element of A to which F assigns no value, and there is no element of A to which F assigns more than one value. We call such a relation a **partial function** from A *into* B. In order to emphasize that a certain function F is not a partial function by satisfying both the existence and uniqueness conditions, we sometimes describe F as a **total function**. If we do not say whether a function is total or partial, then we presuppose that the function is total.

If F is a function from a set B into B, we say that F is a function on B. As we mentioned in Chapter One, functions have places too. An n-place function on B is a function from B^n into B. Such a function takes n-tuples whose coordinates are members of B as arguments and assign single members of B as values. For instance, the function A that assigns to every pair of natural numbers their sum is a 2-place function on $\mathbb{N}$. In other words, A is a function from $\mathbb{N}^2$ into $\mathbb{N}$. We may define A as follows: for every ordered pair $\langle n, m \rangle$, where n and m are members of $\mathbb{N}$, A(n, m) = n+m. Most of the time we do not use the language of ordered tuples; rather we simply define: for all natural numbers n and m, A(n, m) = n+m. The **domain** of a function F that is from D into B is the set D and its **range** is the set of all its values. Thus the range of F is the set that consists of all the members of B that are assigned as values to arguments in D. We denote the domain and range of F, respectively, as "dom(F)" and "ran(F)." It is clear that ran(F) $\subseteq$ B. If F is a function on D, then dom(F) = D and ran(F) $\subseteq$ D.

There are three basic types of functions (whether total or partial): an **onto-function**,[2] a **one-to-one** function,[3] and a **one-to-one correspondence**.[4] F is a function from D *onto* B iff F is a function from D into B and every element of B is the value of some argument; in other words, for every $y \in B$, there exists $x \in D$, such that $F(x) = y$. Using the notion of range, we can define a function F from D into B to be an onto-function just in case ran(F) = B. The 2-place function A on the set of natural numbers ℕ that assigns to any pair of natural numbers their sum is an onto-function since ran(A) = ℕ, that is, every natural number n is the sum of two numbers (e.g., 0 and n). A *one-to-one* function from D into B is a function from D into B that does not assign a single value to more than one argument. More precisely, for all x and y in D, if $F(x) = F(y)$, then $x = y$. In informal terms, a function is one-to-one just in case no two different inputs have the same output. The sum-function A is not a one-to-one function on ℕ since two different pairs of numbers might have the same sum; for example, A(2, 3) = A(1, 4) = 5. On the other hand, the function H on ℕ that assigns to every natural number its square is a one-to-one function but not onto. No two natural numbers have the same square, but there are many natural numbers that are not the squares of any natural numbers. A *one-to-one* function from D *onto* B is called "a one-to-one correspondence between D and B." If we let **S** be the subset of ℕ that consists of all perfect squares, then the function H (as defined above) is a one-to-one correspondence between ℕ and **S**. However, H is not a one-to-one correspondence between the set of the integers ℤ and **S** since two different integers might have the same square, e.g., H(2) = H(–2) = 4. But H is a function from ℤ *onto* **S** since ran(H) = **S**. Finally, H is neither a one-to-one nor an onto-function from ℤ into ℕ; nevertheless, it is still a function from ℤ into ℕ.

A useful operation is **function composition**. If F is a function from A into B and G is a function from B into C, then the composition of F and G, which is denoted as "G∘F," is the function from A into C that assigns to every argument x in A, the value G(F(x)). In other words, G∘F(x) = G(F(x)). ('G∘F' is read 'G circle F'.) We explain function composition by means of an example. Let F be the function from ℤ (the set of the integers) into ℕ such that F(±n) = n, for every natural number n. For example F(0) = 0, F(–2) = F(2) = 2, and F(–7) = F(7) = 7. Let G be the function from ℕ into ℚ (the set of the rational numbers) such that G(0) = 0 and G(n) = 1/n for each n > 0, that is, G assigns to every positive integer its reciprocal; for instance, F(1) = 1, F(2) = 1/2, and F(7) = 1/7. The composition G∘F is the function from ℤ into ℚ that assigns 0 to 0, 1/n to every positive integer n, and –1/n to every negative integer n. For example, if n = –7, G∘F applies first the function F to –7, producing the value 7, and then the function G to 7, producing 1/7; thus G∘F(–7) = G(F(–7)) = G(7) = 1/7. Informally speaking, the composition G∘F assigns values according to the following two-step procedure: the function at the right (i.e., F) is applied first to the argument n, producing a certain value

---

[2] By using the term 'onto-function', we are violating a prohibition against using 'onto' as a modifier. These functions are sometimes called "surjective." However, our usage is in line with a popular mode of oral discussion of functions that is usually invoked in classroom presentations. We are simply producing in a written format what many instructors use for describing surjective functions.

[3] One-to-one functions are sometimes called "injective."

[4] A one-to-one correspondence is sometimes called "bijective."

F(n), and then G is applied to this value F(n), producing a new value G(F(n)). Function composition is a useful operation for generating new functions from other functions — new functions that "combine" the effects of the given functions.

Another useful notion is that of an **inverse function**. If F is a one-to-one function from A onto B (i.e., F is a one-to-one correspondence), then its inverse, which is denoted as "F^{-1}," is the one-to-one function from B onto A that "reverses" the effect of F. F takes an argument x from the domain A and produces a value F(x) in B. F^{-1} takes the value F(x) and produces the original argument x. So in a manner of speaking the inverse function F^{-1} undoes the effect of the function F. Precisely speaking, we define the inverse function as follows: if F is a one-to-one function from A onto B, then F^{-1} is the one-to-one function from B onto A such that $F^{-1} \circ F(x) = F^{-1}(F(x)) = x$. For example, if F is the function from $\mathbb{N}$ onto E (the set of the even natural numbers) that takes every natural number n to its double 2n, then F^{-1} is the function that takes every even natural n number to its half n/2. If we combine the effects of these two functions, we obtain the following two step-procedure: F assigns 2n to n and F^{-1} assigns 2n/2 to 2n; hence the net effect is that n is assigned to n. The function on any nonempty set A that assigns x to x is called the **identity function** on A and is denoted as "I." Therefore I is the function on A such that for every $x \in A$, I(x) = x. We can use this notation to give a compact definition of the inverse function: if F is a one-to-one function from A onto B, then F^{-1} is the function from B onto A, such that $F^{-1} \circ F = I$ on A. Note that if F fails to be a one-to-one onto-function, its inverse would not be a function. For if there are two different arguments x and y such that F(x) = F(y) = z, then $F^{-1}(z)$ cannot return a unique value. Also, if ran(F) ≠ B, which entails that there is $y \in B$ such that y is not assigned as a value to any argument in A, then $F^{-1}(y)$ would not return any value. Said differently, if F fails to be one-to-one, F^{-1} would not satisfy the uniqueness condition, and if F fails to be an onto-function, F^{-1} would not satisfy the existence condition.

2.3.4 Intuitively speaking, the **cardinality** of a set is its size, that is, the number of its elements. In set theory it is possible to define precisely the notion of a cardinal number and, consequently, the notion of the cardinality of a set. However, since this is not a book on set theory, we will not introduce these definitions here. We only note that to every set is attached a unique set-theoretic object that represents the number of its members. This set-theoretic object is called the cardinality of the set and is denoted as "card(A)," where A is any set. As expected, if A is a finite set, card(A) is a natural number; and if A is an infinite set, card(A) is an infinite number. In particular, card($\emptyset$) = 0, card({x}) = 1, card({x, y}) = 2 if x ≠ y, card({$x_1, x_2, x_3, \ldots, x_n$}) = n if $x_i \neq x_j$ for all $i \neq j$, and card($\mathbb{N}$) = $\aleph_0$, where $\mathbb{N}$ is the set of all natural numbers and $\aleph_0$ is the smallest infinite number.

Among the important achievements of the theory of cardinal numbers are the definitions of the relations of identity and less-than between cardinalities. We present these definitions here. Two sets are said to be **equinumerous** iff there is a one-to-one correspondence between them. If A and B are equinumerous sets, we write 'A ≈ B'. Equinumerous sets have the same cardinality. The idea is sufficiently intuitive. If we can match every element of A with a unique element of B and every element of B with a unique element of A, then it seems that A and B ought to have the same number of elements, that is, the same cardinality. Hence equinumerosity is adopted as the condition for identity among cardinalities. We state this condition below. It is commonly called "Hume's Principle."

Hume's Principle: For all sets A and B, card(A) = card(B) iff $A \approx B$, that is, iff there is a one-to-one correspondence between A and B.

Several important definitions can be based on the notion of equinumerosity. Here are the definitions of ordering relations among cardinalities.

Less-than and greater-than: For all sets A and B,

2.3.4a card(A) is less than or equal to card(B) (symbolically, card(A) $\leq$ card(B)) iff there is a set C, such that $C \subseteq B$ and $A \approx C$.

2.3.4b card(A) is less than card(B) (symbolically, card(A) < card(B)) iff card(A) $\leq$ card(B) and card(A) $\neq$ card(B).

2.3.4c card(A) is greater than or equal to card(B) (symbolically, card(A) $\geq$ card(B)) iff card(B) $\leq$ card(A).

2.3.4d card(A) is greater than card(B) (symbolically, card(A) > card(B)) iff card(A) $\geq$ card(B) and card(A) $\neq$ card(B).

We can also define the notions of infinite and finite sets using the notion of equinumerosity.

Infinite and finite sets: For every set A,

2.3.4e A is infinite iff there is a set B, such that $B \subset A$ and $A \approx B$.

2.3.4f A is finite iff it is not infinite.

The definitions above make a good deal of sense. Definition 2.3.4a says, intuitively, that if A is equinumerous with a subset of B, then B must contain at least as many elements as A. Definition 2.3.4b asserts, also intuitively, that if A is equinumerous with a subset of B but it is not equinumerous with B itself, then B must contain strictly more elements than A does. The definitions of the relations $\geq$ and > can be given similar intuitive justifications. The definition of an infinite set is very interesting. It is a profound insight of the father of set theory, Georg Cantor (1845–1918). Since antiquity the concept of an **actual infinity** has been challenged by many mathematicians and philosophers as incoherent. The reason for this charge is that an actually infinite set is of the same size as many of its proper parts; but since the whole is strictly greater than any of its proper parts, no actually infinite sets can exist. The principle that the whole is greater than any of its proper parts, which we shall call "the Whole-Part Principle," was accepted as a self-evident truth. After all, is it not obvious that the set of the natural numbers contains infinitely more members than the set of the even natural numbers? The set ℕ contains the numbers 1, 3, 5, 7, and so on, none of which is a member of the set **E** that consists of the even natural numbers. But "paradoxically," the sets

ℕ and E are equinumerous: every natural number n can be matched with a single even natural number, namely 2n, and every even natural number can be matched with a single natural number, namely n/2. This one-to-one correspondence clearly shows that there are as many natural numbers as there are even natural numbers. This seemed impossible to many philosophers and mathematicians since E is a proper subset of ℕ. The conclusion was that the concept of an actually infinite set is incoherent, and hence no such set can exist. Of course, people were familiar with the set ℕ and its various subsets, as arithmetic is an ancient field of study. Their attitude about such sets, however, was to consider them **potentially infinite**, not actually infinite. The set ℕ is "indefinitely large"; it is "unending"; it is "constantly expanding"; but never "actually completed." The traditional wisdom was that the infinite could not be completed; it is a never-ending process of adding more and more elements. To say that an actual infinity exists entails that an unending process has ended and been completed, which sounds contradictory.

Cantor suggested that there was nothing incoherent about the concept of actual infinity. The problem lies in thinking that the Whole-Part Principle is universally valid for all sets, finite and infinite. Cantor's profound insight was that the defining characteristic of the actually infinite is that it is not subject to the Whole-Part Principle. For Cantor, sets were "static," not "dynamic," objects. Sets just exist with all their elements present: they are all completed. A potentially infinite process might make perfect sense in geometry, such as a moving point in space or a constantly extending line, but sets are not subject to such processes. A set is defined by the presence of its elements. If some of its elements do not exist *yet*, then the set does not exist at all. All the members of a set must exist in order for the set to exist. By challenging the universal validity of the Whole-Part Principle, Cantor challenged the ancient wisdom about the incoherence of the concept of actual infinity. The door was opened for the exploration of actual infinites, their properties, and relations. Actual infinities were called "transfinite sets," their cardinalities "transfinite cardinalities," and their arithmetic "transfinite arithmetic." We will present here a small part of this exploration.

It is known that $\aleph_0$, the cardinality of ℕ, is the smallest infinite cardinality. Hence any cardinality that is less than $\aleph_0$ is finite and any infinite cardinality is greater than or equal to $\aleph_0$. This entails that all infinite subsets of ℕ, such as E, have the same cardinality as ℕ. A set whose cardinality is less than or equal to $\aleph_0$ is called **countable**. All sets whose cardinalities are greater than $\aleph_0$ are called **uncountable**. We gave an intuitive explanation of this distinction in Subsection 1.2.4. An infinite countable set, that is, a set whose cardinality is $\aleph_0$, is called **countably infinite.**

One of the interesting results of the theory of infinite cardinalities is that there are infinitely many cardinalities that are greater than $\aleph_0$, and none of which are equal to each other. This was quite interesting because it also challenged the ancient wisdom that an actually infinite set cannot be made any bigger by adding more elements to it. That was one of the reasons that lead to the rejection of the actually infinite as incoherent. It is true that adding finitely many new elements to an infinite set does not change its cardinality. In fact sometimes adding infinitely many new elements to an infinite set does not change its cardinality. For example, the cardinality of the set **E** (of all the even natural numbers) and of the set **O** (of all the odd natural numbers) is $\aleph_0$, since they are both infinite subsets of ℕ and the cardinality of ℕ is $\aleph_0$, which is the smallest infinite cardinality. The previous remark makes it clear that if we add all the odd natural numbers to **E**, the resulting set, which is ℕ, has the same cardinality as **E**. Cantor proved in a famous theorem that this is not always true. Sometimes if we enlarge an infinite set beyond a certain limit, we reach a genuinely greater cardinality. In fact we can prove the existence of a sequence of infinite cardinalities $\kappa_1, \kappa_2, \kappa_3, \ldots, \kappa_n, \ldots$, such that $\aleph_0 < \kappa_1 < \kappa_2 < \ldots < \kappa_n < \ldots$. This result is an immediate consequence of Cantor's Theorem.

Cantor's Theorem: For every set A, the cardinality of $\mathcal{P}A$ is greater than the cardinality of A.

Proof

1. We first show that $\text{card}(\mathcal{P}A) \geq \text{card}(A)$. By Definition 2.3.4c, we need to show that $\text{card}(A) \leq \text{card}(\mathcal{P}A)$. By Definition 2.3.4a, we must prove that there exists a subset $\mathcal{K}$ of $\mathcal{P}A$, such that $A \approx \mathcal{K}$, that is, there exists a one-to-one correspondence between A and $\mathcal{K}$.
2. By the definition of the powerset, $\mathcal{P}A$ consists of all the subsets of A. By the definition of subset, for every x in A, the singleton of x, {x}, is a subset of A; hence {x} ∈ $\mathcal{P}A$.
3. Let $\mathcal{K}$ be the subset of $\mathcal{P}A$ that consists of all the singletons of the members of A. In other words, $\mathcal{K} = \{Z: Z \in \mathcal{P}A$ and $Z = \{x\}$ for some x in A$\}$.
4. By the definition of subset, it is clear that $\mathcal{K} \subseteq \mathcal{P}A$.
5. Define F to be the function from A into $\mathcal{K}$ that assigns to every element x of A its singleton {x}, which is an element of $\mathcal{K}$. Since every object has exactly one singleton, and since $\mathcal{K}$ consists of *all* the singletons of the elements of A, F is a one-to-one correspondence between A and $\mathcal{K}$, that is, $A \approx K$.
6. From 1 and 5: $\text{card}(\mathcal{P}A) \geq \text{card}(A)$.
7. In order to show that $\text{card}(\mathcal{P}A) > \text{card}(A)$, by Definition 2.3.4d, we need to prove that $\text{card}(\mathcal{P}A) \neq \text{card}(A)$.
8. Reductio Assumption: $\text{card}(\mathcal{P}A) = \text{card}(A)$.
9. From 8 by Hume's Principle: $\mathcal{P}A \approx A$.
10. From 9 by the definition of equinumerosity: there is a one-to-one correspondence F between A and $\mathcal{P}A$.
11. We introduce two terms to simplify the reasoning. Let x be any member of A. Since F is a function from A into $\mathcal{P}A$, there exists a unique member of $\mathcal{P}A$ that is the value of F at x. As usual, we denote this value as "F(x)." By the definition of powerset, $F(x) \subseteq A$. Since x is a member of A and F(x) is a subset of A, it makes sense to ask

whether x belongs to F(x) or not. If x ∈ F(x), we call x "irregular," and if x ∉ F(x), we call x "regular." Thus the irregular elements of A are those elements that belong to the values that F assigns to them, and the regular elements of A are those elements that do not belong to the values that F assigns to them. For example if F(3) = {3, 7, 8}, 3 is irregular because it is a member of F(3); and if F(7) is {0, 13, 9}, 7 is regular because it is not a member of F(7).

12. From 11: define D to be the set that consists of all the regular elements of A, that is, D = {x: x ∈ A and x ∉ F(x)}.

13. From 12: it is clear that D ⊆ A. Thus by the definition of powerset, D ∈ 𝒫A.

14. From 10 and 13: since F is a one-to-one correspondence between A and 𝒫A, it is an onto-function. This entails that D must be the value of F at some argument in A; this entails that there is an element e in A such that F(e) = D.

15. From 11 and 14: either e is a regular or an irregular element of A.

16. Case 1: suppose that e is a regular element of A.

17. From 12 and 16: since D is the set of all the regular elements of A, e ∈ D.

18. From 11, 14, and 17 by substitution: e ∈ F(e). By the definition of an irregular element, e is irregular, which contradicts 16.

19. Case 2: suppose that e is an irregular element of A.

20. From 11 and 19 by the definition of an irregular element: e ∈ F(e).

21. From 14 and 20 by substitution: e ∈ D.

22. From 12 and 21: since D consists of the regular elements of A and e ∈ D, e is regular, which contradicts 19.

23. From 15, 18, and 22: both cases lead to contradictions. Hence the Reductio Assumption entails a contradiction.

24. From 8 through 23: the Reductio Assumption is false, that is, card(𝒫A) ≠ card(A).

25. From 6 and 24 by the definition of "greater-than": the cardinality of 𝒫A is greater than the cardinality of A.

Since the cardinality of ℕ is $\aleph_0$, which is an infinite cardinality, by Cantor's Theorem, card(𝒫ℕ) > $\aleph_0$. Let card(𝒫ℕ) = κ_1. Again, by Cantor's Theorem, card(𝒫𝒫ℕ) > card(𝒫ℕ) = κ_1. Let card(𝒫𝒫ℕ) = κ_2. Once more, Cantor's theorem entails that card(𝒫𝒫𝒫ℕ) > card(𝒫𝒫ℕ) = κ_2. We let κ_3 be card(𝒫𝒫𝒫ℕ). So far, we have obtained the following sequence of infinite cardinalities: $\aleph_0 < \kappa_1 < \kappa_2 < \kappa_3$. Since it is clear that we can iterate the powerset operation in this manner indefinitely, we can generate an increasing sequence of infinite cardinalities all of which are uncountable, that is, $\aleph_0 < \kappa_1 < \kappa_2 < \kappa_3 < \ldots < \kappa_n < \ldots$.

2.3.5   Is there a familiar set that is uncountable? Cantor also proved that the set ℝ of all the real numbers is uncountable, that is, its cardinality, which is usually denoted as "C," is greater than $\aleph_0$. Cantor's proof of this theorem introduced a style of proof that came to be known as **diagonalization**. We will encounter this proof style more than once in later chapters. In this subsection we will describe **Cantor's Diagonal Argument** which establishes that $\mathbb{R}$ is uncountable, and we will also prove that the set of all the rational numbers, $\mathbb{Q}$, is countable. Since $\mathbb{Q}$ is an infinite set and countable, its cardinality is also $\aleph_0$. This is interesting, since $\mathbb{Q}$ seems to be a very large set: while $\mathbb{N}$ is *discrete* in the sense that there are no natural numbers between any consecutive natural numbers, $\mathbb{Q}$ is *dense* in the sense that between any two rational numbers there are infinitely many rational numbers.

We first prove that $\mathbb{R}$ is uncountable. We will show that $C > \aleph_0$, and this immediately entails that $\mathbb{R}$ is uncountable. To be precise, we will prove that a subset of $\mathbb{R}$ is larger than $\mathbb{N}$. Of course, if a subset of A is larger than a set B, then A is larger than B as well. Let **I** be the *half-open unit interval*, that is, **I** is the set that consists of all the real numbers that are greater than or equal to 0 and less than 1. In other words, $\mathbf{I} = \{x : x \in \mathbb{R} \text{ and } 0 \leq x < 1\}$. It is clear that **I** is a proper subset of $\mathbb{R}$.

Cantor's Diagonal Argument: $\text{card}(\mathbf{I}) > \text{card}(\mathbb{N})$.

Proof

1. By Definition 2.3.4d, we need to prove that $\text{card}(\mathbf{I}) \geq \text{card}(\mathbb{N})$ and $\text{card}(\mathbf{I}) \neq \text{card}(\mathbb{N})$. We will think of the real numbers in terms of their infinite decimal expansions. Thus 2 is 2.000…, 1/2 is 0.500…, 10/3 is 3.333…, and so on. Many real numbers have two decimal expansions, e.g., 2 is also 1.999… and 1/2 is also 0.499… . In order to have a unique decimal expansion for each real number, we will disallow decimal expansions ending with infinite sequences of 9's.

2. We first show that $\text{card}(\mathbf{I}) \geq \text{card}(\mathbb{N})$. By Definitions 2.3.4c and 2.3.4a, our goal is to show that there is a set M such that $M \subseteq \mathbf{I}$ and $\mathbb{N} \approx M$. This is trivial. We take M to be the subset of **I** that consists of the real numbers whose decimal expansions are of the form 0.n000…, where n is any natural number. For instance, 0.000…, 0.700…, and 0.24600… are all members of M. The function that matches every natural number n with 0.n000… is clearly a one-to-one correspondence between $\mathbb{N}$ and M. Thus we established that $\text{card}(\mathbf{I}) \geq \text{card}(\mathbb{N})$.

3. It remains to show that $\text{card}(\mathbf{I}) \neq \text{card}(\mathbb{N})$. We employ a Reductio Ad Absurdum proof. The Reductio Assumption is that $\text{card}(\mathbf{I}) = \text{card}(\mathbb{N})$.

4. From 3 by Hume's Principle: $\mathbf{I} \approx \mathbb{N}$.

5. From 4 by the definition of equinumerosity: there is a one-to-one correspondence between **I** and $\mathbb{N}$. Thus the members of **I** can be arranged in an infinite list, $r_0, r_1, r_2, r_3, \ldots, r_n, \ldots$. Every member of **I** appears once and only once in this infinite list. Given that every r_n is an infinite decimal expansion, we can list the members of **I** and their decimal digits in a matrix. Each entry of this matrix is a single digit that is either 0, 1, …, or 9. The symbol a_j^i represent the jth decimal digit of the ith real

number in **I**, that is, r_i. (We assume that the first decimal digit occupies the 0th place.) Here is how this matrix looks.

$r_0 = 0.\ a_0^0\ a_1^0\ a_2^0\ a_3^0\ a_4^0\ a_5^0\ \ldots$

$r_1 = 0.\ a_0^1\ a_1^1\ a_2^1\ a_3^1\ a_4^1\ a_5^1\ \ldots$

$r_2 = 0.\ a_0^2\ a_1^2\ a_2^2\ a_3^2\ a_4^2\ a_5^2\ \ldots$

$r_3 = 0.\ a_0^3\ a_1^3\ a_2^3\ a_3^3\ a_4^3\ a_5^3\ \ldots$

$r_4 = 0.\ a_0^4\ a_1^4\ a_2^4\ a_3^4\ a_4^4\ a_5^4\ \ldots$

$r_5 = 0.\ a_0^5\ a_1^5\ a_2^5\ a_3^5\ a_4^5\ a_5^5\ \ldots$

$\vdots$

6. From 5: the diagonal of this matrix consists of the digits $a_0^0, a_1^1, a_2^2, a_3^3, \ldots, a_n^n, \ldots$. We define the real number $0.b_0b_1b_2b_3\ldots b_n\ldots$ as follows: if $a_0^0 = 1$, $b_0 = 2$, and if $a_0^0 \neq 1$, $b_0 = 1$; if $a_1^1 = 1$, $b_1 = 2$, and if $a_1^1 \neq 1$, $b_1 = 1$; if $a_2^2 = 1$, $b_2 = 2$, and if $a_2^2 \neq 1$, $b_2 = 1$; and in general if $a_n^n = 1$, $b_n = 2$, and if $a_n^n \neq 1$, $b_n = 1$. Let q be the real number just defined. q is sometimes called "the anti-diagonal number," because each of its decimal digits is different from the corresponding digit of the diagonal of the matrix above.

7. From 6: since q begins with either 0.2 or 0.1, it is greater than 0 and less than 1; hence q ∈ **I**.

8. From 5 through 7: given that every member of **I** appears somewhere in the list $r_0, r_1, r_2, r_3, \ldots, r_n, \ldots$, q must be r_k, for some natural number k. Using the notation of the matrix above, r_k may be represented as the infinite decimal expansion $0.\ a_0^k\ a_1^k\ a_2^k\ a_3^k\ \ldots a_k^k\ldots$, where $a_0^k = b_0$, $a_1^k = b_1$, $a_2^k = b_2$, …, and $a_k^k = b_k$, … .

9. From 6 and 8: according to the definition of q, if $a_k^k = 1$, $b_k = 2$, and if $a_k^k \neq 1$, $b_k = 1$. Thus $a_k^k \neq b_k$, which contradicts 8—namely that $a_k^k = b_k$.

10. From 3 through 9: since we obtain a contradiction, the Reductio Assumption must be false, and hence card(**I**) ≠ card(**N**).

11. From 2 and 10 by Definition 2.3.4d: card(**I**) > card(**N**).

We now prove that the set of the rational numbers, ℚ, is countable. We will show that ℚ is equinumerous with a subset of **N**. As explained above, this entails that ℚ is countable. Since ℚ is an infinite set, it is countably infinite. Hence card(ℚ) = $\aleph_0$.

Theorem 2.3.1: ℚ is countable.

Proof

1. We will first prove that the set of all the non-negative rational numbers, ℚ⁺, is countable. Every member of ℚ⁺ can be expressed as the ratio between two natural numbers n and m, such that m is not zero. In set notation, ℚ = {n/m: n and m ∈ **N** and m ≠ 0}. Therefore the members of ℚ⁺ can be represented as the entries of a matrix whose rows supply the numerators of the ratios and whose columns supply

the denominators of the ratios. We should note that these representations are not unique: every rational number has infinitely many representations. For instance, 0/1 = 0/2 = 0/3 = ..., and 6/7 = 12/14 = 18/21, ... Thus the matrix below has many duplicate entries.

	1	2	3	4	5	...
0	0/1 [0th]	0/2 [1st]	0/3 [3rd]	0/4 [6th]	0/5 [10th]	...
1	1/1 [2nd]	1/2 [4th]	1/3 [7th]	1/4 [11th]	1/5	...
2	2/1 [5th]	2/2 [8th]	2/3 [12th]	2/4	2/5	...
3	3/1 [9th]	3/2 [13th]	3/3	3/4	3/5	...
4	4/1 [14th]	4/2	4/3	4/4	4/5	...
5	5/1	5/2	5/3	5/4	5/5	...
⋮	⋮	⋮	⋮	⋮	⋮	...

2. From 1: the arrows between the entries of the matrix describe the counting process. This is an infinite counting process in which every non-negative rational number is counted infinitely many times. For example, 0/1 is counted as the 0th entry, 0/2 as the 1st, 0/3 as the 3rd, 0/4 as the 6th, and so on. But all of these entries are representations of the same rational number. We stipulate that each non-negative rational number is to be counted only the first time it appears in the list and that all subsequent listings of it are to be deleted. This stipulation creates a one-to-one correspondence between $\mathbb{Q}^+$ and a *proper subset* of $\mathbb{N}$. This entails that $\mathbb{Q}^+$ is countable; and since it is infinite, it is countably infinite. Therefore, card($\mathbb{Q}^+$) = $\aleph_0$.

3. From 2: the same counting process can be used to count the non-positive rational numbers $\mathbb{Q}^-$. The only change we need to make in the matrix above is to replace the non-negative natural numbers that enumerate its rows with their negative counterparts. By the same reasoning, as in 2, we establish that $\mathbb{Q}^-$ is countably infinite.

4. From 2 and 3: it is an easy exercise to show that the union of two countable sets is countable. Since $\mathbb{Q} = \mathbb{Q}^+ \cup \mathbb{Q}^-$, $\mathbb{Q}$ is also countably infinite. (We will prove a more general statement in the exercises.)

2.4   An Economical Version of PL

2.4.1   The version of PL that we studied in Chapter One is an "inflated" version. Many aspects of that version can be derived from other aspects of it. For example, we do not need all of the five sentential connectives we included in PL syntax. As we will explain below, the set {¬, →} is **expressively complete**; this means that every binary sentential connective can be, in some sense, expressed in terms of the connectives ¬ and →. We could have included only these connectives in the official syntax of PL, and then introduced as many truth-functional binary connectives as we please as **defined notation**, that is, as symbols that abbreviate other strings of symbols. For instance, we could have defined the notation $X \vee Y$ as an abbreviation for the string $\neg X \rightarrow Y$. Also, we could have adopted the universal quantifier as our only quantifier and defined $(\exists z)X$ as an abbreviation for the string $\neg(\forall z)\neg X$. We know from Chapter One that if the formula $(\exists z)X$ is a component of some PL sentence, then the sentence that results from replacing $(\exists z)X$ with $\neg(\forall z)\neg X$ is logically equivalent to the original sentence. This approach could have greatly limited the inference rules we included in our Natural Deduction System described in Chapter One. Since in the remaining part of this book we will be mostly occupied with proving metatheorems about PL, it is advisable that we work with as few connectives, quantifiers, and inference rules as possible. This would simplify and shorten most of our proofs. In this section, we will develop an economical version of PL that is equivalent to the full version developed in the previous chapter. We will not give this version a new name. Since both versions are equivalent (they validate exactly the same set of inferences), we will call each one of them "First-Order Predicate Logic" or simply "PL." In what remains of this book, when we speak of PL, we intend the economical version, unless we explicitly state otherwise.

We begin our construction of the economical version of PL by explaining the notion of **expressive completeness.** There are certain sets of sentential connectives that can, in some sense, *express* all of the unary and binary truth-functional connectives. We call such sets **expressively complete**. This notion of expressive completeness is different from the proof-theoretic notion of completeness that we studied in Chapter One. It should be clear by now that a truth-functional connective is a connective such that the truth value of a sentence generated by that connective is fully determined by the truth values of its sentential components. Since a unary connective operates on a single sentence and since the maximum number of truth values a sentence may have is two, there are 4 unary truth-functional connectives: one preserves the truth value of the sentence it applies to, one switches its truth value, one assigns only F, and one assigns only T. A binary truth-functional connective operates on two sentences, and it assigns a truth value to the compound sentence that is determined by the truth values of its immediate components. Since there is a maximum of 4 different distributions of truth values to any two PL sentences, and since for every such distribution the compound sentence can either be true or false, there are 16 binary truth-functional connectives. The four binary connectives that are part of the traditional vocabulary of PL have special significance since they are meant to correspond to popular natural-language connectives. However, their choice is partly a historical accident. For instance, there is no compelling reason why we do not have special PL connectives that correspond to the English connectives 'neither-nor' and 'not both'. At any rate, the traditional set of connectives, $\{\neg, \wedge, \vee, \rightarrow, \leftrightarrow\}$, is expressively complete. In fact, it contains more than enough: several of these connectives can be expressed in terms of other connectives. To express a truth-functional connective * in terms of a set S of other truth-functional connectives is to construct a method for rewriting any sentence that contains the connective * as a sentence that consists only of the atomic components of the original sentence and some or all of the connectives in S, such that the two sentences have the same truth conditions. If **X** is a compound sentence that contains the connective *, **Y** is a sentence that contains only connectives from the set S, and **X** and **Y** have identical truth conditions, then we say that **Y** is **an expansion of X in terms of** S. Using this terminology, we can now say that a connective * is **expressible in terms of a set** S of connectives if and only if every sentence that contains * has an expansion in terms of S. We are ready to state the definition of expressive completeness.

Expressive Completeness: A set S of truth-functional connectives is expressively complete if and only if every unary and binary truth-functional connective is expressible in terms of S.

We will give the truth conditions of all the unary and binary truth-functional connectives. We will express these truth conditions as truth tables. We will assign symbols to all of them: some of these symbols are standard and some of them are invented here. We will list all the standard symbols that are widely in use. If the connective can reasonably be considered as corresponding to English connectives, we will also indicate some of these natural connectives. The list below is tedious. We could combine all of these connectives in one large table, but we aim at clarity and thoroughness rather than economy.

Unary connective: **tautology**.

X	⊤ (standard symbol)
T	T
F	T

Unary connective: **contradiction**.

X	⊥ (standard symbol)
T	F
F	F

Unary connective: **logical identity**; 'it is the case that', 'it is true that'.

X	X
T	T
F	F

Unary connective: **negation** or **denial**; 'no', 'not', 'it is not the case that', 'it is false that'.

X	¬X, ~X, –X (standard symbols)
T	F
F	T

Binary connective: **conjunction**; 'and', 'as well as', 'but', 'although', 'in spite of'.

X	Y	(X∧Y), (X&Y), (X•Y) (standard symbols)
T	T	T
T	F	F
F	T	F
F	F	F

Binary connective: the **Sheffer Stroke** (or NAND); 'not both'.

X	Y	(X↑Y), (X|Y) (standard symbols)
T	T	F
T	F	T
F	T	T
F	F	T

Binary connective: **Peirce's Arrow** (or NOR); 'neither-nor'.

X	Y	(X↓Y) (standard symbol)
T	T	F
T	F	F
F	T	F
F	F	T

Binary connective: **inclusive disjunction**; 'or', 'either-or', 'and/or'

X	Y	(X∨Y) (standard symbol)
T	T	T
T	F	T
F	T	T
F	F	F

Binary connective: **exclusive disjunction**; 'either-or but not both'.

X	Y	(X⊕Y), (X+Y) (standard symbols)
T	T	F
T	F	T
F	T	T
F	F	F

Binary connective: **material conditional** or **material implication**; 'if-then', 'only if'.

X	Y	(X→Y), (X⊃Y) (standard symbols)
T	T	T
T	F	F
F	T	T
F	F	T

Binary connective: **converse implication**.

X	Y	(X←Y), (X⊂Y) (standard symbols)
T	T	T
T	F	T
F	T	F
F	F	T

Binary connective: **material nonimplication**.

X	Y	(X↛Y), (X⊅Y) (standard symbols)
T	T	F
T	F	T
F	T	F
F	F	F

Binary connective: **converse nonimplication**.

X	Y	(X↚Y), (X⊄Y) (standard symbols)
T	T	F
T	F	F
F	T	T
F	F	F

Binary connective: **biconditional**; 'if and only if', 'just in case', 'when and only when'.

X	Y	(X↔Y), (X≡Y) (standard symbols)
T	T	T
T	F	F
F	T	F
F	F	T

The remaining six "binary" connectives do not have special symbols,[5] and they are not "really" binary—they are essentially either constant or unary. Syntactically, they can be represented as connecting two PL sentences (a practice that we will follow here), but the truth value of the compound sentence is either independent of the truth values of the immediate components or it depends on the truth values of only one immediate component. We will present the truth conditions of these connectives in a single truth table. The names of these connectives are, respectively, 'tautology', 'contradiction', 'X-negation', 'Y-negation', 'X-projection', and 'Y-projection'. The reader should be warned that the symbolism displayed in this table is nonstandard.

X	Y	(X⊤Y)	(X⊥Y)	(¬XY)	(X¬Y)	(ρXY)	(XρY)
T	T	T	F	F	F	T	T
T	F	T	F	F	T	T	F
F	T	T	F	T	F	F	T
F	F	T	F	T	T	F	F

Let S be the set of all these unary and binary truth-functional connectives. Many subsets of S are expressively complete. In this subsection we will give two examples of these subsets. The exercises contain one more example of such a set of connectives. The set $\{\neg, \rightarrow\}$ is expressively complete. The rules below show how every compound sentence whose main connective is one of the connectives in S can be expanded in terms of the set $\{\neg, \rightarrow\}$. Simple truth tables can show that the sentences of each of the following pairs have the same truth conditions. At the end of the list we will construct an example of such a table. The reader can easily construct the other tables.

Tautology:	⊤	=	(X→X)
Contradiction:	⊥	=	¬(X→X)
Logical identity:	X	=	X
Negation:	¬X	=	¬X
Conjunction:	(X∧Y)	=	¬(X→¬Y)
The Sheffer Stroke:	(X↑Y)	=	(X→¬Y)
Peirce's Arrow:	(X↓Y)	=	¬(¬X→Y)
Inclusive Disjunction:	(X∨Y)	=	(¬X→Y)
Exclusive Disjunction:	(X⊕Y)	=	¬((¬X→Y)→¬(X→¬Y))
Material conditional:	(X→Y)	=	(X→Y)

---

5   In fact, every truth-functional binary connective has a special "prefix." For example, the material conditional of A and B may be written as 'CAB', where the letter 'C' is the prefix of the material conditional, and the exclusive disjunction of A and B may be written as 'JAB', where the letter 'J' is the prefix of exclusive disjunction. However, these prefixes are not widely used in the literature, whether introductory or advanced.

Converse implication: $(X \leftarrow Y)$ = $(Y \rightarrow X)$
Material nonimplication: $(X \not\rightarrow Y)$ = $\neg(X \rightarrow Y)$
Converse nonimplication: $(X \not\leftarrow Y)$ = $\neg(Y \rightarrow X)$
Biconditional: $(X \leftrightarrow Y)$ = $\neg((X \rightarrow Y) \rightarrow \neg(Y \rightarrow X))$
Tautology and contradiction: as above.
X-negation and Y-negation: $(\neg XY)$ = $\neg X$ and $(X\neg Y)$ = $\neg Y$
X-projection and Y-projection: (ρXY) = X and $(X\rho Y)$ = Y

The following table shows that the expansions of Peirce's Arrow and exclusive disjunction in terms of the set $\{\neg, \rightarrow\}$ are correct. The columns of the main connectives are boldfaced.

X	Y	X↓Y	¬(¬X → Y)	X⊕Y	¬((¬X → Y) → ¬(X → ¬Y))
T	T	T F T	**F** F T T T	T F T	**F** F T T T T T T F F T
T	F	T F F	**F** F T T F	T T F	**T** F T T F F F T T T F
F	T	F F T	**F** T F T T	F T T	**T** T F T T F F F T F T
F	F	F T F	**T** T F F F	F F F	**F** T F F F T F F T T F

The columns under Peirce's Arrow and under the main negation sign of the second compound sentence are identical. This indicates that these sentences have identical truth conditions. The same observation applies to the third and fourth compound sentences.

The only single connectives that are expressively complete are the Sheffer Stroke and Peirce's Arrow. In other words, every PL compound sentence that contains no quantifiers and no function symbols and whose connectives are among the 20 connectives discussed above has expansions in terms of $\{\uparrow\}$ and in terms of $\{\downarrow\}$. Since we established the expressive completeness of the set $\{\neg, \rightarrow\}$, it suffices to show that the negation and the material conditional are expressible in terms of these connectives. This would prove that each of $\{\uparrow\}$ and $\{\downarrow\}$ is expressively complete. We will demonstrate here the expressive completeness of the Sheffer Stroke, and we leave the case of Peirce's arrow as an exercise.

Negation: $\neg X$ = $(X \uparrow X)$
Material conditional: $(X \rightarrow Y)$ = $(X \uparrow (Y \uparrow Y))$

We demonstrate the correctness of these expansions by means of the table below.

X	Y	¬X	X↑X	X → Y	X ↑ (Y ↑ Y)
T	T	**F** T	**T** F T	**T** T T	**T** T T F T
T	F	**F** T	**T** F T	**T** F F	**T** F F T F
F	T	**T** F	**F** T F	**F** T T	**F** T T F T
F	F	**T** F	**F** T F	**F** T F	**F** T F T F

The columns headed by the first negation sign and by the first Sheffer Stroke are the same. This establishes that the truth conditions of these sentences are identical. The same is true of the third and fourth sentences. Since the negation and the material conditional are expressible in terms of the Sheffer Stroke and since {¬, →} is expressively complete, the singleton {↑} is also expressively complete.

2.4.2   The syntax and semantics of the economical version of PL is identical with the syntax and semantics of the full version of PL except for the logical symbols. In the economical version, we have only four logical symbols: one unary connective ¬, one binary connective →, one quantifier symbol ∀, and the identity symbol =. Of course, once we limit the syntax of PL, the formation rules and the truth conditions listed in Chapter One should be truncated by deleting the formation rules and the truth conditions of the forms $X \wedge Y$, $X \vee Y$, $X \leftrightarrow Y$, and $(\exists z)X$. One may wish to define the rest of the binary connectives and the existential quantifier on the basis of {¬, →, ∀}, or simply work with the economical version as is. We will choose the latter approach in proving the Soundness and Completeness Theorems (Chapter Three), but we will follow the former approach in outlining the proofs of the Incompleteness Theorems (Chapter Five).

The important task is to select inference rules that are *independent*, *sound*, and *complete*. By 'independent' we mean that none of these rules is derivable from the other rules. We explained in Chapter One what we mean by 'sound' and 'complete'. We will introduce nine inference rules that are sufficient for deriving all the inference rules of the Natural Deduction System (NDS) described in Chapter One. (We are counting Reductio Ad Absurdum as two rules.) We call this economical set of inference rules the **Mini Deduction System** (MDS). This is not a standard name. Indeed, all sets of inference rules that are sound and complete are usually referred to as 'natural deduction systems'. But we want to distinguish the economical set of rules from the system described in Subsection 1.4.5. We begin by listing the rules of MDS.

Reiteration (Reit)
Modus Ponens (MP)
Conditional Proof (CP)
Universal Instantiation (UI)
Universal Generalization (UG)
Identity (Id)
Substitution (Sub)
Both parts of Reductio Ad Absurdum (RAA)

Deriving all the rules of NDS from the rules above is a tedious task, since our NDS consists of thirty-three rules, many of which have two parts. We will simplify matters by invoking a well-known fact about classical inference rules. All of the inference rules we included in NDS can be derived from the Gentzen introduction and elimination rules (GDS), which we described in Subsection 1.4.7. For the convenience of the reader, we relist here the names and symbols of these 17 GDS rules: (1) Reiteration; (2–3) Conjunction Introduction (∧I) and Conjunction Elimination (∧E); (4–5) Conditional Introduction (→I) and Conditional Elimination (→E); (6–7) Universal Introduction (∀I) and Universal Elimination (∀E); (8–9) Existential Introduction (∃I) and Existential Elimination (∃E); (10–11) Identity Introduction (=I) and Identity Elimination (=E); (12–13) Negation Introduction (¬I) and Negation Elimination (¬E); (14–15) Disjunction Introduction (∨I) and Disjunction Elimination (∨E); (16–17) Biconditional Introduction (↔I) and Biconditional Elimination (↔E). Thus it is sufficient to derive the GDS rules from the MDS rules. Since the introduction and elimination rules for the conditional, the negation, the universal quantifiers, and the identity predicate are, respectively, Conditional Proof and Modus Ponens, the two parts of Reductio Ad Absurdum, Universal Generalization and Universal Instantiation, and Identity and Substitution, we only need to derive the introduction and elimination rules for the conjunction, disjunction, biconditional, and the existential quantifier. Before presenting these derivations, we will derive Modus Tollens (MT). It will simplify some of our derivations.

Deriving MT

[n
  ⋮
  h   X→Y   Given
  ⋮
  i   ¬Y   Given
[n+1 i+1   X   RA
     i+2   Y   h, (i+1), MP
n+1] i+3   ¬Y   i, Reit
     i+4   ¬X   (i+1)–(i+3) RAA
  ⋮
n]

∧I and ∧E: X∧Y = ¬(X→¬Y)

Deriving ∧I:

[n
  ⋮
  h   X   Given
  ⋮
  i   Y   Given
[n+1 i+1   X→¬Y   RA
     i+2   ¬Y   h, (i+1), MP
n+1] i+3   Y   i, Reit
     i+4   ¬(X→¬Y)   (i+1)–(i+3), RAA
  ⋮
n]

Deriving ∧E (two parts):

[n
 ⋮
  h ¬(X→¬Y) Given
[n+1 h+1 ¬X RA
[n+2 h+2 X CPA
[n+3 h+3 Y RA
  h+4 X h+2, Reit
n+3] h+5 ¬X h+1, Reit
n+2] h+6 ¬Y (h+3)–(h+5), RAA
  h+7 X→¬Y (h+2)–(h+6), CP
n+1] h+8 ¬(X→¬Y) h, Reit
  h+9 X (h+1)–(h+8), RAA
 ⋮
n]

[n
 ⋮
  h ¬(X→¬Y) Given
[n+1 h+1 ¬Y RA
[n+2 h+2 X CPA
n+2] h+3 ¬Y h+1, Reit
  h+4 X→¬Y (h+2)–(h+3), CP
n+1] h+5 ¬(X→¬Y) h, Reit
  h+6 Y (h+1)–(h+5), RAA
 ⋮
n]

∨I and ∨E: X∨Y = ¬X→Y

Deriving ∨I (two parts):

[n
 ⋮
  h X Given
[n+1 h+1 ¬X CP
[n+2 h+2 ¬Y RA
  h+3 X h, Reit
n+2] h+4 ¬X h+1, Reit
n+1] h+5 Y (h+2)–(h+4), RAA
  h+6 ¬X→Y (h+1)–(h+5), CP
 ⋮
n]

[n
    ⋮
    h   Y   Given

[n+1] h+1   ¬X   CP
n+1]   h+2   Y   h, Reit
        h+3   ¬X→Y   (h+1)–(h+2), CP
    ⋮
n]

Deriving ∨E:

[n
    ⋮
    h   ¬X→Y   Given
[n+1] h+1   X   Assumption
        ⋮
n+1]   m   Z   (the subderivation (h+1)–m is given)
[n+j]   m+1   Y   Assumption
        ⋮
n+j]   k   Z   (the subderivation (m+1)–k is given)
        k+1   X→Z   (h+1)–m, CP
        k+2   Y→Z   (m+1)–k, CP
[n+j+1 k+3   ¬Z   RA
        k+4   ¬X   k+1, k+3, MT
        k+5   ¬Y   k+2, k+3, MT
n+j+1] k+6   Y   h, k+4, MP
        k+7   Z   (k+3)–(k+6), RAA
    ⋮
n]

↔I and ↔IE: $X \leftrightarrow Y = \neg((X \to Y) \to \neg(Y \to X))$

Deriving ↔I:

[n			
	⋮		
[n+1	h	X	Assumption
	⋮		
n+1]	m	Y	(subderivation h–m is given)
[n+j	m+1	Y	Assumption
	⋮		
n+j]	k	X	(subderivation (m+1)–k is given)
	k+1	X→Y	h–m, CP
	k+2	Y→X	(m+1)–k, CP
[n+j+1	k+3	(X→Y)→¬(Y→X)	RA
	k+4	¬(Y→X)	k+1, k+3, MP
n+j+1]	k+5	Y→X	k+2, Reit
	k+6	¬((X→Y)→¬(Y→X))	(k+3)–(k+5), RAA
	⋮		
n]			

Deriving ↔E (two parts)

[n			
	⋮		
	h	¬((X→Y)→¬(Y→X))	Given
	⋮		
	i	X	Given
[n+1	i+1	¬Y	RA
[n+2	i+2	X→Y	CPA
[n+3	i+3	Y→X	RA
	i+4	Y	i, i+2, MP
n+3]	i+5	¬Y	i+1, Reit
n+2]	i+6	¬(Y→X)	(i+3)–(i+5), RAA
	i+7	(X→Y)→¬(Y→X)	(i+2)–(i+6), CP
n+1]	i+8	¬((X→Y)→¬(Y→X))	h, Reit
	i+9	Y	(i+1)–(i+8), RAA
	⋮		
n]			

```
[n
    ⋮
    h    ¬((X→Y)→¬(Y→X))        Given
    ⋮
    i    Y                       Given
[n+1    i+1  ¬X                  RA
[n+2    i+2  X→Y                 CPA
[n+3    i+3  Y→X                 RA
        i+4  X                   i, i+2, MP
 n+3]   i+5  ¬X                  i+1, Reit
 n+2]   i+6  ¬(Y→X)              (i+3)–(i+5), RAA
        i+7  (X→Y)→¬(Y→X)        (i+2)–(i+6), CP
 n+1]   i+8  ¬((X→Y)→¬(Y→X))     h, Reit
        i+9  X                   (i+1)–(i+8), RAA
    ⋮
 n]
```

∃I and ∃E: (∃z)Y = ¬(∀z)¬Y

Deriving ∃I: **X** is a PL sentence, **s** is a PL *singular term* that occurs in **X**, and **z** is a PL variable that does not occur in **X**. **X[z, s]** is a PL *formula* formed by replacing *one or more* of the occurrences of **s** in **X** by **z**.

```
[n
        ⋮
        h    X                    Given
[n+1    h+1  (∀z)¬X[z, s]         RA
        h+2  ¬X[s, s]             h+1, UI    (X[s, s] is obtained from X[z, s] by
                                              replacing all the occurrences of the
                                              variable z in X[z, s] by the singular term s.
                                              Since X[z, s] is formed by replacing one or
                                              more of the occurrences of s in X by z, X[s,
                                              s] is simply X.)
 n+1]   h+3  X                    h, Reit
        h+4  ¬(∀z)¬X[z, s]        (h+1)–(h+3), RAA
        ⋮
 n]
```

RESOURCES OF THE METATHEORY

Deriving ∃E: **s** is a PL *name*, **z** is a PL variable, and **X** is a PL formula that contains occurrences of **z** but no **z**-quantifiers. **s** satisfies three conditions: (1) it does not occur in any premise or undischarged assumption prior to line h, (2) it does not occur in $\neg(\forall z)\neg X$, and (3) it does not occur in **Y**. **X[s]** is the PL sentence formed by replacing *all* the occurrences of **z** in **X** by **s**.

```
[n
        ⋮
        h–1     ¬(∀z)¬X          Given
[n+1    h       X[s]             Assumption
        ⋮
n+1]    k       Y                (the subderivation h–k is given)
        k+1     X[s]→Y           h–k, CP
[n+2    k+2     ¬Y               RA
        k+3     ¬X[s]            k+1, k+2, MT
        k+4     (∀z)¬X[z]        k+3, UG   (X[z] is obtained from X[s] by replacing all the
                                            occurrences of the name s with the variable z.
                                            Given that s does not occur in X (second
                                            condition) and that X[s] is formed by
                                            replacing all the occurrences of z in X by s, the
                                            formula X[z] is simply X. s is arbitrary at line
                                            k+3 since s does not occur in any premise or
                                            undischarged assumption prior to line h (the
                                            first condition), s occurs in the assumption at
                                            line h but this assumption is discharged after
                                            line k, and s does not occur in the assumption
                                            at line k+2 (third condition).)
n+2]    k+5     ¬(∀z)¬X          h–1, Reit
        k+6     Y                (k+2)–(k+5), RAA
  ⋮
n]
```

The above derivations establish our claim that NDS is derivable from MDS. All the rules of MDS are included in NDS. Hence MDS and NDS are "equivalent" systems: they validate exactly the same inferences. We state this claim precisely as the metatheorem below. But before we state the theorem, we need to introduce some notation and an expression. Let Ω be the set $\{\neg, \rightarrow, \forall\}$. A PL sentence all of whose logical operators belong to Ω is written as Y^Ω. If **X** is a PL sentence whose logical operators belong to the set $\{\neg, \wedge, \vee, \rightarrow, \leftrightarrow, \forall, \exists\}$, then X^Ω is the sentence that is obtained from **X** by expressing all the logical operators of **X** in terms of the members of Ω, using the abbreviations introduced in the derivations above. For example, if **X** is $(\exists w)(\forall x)(Px \vee Gxw) \wedge (Gra \rightarrow (\forall z)(Dz \leftrightarrow Ds))$, X^Ω can be obtained from **X** via the following sequence of transformations:

1. $(\exists w)(\forall x)(Px \vee Gxw) \Rightarrow \neg(\forall w)\neg(\forall x)(Px \vee Gxw) \Rightarrow \neg(\forall w)\neg(\forall x)(\neg Px \to Gxw)$
2. $(Gra \to (\forall z)(Dz \leftrightarrow Ds)) \Rightarrow (Gra \to (\forall z)\neg((Dz \to Ds) \to \neg(Ds \to Dz)))$
3. $\mathbf{X} = (\exists w)(\forall x)(Px \vee Gxw) \wedge (Gra \to (\forall z)(Dz \leftrightarrow Ds)) \Rightarrow \neg((\exists w)(\forall x)(Px \vee Gxw) \to \neg(Gra \to (\forall z)(Dz \leftrightarrow Ds)))$
4. Substituting 1 and 2 in 3, we have: $\neg((\exists w)(\forall x)(Px \vee Gxw) \to \neg(Gra \to (\forall z)(Dz \leftrightarrow Ds))) \Rightarrow \neg(\neg(\forall w)\neg(\forall x)(\neg Px \to Gxw) \to \neg(Gra \to (\forall z)\neg((Dz \to Ds) \to \neg(Ds \to Dz)))) = \mathbf{X}^{\Omega}$.

We call $\mathbf{X}^{\Omega}$ "the Ω expansion of $\mathbf{X}$." We extend this notation to sets of PL sentences: if Γ is a set of PL sentences, Γ^{Ω} is the set that consists of all the Ω expansions of the members of Γ. It is clear that if the members of Γ are all expressed in terms of the operators in Ω, then $\Gamma^{\Omega} = \Gamma$. Now we state the metatheorem.

Theorem 2.4.1: For every PL sentence $\mathbf{X}$ and for every set Γ of PL sentences, $\mathbf{X}$ is derivable from Γ in NDS iff $\mathbf{X}^{\Omega}$ is derivable from Γ^{Ω} in MDS—that is, $\Gamma \vdash_{\text{NDS}} \mathbf{X}$ iff $\Gamma^{\Omega} \vdash_{\text{MDS}} \mathbf{X}^{\Omega}$, where $\vdash_{\text{NDS}}$ and $\vdash_{\text{MDS}}$ denote the relation of derivability in NDS and MDS, respectively.

The derivations we gave above constitute a proof of this theorem.

It is obvious that every sentence is logically equivalent with its Ω expansion. This entails that the following metatheorem is true.

Theorem 2.4.2: For every PL sentence $\mathbf{X}$ and for every set Γ of PL sentences, $\mathbf{X}$ is a logical consequence of Γ iff $\mathbf{X}^{\Omega}$ is a logical consequence of Γ^{Ω}—that is, $\Gamma \vDash \mathbf{X}$ iff $\Gamma^{\Omega} \vDash \mathbf{X}^{\Omega}$.

We leave to the reader the proof of this theorem as an easy exercise (see Exercise 2.5.9). We will prove the soundness and completeness of MDS in the next chapter. Given Theorems 2.4.1 and 2.4.2, the soundness and completeness of MDS entails that NDS is sound and complete, and vice versa. We state and prove this corollary below.

Corollary 2.4.1: NDS is sound and complete iff MDS is sound and complete.

Proof

1. First, suppose that NDS is sound and complete.
2. From 1: for every PL sentence $\mathbf{X}$ and every PL set Γ, $\Gamma \vDash \mathbf{X}$ iff $\Gamma \vdash_{\text{NDS}} \mathbf{X}$.
3. Our goal now is to establish that MDS is sound and complete. In other words, we want to show that for every PL sentence $\mathbf{Y}^{\Omega}$ and every PL set Σ^{Ω}, $\Sigma^{\Omega} \vDash \mathbf{Y}^{\Omega}$ iff $\Sigma^{\Omega} \vdash_{\text{MDS}} \mathbf{Y}^{\Omega}$. We begin by assuming that $\Sigma^{\Omega} \vDash \mathbf{Y}^{\Omega}$.
4. From 2 and 3: $\Sigma^{\Omega} \vdash_{\text{NDS}} \mathbf{Y}^{\Omega}$.
5. From 4 by Theorem 2.4.1: $\Sigma^{\Omega} \vdash_{\text{MDS}} \mathbf{Y}^{\Omega}$.
6. From 3 through 5: if $\Sigma^{\Omega} \vDash \mathbf{Y}^{\Omega}$, then $\Sigma^{\Omega} \vdash_{\text{MDS}} \mathbf{Y}^{\Omega}$. Hence MDS is complete.
7. Now we assume that $\Sigma^{\Omega} \vdash_{\text{MDS}} \mathbf{Y}^{\Omega}$.
8. From 7 by Theorem 2.4.1: $\Sigma^{\Omega} \vdash_{\text{NDS}} \mathbf{Y}^{\Omega}$.

9. From 2 and 8: $\Sigma^\Omega \vDash Y^\Omega$.
10. From 7 through 9: if $\Sigma^\Omega \vdash_{MDS} Y^\Omega$, then $\Sigma^\Omega \vDash Y^\Omega$. Hence MDS is sound.
11. From 1 through 10: if NDS is sound and complete, then MDS is sound and complete.
12. Second, suppose that MDS is sound and complete.
13. From 12: for every PL sentence Y^Ω and every PL set Σ^Ω, $\Sigma^\Omega \vDash Y^\Omega$ iff $\Sigma^\Omega \vdash_{MDS} Y^\Omega$.
14. We want to prove now that NDS is sound and complete; that is, for every PL sentence X and every PL set Γ, $\Gamma \vDash X$ iff $\Gamma \vdash_{NDS} X$. We show first the completeness of NDS. Thus assume that $\Gamma \vDash X$.
15. From 14 by Theorem 2.4.2: $\Gamma^\Omega \vDash X^\Omega$.
16. From 13 and 15: $\Gamma^\Omega \vdash_{MDS} X^\Omega$.
17. From 16 by Theorem 2.4.1: $\Gamma \vdash_{NDS} X$.
18. From 14 through 17: if $\Gamma \vDash X$, then $\Gamma \vdash_{NDS} X$. This means that NDS is complete.
19. To show that NDS is sound, we assume that $\Gamma \vdash_{NDS} X$.
20. From 19 by Theorem 2.4.1: $\Gamma^\Omega \vdash_{MDS} X^\Omega$.
21. From 13 and 20: $\Gamma^\Omega \vDash X^\Omega$.
22. From 21 by Theorem 2.4.2: $\Gamma \vDash X$.
23. From 19 through 22: if $\Gamma \vdash_{NDS} X$, then $\Gamma \vDash X$. Hence NDS is sound.
24. From 12 through 23: if MDS is sound and complete, then NDS is sound and complete.
25. From 11 and 24: NDS is sound and complete iff MDS is sound and complete.

Given Theorems 2.4.1 and 2.4.2 and Corollary 2.4.1, in the remainder of this book we will only prove metatheorems about the economical version of PL, whose logical symbols are $\neg$, $\rightarrow$, $\forall$, and $=$, and whose deduction system is MDS. All of these theorems are equally true of the full version of PL, whose logical symbols are $\neg$, $\wedge$, $\vee$, $\rightarrow$, $\leftrightarrow$, $\forall$, $\exists$, and $=$, and whose deduction system is NDS. We will no longer speak of economical versus full versions of PL, nor do we write '$\vdash_{MDS}$' for the derivability relation in MDS. It should be understood that, unless it is stated otherwise, all our references to PL and its deduction system in the following chapters are to the economical version of PL and to MDS.

2.5 Exercises

Note: All answers must be adequately justified.

2.5.1 A binary relation R is an **equivalence relation** iff R is reflexive, symmetric, and transitive. Prove that the relation of equinumerosity, ≈, is an equivalence relation between sets.

2.5.2 Use PMI or PCI to prove the following.

    2.5.2a   For every natural number $n > 0$, $1 + 3 + 5 + \ldots + (2n-1) = n^2$.
    2.5.2b*   For every natural number $n > 0$, $3 + 6 + 9 + \ldots + 3n = 3n(n+1)/2$.
    2.5.2c   For every natural number $n > 0$, $1^3 + 2^3 + 3^3 + \ldots + n^3 = (1 + 2 + 3 + \ldots + n)^2$.
    2.5.2d   For every compound PL formula **X** that contains parentheses, the following counting process ρ always terminates with 0, and it never assigns any negative number to any parenthesis: ρ proceeds from left to right, assigning 1 to the first parenthesis and adding 1 for every subsequent left parenthesis and subtracting 1 for every subsequent right parenthesis.
    2.5.2e*   Every quantifier-free PL sentence has an expansion in terms of {¬, ∧}. (Use the full version of PL.)
    2.5.2f*   Let $\mathcal{F}$ be a nonempty, pairwise disjoint family of nonempty sets. A set C is called a **choice set** for $\mathcal{F}$ iff for every A in $\mathcal{F}$, the intersection of C and A is a singleton (i.e., C shares a single element with each member of $\mathcal{F}$). Every nonempty, finite, pairwise disjoint family of nonempty sets has a choice set.
    2.5.2g   For every nonempty, finite family $\mathcal{F}$ of countably infinite sets, $\cup\mathcal{F}$ is countably infinite.

2.5.3 Let $\mathcal{F}$ be a countably infinite family of countably infinite sets. Prove that $\cup\mathcal{F}$ is countably infinite.

2.5.4 For every set A, define the **successor** of A, which is denoted as "SA," to be the set $A \cup \{A\}$. For example, if A = {e, d}, S{e, d} = {e, d}∪{{e, d}} = {e, d, {e, d}}; if A = ∅, S∅ = ∅∪{∅} = {∅}; and if A = {∅}, S{∅} = {∅}∪{{∅}} = {∅, {∅}}. Furthermore, for any set A, define $S^0 A$ to be A and $S^{n+1}A$, where n is any natural number, to be the successor of $S^n A$, that is, $SS^n A$. Informally speaking, if $n \geq 1$, $S^n A$ is obtained from A by iterating the successor operation n times. Prove the following statements.

    2.5.4a   For every natural number n, $S^n\emptyset = \{S^q\emptyset$: q is a natural number $< n\}$.
    2.5.4b*   For all natural numbers n and m, $m = n$ iff $S^m\emptyset = S^n\emptyset$.
    2.5.4c   For all natural numbers n and m, $m < n$ iff $S^m\emptyset \subset S^n\emptyset$.

2.5.5 A family $\mathcal{F}$ of sets is said to be **inductive** iff $\emptyset \in \mathcal{F}$, and for every set A, if $A \in \mathcal{F}$, SA $\in \mathcal{F}$ (recall that SA is the successor of A as defined above). The set ω is defined as the "smallest" inductive set, that is, ω is inductive and for every inductive set $\mathcal{F}$, ω ⊆ $\mathcal{F}$. Establish the following statements.

    2.5.5a   For every β ⊆ ω, if ∅ ∈ β, and SA ∈ β whenever A ∈ β, then β = ω.
    2.5.5b* For every A, A ∈ ω iff there is a natural number n such that A = $S^n\emptyset$.
    2.5.5c   For all A and B in ω, A ∈ B iff A ⊂ B.
    2.5.5d   The relation ∈ on ω is irreflexive, asymmetric, transitive, connex, and has a minimal element.
    2.5.5e* Every nonempty subset of ω has an ∈-minimal element.
    2.5.5f   ω ≈ ℕ.

2.5.6*   Prove that for any finite set A, if card(A) = n, card($\mathcal{P}$A) = 2^n.

2.5.7   A binary relation R on a set A is a **linear ordering** of A iff R is asymmetric, transitive, and connex. R is a **well-founded** relation on A iff every nonempty subset of A has an R-minimal element. R is a **well-ordering** of A iff R is a well-founded linear ordering of A. If R is a well-ordering of a set A, we also say that A *is well-ordered by* R. Furthermore, we say that a set A *can be well-ordered* iff there is a well-ordering of A. Prove the following statements.

    2.5.7a   ∈ is a well-ordering of ω.
    2.5.7b   For all sets A and B, if A can be well-ordered and B ≈ A, then B can be well-ordered.
    2.5.7c   For all sets A and B, if A and B can be well-ordered, then A∪B can be well-ordered.
    2.5.7d* For all sets A and B, if A and B can be well-ordered, then their Cartesian product, A×B, can be well-ordered.

2.5.8   A set $\mathcal{A}$ is **transitive** iff $\mathcal{A}$ is a family of sets that contains the members of its members, that is, for all x and B, if B ∈ $\mathcal{A}$ and x ∈ B, then x ∈ $\mathcal{A}$. (Observe that this notion of a transitive set is different from the notion of a transitive relation.) An **ordinal** is a transitive set that is well-ordered by ∈. We use lowercase Greek letters to denote ordinals. Let **Ord** be the *class* of all ordinals. Prove the following statements. You may invoke these two facts without a proof: for every set D, D ∉ D; and for all ordinals α and β, α ∈ β or β ∈ α or α = β.

    2.5.8a   ω is an ordinal.
    2.5.8b*   Every member of an ordinal is an ordinal.
    2.5.8c   **Ord** is not a set.
    2.5.8d   For every ordinal α, Sα is an ordinal; and for any ordinal β, if α ∈ β, β ∉ Sα.
    2.5.8e*   For all ordinals α and β, β ∈ α iff β ⊂ α.
    2.5.8f   For every set K of ordinals, (1) ∪K is an ordinal; (2) every α in K belongs or is equal to ∪K; and (3) for any ordinal β, if each ordinal in K belongs to or is equal to β, then β ∉ ∪K.

2.5.9   Prove Theorem 2.4.2.

2.5.10   Using only the rules of MDS, show that Eg^2cc is derivable from the following set:
S1 $(\forall x)(\forall y)(Axy \rightarrow (\neg Bg^2xy \rightarrow Eg^2cc))$
S2 $\neg(\forall x)(Rx \rightarrow \neg Abx)$
S3 $(\forall x)(\forall y)\neg Bg^2xy$

2.5.11   Let Ω be $\{\neg, \wedge, \exists, =\}$. Suppose that we construct another economical version of PL that is based on these connectives, quantifier, and the identity predicate. Let us call this version PL^Ω. Define a deduction system DS, that is minimal, sound, and complete for PL^Ω. Assuming that MDS is complete, prove that your deduction system is also complete by deriving the rules of MDS from it.

2.5.12* Prove that for all PL singular terms **s** and **t**, for any PL sentence **X[s]** in which **s** occurs, and for every PL interpretation J that is relevant to **X[s]** and **t** (i.e., J interprets **X[s]** and **t**), if **s** = **t** is true on J, then the PL sentences **X[s]** and **X[s, t]** have the same truth value on J, where **X[s, t]** is obtained from **X[s]** by replacing one or more occurrences of **s** with **t**. You may invoke without proof the following fact: if J(**s**) is the same as J(**t**), then replacing one or more of the occurrences of **s** in some PL singular term **q** with **t** does not change J(**q**). (Recall that J(**r**) is the referent that J assigns to the PL singular term **r**.)

2.5.13   Suppose that we have a logical system that allows us to form infinite conjunctions of any sentences. For instance, in this logic the infinite conjunctions $Pa \wedge \neg Pa \wedge Pa \wedge \neg Pa \wedge Pa \wedge \neg Pa \wedge \ldots$ and $Pa \wedge Pa \wedge \neg Pa \wedge Pa \wedge Pa \wedge \neg Pa \wedge Pa \wedge Pa \wedge \neg Pa \wedge \ldots$ are well-formed sentences. Use diagonalization to prove that the set whose members are all the infinite conjunctions that are made of any combinations of Pa and $\neg Pa$ is uncountable.

2.5.14   Let $\mathcal{F}$ be a pairwise disjoint family of nonempty subsets of a set S. Prove that if S is countable, then so is $\mathcal{F}$.

2.5.15* Construct a PL sentence **X** from the 1-place predicates P and Q, the 1-place function symbol g, and the logical vocabulary of PL, such that for every positive integer n, **X** has a model of size 2n, but **X** has no models of odd sizes.

Solutions to the Starred Exercises

SOLUTIONS TO 2.5.2

2.5.2b Let E(n) be the formula: $3 + 6 + 9 + \ldots + 3n = 3n(n+1)/2$. We want to prove that for every natural number $n \geq 1$, E(n) is true.

Proof by PMI

1. The Base Step: Let $n = 1$. Left-hand side of E(1) = 3; right-hand side of E(1) = $3 \times 1(1+1)/2 = 3$. Therefore E(1) is true.

2. The Inductive Step: Assume that for some natural number $k \geq 1$, E(k) is true, that is, $3 + 6 + 9 + \ldots + 3k = 3k(k+1)/2$. This is the Induction Hypothesis. We want to prove that the statement is true for the successor of k. In other words, we want to prove that E(k+1) is true.

   Left-hand side of E(k+1) = $3 + 6 + 9 + \ldots + 3k + 3(k+1) = 3k(k+1)/2 + 3(k+1) = (3k(k+1) + 6(k+1))/2 = 3(k+1)(k+2)/2$ = right-hand side of E(k+1). Hence E(k+1) is true.

3. From 1 and 2 by PMI: for every natural number $n \geq 1$, E(n) is true.

2.5.2e We will show that every PL sentence that contains no quantifiers has an expansion in terms of the connectives $\neg$ and $\wedge$ by using mathematical induction on the number of sentential connectives in a sentence. We define the *complexity* of a PL sentence as the number of its sentential connectives. Precisely speaking, we will prove that the following formula is true of all natural numbers n:

E(n)   For every quantifier-free PL sentence **X**, if the complexity of **X** is n, then **X** has an expansion in terms of $\{\neg, \wedge\}$.

Proof by PCI

1. The Base Step: Let $n = 0$. Hence **X** contains no sentential connectives. **X** is its own expansion.

2. The Inductive Step: Suppose that for some natural number $k > 0$, E(m) is true for all natural numbers $m < k$. This is the Induction Hypothesis. Our goal now is to prove that E(k) is true. Thus let **X** be a quantifier-free PL sentence whose complexity is k. Since $k > 0$, **X** has at least one sentential connective. **X** can be $\neg$**Y**, **Y**$\wedge$**Z**, **Y**$\vee$**Z**, **Y**$\to$**Z**, or **Y**$\leftrightarrow$**Z**. We will consider each case in turn.

2.1. If X is $\neg Y$, the complexity of $Y < k$ since the complexity of $\neg Y$ is k. So the Induction Hypothesis is applicable to Y. Thus Y has an expansion in terms of $\{\neg, \wedge\}$. Let this expansion be Y^*. $\neg Y^*$ is, therefore, an expansion of $\neg Y$ in terms of $\{\neg, \wedge\}$.

2.2. If X is $Y \wedge Z$, the complexity of each of Y and $Z < k$. We apply the Induction Hypothesis to both Y and Z. Hence Y and Z have expansions in terms of $\{\neg, \wedge\}$. Let these expansions be Y^* and Z^*. It follows that $Y^* \wedge Z^*$ is an expansion of $Y \wedge Z$ in terms of $\{\neg, \wedge\}$.

2.3. If X is $Y \vee Z$, the complexity of each of Y and $Z < k$. The Induction Hypothesis is applicable to Y and Z. Therefore Y and Z have expansions in terms of $\{\neg, \wedge\}$. Let these expansions be Y^* and Z^*. Since $Y \vee Z$ is logically equivalent to $\neg(\neg Y \wedge \neg Z)$, $\neg(\neg Y^* \wedge \neg Z^*)$ is an expansion of $Y \vee Z$ in terms of $\{\neg, \wedge\}$.

2.4. If X is $Y \rightarrow Z$, the Induction Hypothesis is applicable to Y and Z because the complexity of each of Y and $Z < k$. Hence there are expansions of Y and Z in terms of $\{\neg, \wedge\}$. As above, we let these expansions be Y^* and Z^*. We know that $Y \rightarrow Z$ is logically equivalent to $\neg(Y \wedge \neg Z)$. Therefore $\neg(Y^* \wedge \neg Z^*)$ is an expansion of $Y \rightarrow Z$ in terms of $\{\neg, \wedge\}$.

2.5. Finally, if X is $Y \leftrightarrow Z$, we apply the Induction Hypothesis to Y and Z since their complexities $< k$. It follows that Y and Z have expansions in terms of $\{\neg, \wedge\}$. As usual, we take these expansions to be Y^* and Z^*. Since $Y \leftrightarrow Z$ is logically equivalent to $\neg(Y \wedge \neg Z) \wedge \neg(Z \wedge \neg Y)$, $\neg(Y^* \wedge \neg Z^*) \wedge \neg(Z^* \wedge \neg Y^*)$ is an expansion of $Y \leftrightarrow Z$ in terms of $\{\neg, \wedge\}$. This establishes the Inductive Step.

3. From 1 and 2 by PCI: for all natural numbers n, E(n) is true.

2.5.2f We will apply PMI to the cardinality of $\mathcal{F}$ to show that the formula E(n) is true of all natural numbers $n \geq 1$.

E(n) For every pairwise disjoint family $\mathcal{F}$ of nonempty sets, if card($\mathcal{F}$) = n, then $\mathcal{F}$ has a choice set.

Proof by PMI

1. The Base Step: Let n = 1. Since card($\mathcal{F}$) = 1, $\mathcal{F}$ contains only one element. By the definition of $\mathcal{F}$, this element is a nonempty set. Hence there is an object e that belongs to it. The singleton of e, {e}, is a choice set for $\mathcal{F}$.

2. The Inductive Step: Assume that for some natural number $k \geq 1$, E(k) is true—that is, every pairwise disjoint family $\mathcal{F}$ of nonempty sets whose cardinality is k has a choice set. This is the Induction Hypothesis. We want

RESOURCES OF THE METATHEORY 129

to prove that E(k+1) is true. So let $\mathcal{F}$ be a pairwise disjoint family of nonempty sets whose cardinality is k+1. Since $\mathcal{F}$ has k+1 elements, we can represent it as follows: $\mathcal{F} = \{A_1, A_2, A_3, \ldots, A_k, A_{k+1}\} = \{A_1, A_2, A_3, \ldots, A_k\} \cup \{A_{k+1}\}$. Since the family $\{A_1, A_2, A_3, \ldots, A_k\}$ has k elements, the Induction Hypothesis is applicable to it. Thus there is a choice set C such that $C \cap A_m = \{e_m\}$ for every m = 1, 2, 3, ..., k. By the definition of $\mathcal{F}$, A_{k+1} is nonempty and shares no common elements with any of the sets A_1, A_2, A_3, ..., and A_k. It follows that there is an object e_{k+1} that belongs to A_{k+1} but is not a member of any of the sets A_1, A_2, A_3, ..., and A_k. We let C* be the set $C \cup \{e_{k+1}\}$ — that is, we add e_{k+1} to C. The new set C* is a choice set for $\mathcal{F}$, since it shares a single element with each member of $\mathcal{F}$. This establishes the Inductive Step.

3. From 1 and 2 by PMI, for every natural number n ≥ 1, E(n) is true.

SOLUTION TO 2.5.4

2.5.4b Our goal is to prove that for all natural numbers n and m, m = n iff $S^m\emptyset = S^n\emptyset$. The 'only-if' part might appear too obvious to require any proof. After all, is it not clear that $S^n\emptyset = S^n\emptyset$? In fact, there is a point in proving that if n = m, $S^m\emptyset = S^n\emptyset$. The point is to show that applying the successor operation to a set any number of times always generates a *unique* set. There are operations that when applied to a single individual generate two outcomes. For example if we apply the operation "the square root of" to the number 4, we obtain two numbers: 2 and −2 (unless, of course, 'the square root of' means 'the positive square root of'). The 'only-if' part of the theorem we wish to prove ensures that '$S^n(\emptyset)$' denotes exactly one set. The proof is below.

 1. We first prove the 'only-if' part. So let n and m be any identical natural numbers. We want to show that $S^m\emptyset = S^n\emptyset$. We will prove this using PMI.

 2. The Base Step. Let n = 0. $S^0\emptyset = \emptyset$. Since there is exactly one empty set, $S^0\emptyset$ is a unique set.

 3. The Inductive Step. Now suppose that $S^k\emptyset$ is a unique set, where k is some natural number ≥ 0. This is the Induction Hypothesis. We wish to prove that $S^{k+1}\emptyset$ is a unique set. By definition, $S^{k+1}\emptyset = S^k\emptyset \cup \{S^k\emptyset\}$. Hence, for every object x, $x \in S^{k+1}\emptyset$ iff $x \in S^k\emptyset$ or $x = S^k\emptyset$. By the Induction Hypothesis, '$S^k(\emptyset)$' designates one and only one set. It follows, by the Principle of Extensionality, that '$S^{k+1}\emptyset$' designates one and only one set as well.

 4. From 2 and 3 by PMI: for every natural number n, $S^n\emptyset$ is a unique set — that is, for all natural numbers n and m, if m = n, $S^m\emptyset = S^n\emptyset$.

5. Now we prove the 'if' part. Thus let $S^m\emptyset = S^n\emptyset$. We want to show that m = n. We will apply PCI to n.

6. The Base Step. Let n = 0. We have $S^m\emptyset = S^0(\emptyset) = \emptyset$. If m = q+1 for some natural number q, $S^m\emptyset = S^{q+1}\emptyset = S^q\emptyset \cup \{S^q\emptyset\}$. Since $S^q\emptyset \in S^{q+1}\emptyset$, $S^{q+1}(\emptyset) \neq \emptyset$. Given that $S^m\emptyset = \emptyset$, it follows that $S^m\emptyset \neq S^{q+1}(\emptyset)$; and hence m ≠ q+1, for any natural number q. Therefore, m = 0 = n. This establishes the Base Step.

7. The Inductive Step. Suppose that for some natural number k > 0 and for every natural number q < k, if $S^m\emptyset = S^q\emptyset$, then m = q. This is the Induction Hypothesis. We wish to show that if $S^m\emptyset = S^k\emptyset$, then m = k. So let $S^m\emptyset = S^k\emptyset$. By Exercise 2.5.4a, $S^m\emptyset = \{S^q(\emptyset): q \leq m-1\} = S^k\emptyset = \{S^r\emptyset: r \leq k-1\}$. Since $S^{m-1}\emptyset \in S^m\emptyset$, $S^{m-1}\emptyset \in S^k\emptyset$. This entails that $S^{m-1}\emptyset = S^r\emptyset$ for some natural number r ≤ k–1. The Induction Hypothesis is applicable to r. It follows that m–1 = r ≤ k–1. Similarly, since $S^{k-1}\emptyset \in S^k\emptyset$, $S^{k-1}\emptyset \in S^m\emptyset$. Therefore there is a natural number q ≤ m–1 such that $S^{k-1}\emptyset = S^q\emptyset$. Again, the Induction Hypothesis is applicable to k–1. We obtain that k–1 = q ≤ m–1. So far, we have m–1 ≤ k–1 and k–1 ≤ m–1. By the antisymmetry of the relation ≤, m–1 = k–1, which implies that m = k. We conclude that if $S^m\emptyset = S^k\emptyset$, then m = k. This establishes the Inductive Step.

8. From 6 and 7 by PCI: for all natural numbers n and m, if $S^m\emptyset = S^n\emptyset$, then m = n.

9. From 4 and 8: for all natural numbers n and m, m = n iff $S^m\emptyset = S^n\emptyset$.

SOLUTIONS TO 2.5.5

2.5.5b Our goal is to prove that for every A, A ∈ ω iff there is a natural number n such that $A = S^n\emptyset$.

1. We first prove the 'if' part. We want to show that if $A = S^n\emptyset$ for some natural number n, then A ∈ ω. This is equivalent to saying that for every natural number n, $S^n\emptyset \in \omega$. We will establish the latter version using PMI.

2. The Base Step. Let n= 0. By definition, $S^0\emptyset = \emptyset$. Since ω is inductive, ∅ ∈ ω; hence $S^0\emptyset \in \omega$.

3. The Inductive Step. Our Induction Hypothesis is that for some natural number k, $S^k\emptyset \in \omega$. We wish to show that $S^{k+1}\emptyset \in \omega$. This immediately follows from the fact that ω is inductive.

4. From 2 and 3 by PMI: for any natural number n, $S^n\emptyset \in \omega$. Note that we proved the 'if' part without making use of the "minimality" of ω, that is, the fact that ω is a subset of every inductive set. Our proof actually

shows that for every natural number n, $S^n\emptyset$ belongs to every inductive set.

5. The 'only-if' part makes use of the fact that ω is the "smallest" inductive set. We want to show that if A ∈ ω, A = $S^n\emptyset$ for some natural number n. Define $\mathcal{K}$ to be the family {$S^n\emptyset$: n is a natural number}. It is clear that $\mathcal{K}$ is an inductive set; for $\emptyset = S^0\emptyset \in \mathcal{K}$, and $SS^n\emptyset = S^{n+1}\emptyset \in \mathcal{K}$ whenever $S^n\emptyset \in \mathcal{K}$. Since ω is a subset of every inductive set, $\omega \subseteq \mathcal{K}$. This entails that every member of ω is identical with $S^n\emptyset$ for some natural number n.

6. From 4 and 5: for every A, A ∈ ω iff there is a natural number n such that A = $S^n\emptyset$.

2.5.5e We want to prove that for every β, if β ⊆ ω and β is nonempty, then there is A ∈ β such that for each D ∈ β, D ∉ A—that is, A is an ∈-minimal element in β.

1. Let β be any nonempty subset of ω. By exercise 2.5.5b, every member of β is $S^n\emptyset$ for some natural number n.

2. Define E to be the set {n: $S^n\emptyset \in \beta$}. For example, if β = { $S^3\emptyset$, $S^9\emptyset$, $S^{11}\emptyset$, $S^{16}\emptyset$}, then E = {3, 9, 11, 16}. It is clear that E ⊆ ℕ.

3. Since β is nonempty, E is nonempty as well. Hence E has a <-minimal element—that is, there is a number m in E such that for every k in E, not-(k < m). We will show that $S^m\emptyset$ is an ∈-minimal element in β.

4. Let $S^n\emptyset$ be any member of β. By the definition of E, n ∈ E. By the minimality of m, not-(n < m). Thus m ≤ n.

5. If m < n, then, by Exercise 2.5.4c, $S^m\emptyset \subset S^n\emptyset$. In this case, by Exercise 2.5.5c, $S^m\emptyset \in S^n\emptyset$. By Exercise 2.5.5d, ∈ is asymmetric on ω. Therefore $S^n\emptyset \notin S^m\emptyset$.

6. If m = n, then, by Exercise 2.5.4b, $S^m\emptyset = S^n\emptyset$. By Exercise 2.5.5d, ∈ is irreflexive on ω. Hence $S^n\emptyset \notin S^m\emptyset$.

7. From 4 through 6, for every $S^n\emptyset$ in β, $S^n\emptyset \notin S^m\emptyset$. This entails that $S^m\emptyset$ is an ∈-minimal element in β.

8. From 1 and 7: every nonempty subset of ω has an ∈-minimal element.

Solution to 2.5.6

If A = ∅, $\mathcal{P}A$ = {∅}. This is consistent with the theorem, since $2^0 = 1$. The general case is when A is nonempty. Since A is finite and its cardinality is n, we can represent it as the set {e_1, e_2, e_3, ..., e_n}. Every subset B of A can be defined in terms of an n-tuple $\langle z_1, z_2, z_3, ..., z_n \rangle$ where each z_i is either 0 or 1; if z_i is 1, $e_i \in B$, and if z_i is 0, $e_i \notin B$. The table below shows a few subsets of A and their corresponding n-tuples.

Subsets of A	e_1	e_2	e_3	e_4	e_5	...	e_n
∅	⟨0	0	0	0	0	0	0⟩
{e_1, e_2}	⟨1	1	0	0	0	0	0⟩
{e_1, e_3, e_5}	⟨1	0	1	0	1	0	0⟩
{e_1, e_2, e_5, e_n}	⟨1	1	0	0	1	0	1⟩
{e_2, e_3, e_4, e_5, e_n}	⟨0	1	1	1	1	0	1⟩
A = {e_1, e_2, e_3, ..., e_n}	⟨1	1	1	1	1	1	1⟩

Every subset of A corresponds to exactly one n-tuple. Since each n-tuple has n coordinates and each coordinate can be either 0 or 1, there are 2×2×2×...×2 (n times) n-tuples. Thus there are 2^n subsets of A. Therefore card($\mathcal{P}$A) = 2^n.

SOLUTION TO 2.5.7

2.5.7d   Let A and B be any well-ordered sets. We want to prove that A×B can be well-ordered.

   1. Let R be a well-ordering of A and Q be a well-ordering of B.

   2. By the definition of Cartesian product, A×B is the set of all the ordered pairs ⟨x, y⟩ whose first coordinates are members of A and whose second coordinates are members of B. We define the following binary relation L on A×B. For all ⟨x, y⟩ and ⟨u, v⟩ in A×B, ⟨x, y⟩L⟨u, v⟩ iff xRu, or x = u and yQv. L is called a "lexicographical ordering" or "dictionary ordering" because it is similar to the alphabetical ordering of a dictionary.

   3. We first prove that L is a linear ordering of A×B.

      3.1. Let ⟨x, y⟩L⟨u, v⟩. The definition of L entails that xRu, or x = u and yQv. If we assume that xRu, then since R is asymmetric, not-(uRx) and u ≠ x. Hence not-(⟨u, v⟩L⟨x, y⟩). (Observe that an asymmetric relation P is irreflexive, that is, for all x, not-(xPx).) If we assume that x = u and yQv, then since R and Q are asymmetric, not-(uRx) and not-(vQy). Thus not-(⟨u, v⟩L⟨x, y⟩). This establishes that L is asymmetric.

      3.2. Now let ⟨x, y⟩L⟨u, v⟩ and ⟨u, v⟩L⟨w, z⟩. By the definition of L, xRu, or x = u and yQv, and uRw, or u = w and vQz. Assume that xRu. There are two cases. (1) uRw; since R is transitive, xRw, which entails that ⟨x, y⟩L⟨w, z⟩. (2) u = w and vQz; by substitution, we get xRw, which, again, entails that ⟨x, y⟩L⟨w, z⟩. Now assume that x = u and yQv. We have the same two cases as before. (1) uRw; by substitution, xRw, which implies that ⟨x, y⟩L⟨w, z⟩. (2) u = w and vQz; by substitution and the transitivity of Q, we get x = w and yQz, which implies that ⟨x, y⟩L⟨w, z⟩. We conclude that L is transitive.

3.3. To show that L is connex we let $\langle x, y \rangle$ and $\langle u, v \rangle$ be any two members of A×B. Since R and Q are connex we have the following disjunctions: xRu or uRx or x = u, and yQv or vQy or y = v. If xRu, $\langle x, y \rangle L \langle u, v \rangle$. If uRx, $\langle u, v \rangle L \langle x, y \rangle$. If x = u, we have three cases. (1) yQv; this entails that $\langle x, y \rangle L \langle u, v \rangle$. (2) vQy; this entails that $\langle u, v \rangle L \langle x, y \rangle$. (3) y = v; by the Principle of Ordered Tuples, $\langle x, y \rangle = \langle u, v \rangle$. Thus in all cases $\langle x, y \rangle L \langle u, v \rangle$ or $\langle u, v \rangle L \langle x, y \rangle$ or $\langle x, y \rangle = \langle u, v \rangle$. Hence L is connex. Since L is asymmetric, transitive, and connex, it is a linear ordering of A×B.

3.4. Second, we prove that L is well-founded on A×B. Let D be any nonempty subset of A×B. Define D_A to be the set of all the first coordinates of the ordered pairs in D. Since D is nonempty, D_A is nonempty. Given that R is well-founded on A, D_A has an R-minimal element m. Now define D_m to be the set $\{y: y \in B$ and $\langle m, y \rangle \in D\}$. In other words, we collect in D_m all the second coordinates of the ordered pairs $\langle m, y \rangle$ that are in D. Since Q is well founded on B, and D_m is a nonempty subset of B, D_m has a Q-minimal element e. By the definition of D_m, the ordered pair $\langle m, e \rangle$ belongs to D. We claim that $\langle m, e \rangle$ is an L-minimal element in D. To see this let $\langle u, v \rangle$ be any member of D. Since m is R-minimal in D_A, not-(uRm), which implies that mRu or m = u (R is connex). Assume that mRu. It follows that $\langle m, e \rangle L \langle u, v \rangle$. By the asymmetry of L, not-($\langle u, v \rangle L \langle m, e \rangle$). Now assume that m = u. Since e is Q-minimal in D_m, not-(vQe). By the connexity of Q, either eQv or e = v. If eQv, $\langle m, e \rangle L \langle u, v \rangle$. By the asymmetry of L, not-($\langle u, v \rangle L \langle m, e \rangle$). If e = v, given that m = u, $\langle u, v \rangle = \langle m, e \rangle$ (by the Principle of Ordered Tuples). We know that every asymmetric relation is irreflexive; hence not-($\langle u, v \rangle L \langle m, e \rangle$). So in all cases, not-($\langle u, v \rangle L \langle m, e \rangle$). This proves that $\langle m, e \rangle$ is an L-minimal element in D, which establishes that L is well-founded on A×B.

3.5. From 3.3 and 3.4: the lexicographical ordering L is a well-ordering of the product A×B.

Solutions to 2.5.8

2.5.8b Let α be any ordinal and β any member of α. Our goal is to prove that β is an ordinal. Given the definition of an ordinal, we must show that β is a transitive set that is well ordered by $\in$.

1. We first show that β is transitive. We need to prove that β contains the members of its members. So we take δ to be any member of β and γ any member of δ. We wish to show that γ belongs to β. We have the following sequence of membership: $\gamma \in \delta \in \beta \in \alpha$. Since α is an ordinal, it is transitive. Hence the members of β belong to α, i.e., $\delta \in \alpha$. Again

since δ is a member of α and α is transitive, the members of δ belong to α. It follows that γ ∈ α. We obtain that γ, δ, and β all belong to α, and that γ ∈ δ ∈ β. Since ∈ is a transitive relation on α (because ∈ is a well-ordering of α), γ ∈ β, which is the desired conclusion.

2. Second, we prove that ∈ is a linear ordering of β. Let δ and γ be any members of β such that δ ∈ γ. Since β is a member of α and α contains the members of its members, δ and γ belong to α. But δ ∈ γ and ∈ is asymmetric on α; therefore γ ∉ δ. This shows that ∈ is asymmetric on β. Now we take δ, γ, and σ to be any members of β such that δ ∈ γ ∈ σ. Again by the transitivity of α, each of δ, γ, and σ belongs to α. Given that δ ∈ γ ∈ σ and that ∈ is a transitive relation on α, it follows that δ ∈ σ, which establishes that ∈ is a transitive relation on β. Finally we show that ∈ is connex on β. So let δ and γ be any members of β. By the transitivity of α, δ and γ belong to α. Since ∈ is connex on α, δ ∈ γ or γ ∈ δ or δ = γ. We conclude that ∈ is a linear ordering of β.

3. Third, we establish that ∈ is well-founded on β. Let K be any nonempty subset of β. By the definition of subset, every member of K is a member of β. Since β belongs to α and α is a transitive set, α contains all the members of β; and hence it contains all the members of K. This means that K is a nonempty subset of α. Given that ∈ is well-founded on α, K has an ∈-minimal element, which shows that ∈ is well-founded on β.

4. From 1 through 3: β is a transitive set that is well-ordered by ∈. Therefore β is an ordinal.

2.5.8e Our goal is to prove that for all ordinals α and β, β ∈ α iff β ⊂ α. The 'only-if' direction is trivial. If β ∈ α, then, since α is a transitive set, β ⊆ α. However, β cannot be identical with α if β ∈ α; otherwise, α ∈ α, which is impossible. Hence if β ∈ α, then β ⊂ α.

1. Now we prove the 'if' part, which is more involved. So let α and β be any ordinals such that β ⊂ α.

2. By the definition of proper subset, there are members in α that do not belong to β. Thus the set α–β is not empty.

3. Given that α–β is a nonempty subset of α and that ∈ is well-founded on α, α–β has an ∈-minimal element. Let this element be τ. We will prove that τ = β. First we make a few observations. τ ∈ α and τ ∉ β (definition of α–β). τ is an ordinal since it is a member of an ordinal (Exercise 2.5.8b). By the definition of ordinal, τ, β, and α are families of sets. Finally, by the minimality of τ, there is no member x of α–β such that x ∈ τ. We will establish that τ = β by proving that for all sets A, A ∈ τ iff A ∈ β, and then invoking the Principle of Extensionality to conclude that τ

= β. The final conclusion follows immediately: since $\tau \in \alpha$ and $\tau = \beta$, by substitution $\beta \in \alpha$.

4. We first show that for every set A, if $A \in \tau$, then $A \in \beta$. We make a Reductio Assumption: there is a set A such that $A \in \tau$ and $A \notin \beta$.
5. From 3 and 4: $A \in \tau \in \alpha$.
6. From 5 by the transitivity of α: $A \in \alpha$.
7. From 4 through 6: $A \in \alpha-\beta$ and $A \in \tau$, which contradicts 3—namely that τ is an $\in$-minimal element of $\alpha-\beta$.
8. From 4 through 7: we reject the Reductio Assumption and conclude that for every set A, if $A \in \tau$, then $A \in \beta$.
9. Now we show that for every set A, if $A \in \beta$, then $A \in \tau$. Assume for reductio that there is a set A such that $A \in \beta$ and $A \notin \tau$.
10. From 1 and 9: since $\beta \subset \alpha$, $A \in \alpha$. So we have that τ and A belong to α.
11. From 10: given that α is well-ordered by $\in$, it follows that either $A \in \tau$ or $\tau \in A$ or $\tau = A$.
12. From 9 and 11: since $A \notin \tau$, $\tau \in A$ or $\tau = A$.
13. Assume that $\tau \in A$. From 9, we know that $A \in \beta$. Since β is a transitive set, $\tau \in \beta$, which contradicts 3—namely that $\tau \in \alpha-\beta$. So this option is closed.
14. Now assume that $\tau = A$. Again, since $A \in \beta$, $\tau \in \beta$, which contradicts the fact that $\tau \notin \beta$. This option is also closed.
15. From 9 through 14: the Reductio Assumption is false. Therefore, for every set A, if $A \in \beta$, then $A \in \tau$.
16. From 8 and 15: for all sets A, $A \in \tau$ iff $A \in \beta$. By the Principle of Extensionality, $\tau = \beta$.
17. From 3 and 16: since $\tau \in \alpha$, $\beta \in \alpha$.
18. From 1 through 17: if $\beta \subset \alpha$, then $\beta \in \alpha$.

Solution to 2.5.12

We take J to be any PL interpretation that is relevant to **X[s]** and **t**, where **X[s]** is a PL sentence in which the PL singular term **s** occurs and **t** is any PL singular term. We assume that **s = t** is true on J. Our goal is to prove that **X[s]** and **X[s, t]** have the same truth value on J, where **X[s, t]** is obtained from **X[s]** by replacing one or more occurrences of **s** with **t**. We are given the fact that if J(**s**) is the same as J(**t**), then replacing one or more occurrences of **s** in some PL singular term **q** with **t** does not change J(**q**). We define the complexity of a PL sentence to be the number of connectives or quantifiers in it. We will use PCI on the complexity of **X[s]** to establish this theorem.

Base Step: Let the complexity of **X[s]** be 0. Hence **X[s]** is an atomic sentence. There are two types of atomic sentences in PL. Either **X[s]** is an identity sentence or a subject-predicate sentence. Assume first that **X[s]** is an identity sentence. So **X[s]** is of the form **q** = **r**, where **q** and **r** are some PL singular terms such that the singular term **s** occurs in **q**, **r**, or both. In this case, replacing **s** with **t** in *zero* or more occurrences of **s** in **r** or **q** does not change the referent of **r** or of **q** on J (because, by assumption, J(**s**) and J(**t**) are the same individual and because of the fact given above). Let **r*** and **q*** be the terms that result from such replacement. We said that J(**r**) is identical with J(**r***), and that J(**q**) is identical with J(**q***). It follows, by the truth conditions of the identity predicate, that **q** = **r** and **q*** = **r*** have the same truth value on J. Now assume that **X[s]** is of the form $P^n q_1 q_2 ... q_n$, where P^n is an n-place PL predicate and q_1, q_2, ..., and q_n are some PL singular terms, some of which might be **s** or contain occurrences of **s**. Let q_1^*, q_2^*, ..., and q_n^* be the terms that result from q_1, q_2, ..., and q_n by replacing **s** with **t** in zero or more occurrences of **s** in q_1, q_2, ..., or q_n. (Note that every PL expression occurs in itself; thus if q_1 is **s**, then **s** occurs in q_1.) Since J(**s**) and J(**t**) are identical, the referents of q_1^*, q_2^*, ..., and q_n^* on J are the same as the referents of q_1, q_2, ..., and q_n, respectively. By the truth conditions of $P^n q_1 q_2 ... q_n$, the truth value of $P^n q_1 q_2 ... q_n$ and $P^n q_1^* q_2^* ... q_n^*$ is the same on J. Therefore the Base Step is established.

The Inductive Step: Suppose that the theorem is true for every PL sentence **X[s]** whose complexity is m, where m is any natural number that is less than some natural number k (so k > 0). This is the Induction Hypothesis. We want to show that the theorem is true for any sentence with complexity k. So let **X[s]** be a PL sentence with complexity k. Since k > 0, **X[s]** must be a compound sentence. Hence it is of the form ¬**Y**, **Y**→**W**, or (∀z)**Y**. Consider first the case ¬**Y**. **s** occurs in **Y**. Let **Y*** be the sentence obtained from **Y** by replacing one or more of the occurrences of **s** with **t**. It is clear that **Y** has complexity less than k. Thus the Induction Hypothesis applies to **Y**—that is, **Y** and **Y*** have the same truth value on J. Hence, ¬**Y** and ¬**Y*** have the same truth value on J as well. Consider second the case of **Y**→**W**. **s** occurs in **Y** or in **W** (or both). Let **Y*** and **W*** be the sentences obtained from **Y** and **W**, respectively, by replacing zero or more occurrences of **s** with **t**. **Y** and **W** have complexities less than k. Therefore, by the Induction Hypothesis, **Y** and **Y*** have the same truth value on J, and so do **W** and **W***. It follows that **Y**→**W** and **Y***→**W*** have the same truth value on J. Finally, we consider the case of (∀z)**Y**. **s** occurs in the *formula* **Y**. Let (∀z)**Y*** be the sentence obtained from (∀z)**Y** by replacing one or more of **s** with **t**. It is obvious that for every PL name **n** in LN of J, **Y[n]** is a basic substitutional instance of (∀z)**Y** iff **Y*[n]** is a basic substitutional instance of (∀z)**Y***. Since **Y[n]** has complexity less than k, the Induction Hypothesis applies to **Y[n]**. Thus **Y[n]** and **Y*[n]** have the same truth value on J. In other words, all the basic substitutional instances of (∀z)**Y** have the same truth values as the corresponding basic substitutional instances of (∀z)**Y***. By the truth conditions of the universal quantifier, (∀z)**Y** and (∀z)**Y*** have the same truth value on J. Therefore the Inductive Step is established.

By PCI the theorem is true for all complexities of **X[s]**. This means that the theorem is true for every PL sentence **X[s]**.

Solution to 2.5.15

Let **X** be the conjunction of the following four PL sentences:

S1 $(\forall x)(Px \vee Qx) \wedge \neg(\exists y)(Py \wedge Qy)$
S2 $(\forall x)(Px \rightarrow Qgx)$
S3 $(\forall y)(Qy \rightarrow (\exists x)(Px \wedge y = gx))$
S4 $(\forall x)(\forall y)(gx = gy \rightarrow x = y)$

For any positive even number 2n, consider the following PL interpretation of size 2n. (n is any positive integer.)

The PL interpretation I

UD: $\{1, 2, 3, \ldots, 2n\}$
LN: $a_1, a_2, a_3, \ldots, a_{2n}$

Semantical Assignments

I(a_j): j, where j is 1, 2, 3, …, or 2n
I(P): $\{1, 2, \ldots, n\}$; I(Q): $\{n+1, n+2, \ldots, 2n\}$
I(g): $\{<1, n+1>, <2, n+2>, \ldots, <n, 2n>, <n+1, 1>, <n+2, 2>, \ldots, <2n, n>\}$

Here are the natural (or PL-English) readings of S1–S4 on I.

S1 asserts on I that every individual either belongs to the extension of P or the extension of Q but not both.

S2 asserts on I that for every x that is a member of the extension of P, its g value, gx, belongs to the extension of Q.

S3 asserts on I that for every y that is a member of the extension of Q, y is the g value of some member of the extension of P.

S4 asserts on I that for any two individuals x and y, if their g values are identical, then $x = y$.

It is clear that S1–S4 are true on I. Therefore, I is a model of **X**.

Furthermore, no finite interpretation of an odd size can be a model of **X**. Suppose that J is a model of **X** and that J is of some finite size. Since S1 is true on J, the UD of J must be partitioned by the extensions of P and Q. That is, every individual in UD either belongs to J(P) or to J(Q), but not both. The truth of S2–S4 implies that the function that I assigns to g, that is, I(g), is a one-to-one correspondence between J(P) and J(Q). Hence, by Hume's Principle, J(P) and J(Q) have the same cardinality. This entails that the finite UD of J is divided into two equal halves, J(P) and J(Q), which implies that J cannot be of an odd size.

Chapter Three

The Soundness and Completeness Theorems

3.1 The Soundness Theorem

The goal of this section is to prove that MDS is **sound**. More precisely, we will prove that for every set Γ of PL sentences and every PL sentence **X**, if $\Gamma \vdash \mathbf{X}$, then $\Gamma \vDash \mathbf{X}$. Thus we start by taking Γ to be any arbitrary set of PL sentences and **X** any PL sentence that is derivable from Γ. By the definition of derivability, there is a PL derivation D of **X** from Γ. Let Σ_D be the set of the members of Γ that are invoked in D. Using set-theoretic notation, $\Sigma_D = \{\mathbf{Y}: \mathbf{Y} \in \Gamma$ and **Y** appears in D$\}$. Since all PL derivations are finite sequences of PL sentences, it is clear that Σ_D is a finite subset of Γ. Also, by definition, the last sentence of D is **X**. If $\Sigma_D \vDash \mathbf{X}$, then $\Gamma \vDash \mathbf{X}$. To see this, assume that $\Sigma_D \vDash \mathbf{X}$ and let M be any model of Γ that is relevant to **X**. Since any model of Γ is also a model of any subset of Γ, M is a model of Σ_D; hence **X** is true on M. So every model of Γ that is relevant to **X** makes **X** true. This establishes that **X** is a logical consequence of Γ if it is a logical consequence of Σ_D. Thus to show that $\Gamma \vDash \mathbf{X}$, it is sufficient to show that $\Sigma_D \vDash \mathbf{X}$.

We will apply mathematical induction to the number that designates a line of D. Let n be the number of a line of D. If D consists of j lines, then $1 \leq n \leq j$. We take $\mathbf{Z}_n$ to be the sentence that appears in D at line n, and Σ_n to be the set of all the premises and undischarged assumptions that occur in D at line n or prior to line n. Since j, i.e., the length of D, can be any positive natural number, we will prove that for every natural number $n \geq 1$, $\Sigma_n \vDash \mathbf{Z}_n$. This entails that $\Sigma_D \vDash \mathbf{X}$. To see this, observe first that, by the definition of derivation, **X** is the last sentence of D, that is, $\mathbf{X} = \mathbf{Z}_j$, where j is the number of the last line of D. Second, given the rules of blocks, all the subblocks that might be initiated in D must be closed before the main block can be closed. In other words, all the assumptions that might be introduced in D by hypothetical inference rules must be *discharged* before the conclusion of D, which is **X**, is introduced. This means that the set Σ_j consists solely of the premises that are invoked in D—that is, $\Sigma_j = \Sigma_D$. Thus proving that $\Sigma_n \vDash \mathbf{Z}_n$, for every natural number $n \geq 1$, is sufficient for establishing that $\Sigma_D \vDash \mathbf{X}$, which is our desired conclusion.

We will use the Principle of Complete Induction (PCI) to prove that $\Sigma_n \vDash \mathbf{Z}_n$, for every natural number n that is greater than or equal to 1.

The Base Step: Let n = 1. The first line of any derivation has no antecedents. So the sentence Z_1 is either a premise, an assumption of a hypothetical rule, or an identity statement of the form **s** = **s**, which is introduced by the MDS rule Identity. If Z_1 is a premise or an assumption, then $\Sigma_1 = \{Z_1\}$. It is obvious that, in this case, $\Sigma_1 \vDash Z_1$. If Z_1 is of the form **s** = **s** (where **s** is any PL singular term), then Σ_1 is empty. But **s** = **s** is a valid sentence, that is, it is true on all the PL interpretations that are relevant to it. By definition, all PL interpretations are models of the empty set, since an interpretation would fail to be a model of the empty set only if the empty set were to contain a sentence that is false on that interpretation; given that the empty set contains no sentences, no PL interpretation fails to be a model of it. Hence **s** = **s** is true on every model of the empty set that is relevant to it. In other words, **s** = **s** is a logical consequence of the empty set. Therefore, in all cases, $\Sigma_1 \vDash Z_1$. This establishes the Base Step.

The Inductive Step: Let k > 1, and assume that for every m such that $1 \leq m < k$, $\Sigma_m \vDash Z_m$. This is the Induction Hypothesis. Our goal is to prove that $\Sigma_k \vDash Z_k$. Since k > 1, line k might have antecedents. There are three main cases to consider: (1) Z_k is a premise or an assumption introduced by some hypothetical rule, (2) Z_k is an identity sentence of the form '**s** = **s**', which is introduced by the rule Identity, and (3) Z_k is the conclusion of one of the MDS rules other than Identity. Since MDS contains eight rules in addition to Identity, the third case divides into eight subcases. So we need to consider a total of ten cases.

Case 1: Z_k is a premise or an assumption introduced by some hypothetical rule. By the definition of Σ_k, $Z_k \in \Sigma_k$. It is obvious that $\Sigma_k \vDash Z_k$.

Case 2: Z_k is introduced by the rule Identity. Thus Z_k is a sentence of the form '**s** = **s**'. But '**s** = **s**' is valid. Hence Z_k is a logical consequence of every set of PL sentences; in particular, $\Sigma_k \vDash Z_k$.

Case 3: Z_k is introduced by the rule Reiteration (Reit). According to Reit, Z_k must occur on a line p that is prior to line k. Since p < k, the Induction Hypothesis applies to p. Thus $\Sigma_p \vDash Z_p$, where Z_p is Z_k. As any normal rule, Reit must be applied in an open block. This implies that if Σ_p contains undischarged assumptions, which were introduced by hypothetical rules, then none of these assumptions can be discharged at line k or prior to line k (otherwise Z_p would occur in a closed block, and hence it could not be reiterated at line k). However, it is possible that some subblocks are initiated between lines p and k. Therefore Σ_p is either the same set as Σ_k or a proper subset of it. In other words, $\Sigma_p \subseteq \Sigma_k$. Since $\Sigma_p \vDash Z_k$ and $\Sigma_p \subseteq \Sigma_k$, $\Sigma_k \vDash Z_k$.

Case 4: Z_k is the conclusion of the rule Modus Ponens (MP). The antecedents of this rule are two sentences: **Y** and **Y**→Z_k. Both of these sentences occur in the derivation D prior to line k. Assume that they occur on lines p and q, respectively, and that p < q. Since p and q are less than k, the Induction Hypothesis applies to them. So we have that $\Sigma_p \vDash \mathbf{Z}_p$ and $\Sigma_q \vDash \mathbf{Z}_q$. But $\mathbf{Z}_p$ is $\mathbf{Y}$ and $\mathbf{Z}_q$ is $\mathbf{Y} \rightarrow \mathbf{Z}_k$; hence $\Sigma_p \vDash \mathbf{Y}$ and $\Sigma_q \vDash \mathbf{Y} \rightarrow \mathbf{Z}_k$. MP must be applied in an open block. Following the same reasoning as in Case 3, this implies that Σ_p and Σ_q are subsets of Σ_k. It follows that $\Sigma_k \vDash \mathbf{Y}$ and $\Sigma_k \vDash \mathbf{Y} \rightarrow \mathbf{Z}_k$. Let M be any PL model of Σ_k that is also relevant to $\mathbf{Z}_k$. If M is not relevant to $\mathbf{Y}$, we expand M into M*, which is identical with M except it interprets the additional PL vocabulary in $\mathbf{Y}$. It is clear that M* is a model of Σ_k since its interpretations of the vocabulary of Σ_k are identical with those of M. Given that $\mathbf{Y}$ and $\mathbf{Y} \rightarrow \mathbf{Z}_k$ are logical consequences of Σ_k, they are true on M*. By the truth conditions of the conditional, $\mathbf{Z}_k$ is true on M*. But M and M* agree on their interpretations of the PL vocabulary in $\mathbf{Z}_k$; hence $\mathbf{Z}_k$ is true on M. We conclude that any model of Σ_k that is relevant to $\mathbf{Z}_k$ makes $\mathbf{Z}_k$ true. Therefore $\Sigma_k \vDash \mathbf{Z}_k$.

Case 5: $\mathbf{Z}_k$ is the conclusion of the rule Conditional Proof (CP). According to this rule, a CP block precedes line k. This CP block is initiated by introducing an assumption $\mathbf{Y}$ at some line p prior to line k–1 and is exited at line k. Line k–1 is the last line of the CP block. Let $\mathbf{W}$ be the sentence that appears on line k–1. $\mathbf{Z}_k$ is thus the sentence $\mathbf{Y} \rightarrow \mathbf{W}$. The CP Assumption $\mathbf{Y}$ is discharged at line k. By the rules of blocks, any subblock that is initiated after the CP block is opened must be exited before the CP block is exited. Hence all the assumptions that are introduced after line p must be discharged at line k–1 or prior to it. Since the CP Assumption is introduced at line p and is discharged at line k, the only difference between Σ_k and Σ_{k-1} is that the Σ_{k-1} contains in addition to the members of Σ_k the CP Assumption $\mathbf{Y}$. In other words, $\Sigma_{k-1} = \Sigma_k \cup \{\mathbf{Y}\}$. The Induction Hypothesis applies to k–1. Thus $\Sigma_{k-1} \vDash \mathbf{Z}_{k-1}$. $\mathbf{Z}_{k-1}$ is $\mathbf{W}$. We obtain that $\Sigma_k \cup \{\mathbf{Y}\} \vDash \mathbf{W}$. Let M be any model of Σ_k that is also relevant to $\mathbf{Y}$ and $\mathbf{W}$. Either $\mathbf{Y}$ is true on M or not. If it is false, $\mathbf{Y} \rightarrow \mathbf{W}$ is true on M. If it is true on M, then M is a model of $\Sigma_k \cup \{\mathbf{Y}\}$. In this case, $\mathbf{W}$ is true on M, which entails that $\mathbf{Y} \rightarrow \mathbf{W}$ is true on M. Therefore, $\mathbf{Y} \rightarrow \mathbf{W}$, which is $\mathbf{Z}_k$, is true on every model of Σ_k that is relevant to $\mathbf{Z}_k$. It follows that $\Sigma_k \vDash \mathbf{Z}_k$.

Cases 6 and 7: $\mathbf{Z}_k$ is the conclusion of the rule Reductio Ad Absurdum (RAA). This rule has two parts. Since the proofs of these parts are almost identical, we will present only one of them. The RAA block is initiated at some line prior to line k–1 with the introduction of the Reductio Assumption $\neg \mathbf{Z}_k$, and is exited at line k with the discharge of the Reductio Assumption. The last line of the RAA block is line k–1. $\mathbf{Z}_{k-1}$ is $\neg \mathbf{Y}$, where $\mathbf{Y}$ is some PL sentence, and $\mathbf{Y}$ appears in the RAA Block on line k–2. As in Case 5, the only difference between Σ_k and Σ_{k-1} is that Σ_{k-1} contains, in addition to the members of Σ_k, the Reductio Assumption $\neg \mathbf{Z}_k$—that is, $\Sigma_{k-1} = \Sigma_k \cup \{\neg \mathbf{Z}_k\}$. We also note that since $\mathbf{Y}$ and $\neg \mathbf{Y}$ must occur in an open block and since $\mathbf{Z}_{k-1}$ might be a premise, $\Sigma_{k-2} \subseteq \Sigma_{k-1}$. The Induction Hypothesis applies to k–1 and k–2; and hence $\Sigma_{k-1} \vDash \mathbf{Z}_{k-1}$, which is $\neg \mathbf{Y}$, and $\Sigma_{k-2} \vDash \mathbf{Z}_{k-2}$, which is $\mathbf{Y}$. Given that $\Sigma_{k-2} \subseteq \Sigma_{k-1}$, it follows that $\Sigma_{k-1} \vDash \mathbf{Y}$ and $\Sigma_{k-1} \vDash \neg \mathbf{Y}$. Since $\Sigma_{k-1} = \Sigma_k \cup \{\neg \mathbf{Z}_k\}$, we have that $\Sigma_k \cup \{\neg \mathbf{Z}_k\} \vDash \mathbf{Y}$ and $\Sigma_k \cup \{\neg \mathbf{Z}_k\} \vDash \neg \mathbf{Y}$. Let M be any PL model of Σ_k that is relevant to $\mathbf{Z}_k$. If M is not relevant to $\mathbf{Y}$, we expand M into M*, which is identical with M except M* interprets the additional PL vocabulary in **Y**. Since M and M* agree on their interpretations of the vocabulary of Σ_k, M* is a model of Σ_k. If $\neg \mathbf{Z}_k$ is true on M*, then M* is a model of $\Sigma_k \cup \{\neg \mathbf{Z}_k\}$, which implies that **Y** and $\neg$**Y** are true on M*. Since the latter is impossible, $\neg \mathbf{Z}_k$ is false on M*; and hence $\mathbf{Z}_k$ is true on M*. Given that M interprets the vocabulary in $\mathbf{Z}_k$ exactly as M* does, it follows that $\mathbf{Z}_k$ is true on M. Therefore, $\mathbf{Z}_k$ is true on every model of Σ_k that is relevant to $\mathbf{Z}_k$. In other words, $\Sigma_k \vDash \mathbf{Z}_k$.

Case 8: $\mathbf{Z}_k$ is the conclusion of the rule Universal Instantiation (UI). The antecedent of this rule is a sentence of the form $(\forall \mathbf{z})\mathbf{Y}$, which occurs on line p that is prior to line k, and its conclusion is **Y[t]**, where **Y[t]** is obtained from **Y** by replacing every occurrence of the variable **z** with the singular term **t**. $\mathbf{Z}_k$ is **Y[t]**. The Induction Hypothesis applies to p; hence $\Sigma_p \vDash \mathbf{Z}_p$. UI must be applied in an open block. As in Cases 3 and 4, $\Sigma_p \subseteq \Sigma_k$ and $\Sigma_k \vDash \mathbf{Z}_p$. $\mathbf{Z}_p$ is $(\forall \mathbf{z})\mathbf{Y}$. We obtain that $\Sigma_k \vDash (\forall \mathbf{z})\mathbf{Y}$. Let M be any PL model of Σ_k that is relevant to **Y[t]**. Since $(\forall \mathbf{z})\mathbf{Y}$ is a logical consequence of Σ_k, it is true on M.[1] By the truth conditions of the universal quantifier, all the basic substitutional instances of $(\forall \mathbf{z})\mathbf{Y}$ are true on M. By Theorem 1.2.2, every substitutional instance of $(\forall \mathbf{z})\mathbf{Y}$ is true on M; in particular, **Y[t]**, which is $\mathbf{Z}_k$, is true on M. Hence every model of Σ_k that is relevant to $\mathbf{Z}_k$ makes $\mathbf{Z}_k$ true. Therefore $\Sigma_k \vDash \mathbf{Z}_k$.

Case 9: $\mathbf{Z}_k$ is the conclusion of the rule Universal Generalization (UG). According to this rule, all of the following conditions are true: $\mathbf{Z}_k$ is of the form $(\forall \mathbf{z})\mathbf{Y[z]}$; on some line p, which is prior to line k, **Y** appears; a name **s** occurs in **Y**; the variable **z** does not occur in **Y**; and the formula **Y[z]** is obtained from the sentence **Y** by replacing *every* occurrence of **s** with **z**. Furthermore, the rule places a very important condition on the name **s**: **s** must be *arbitrary* at line p—that is, **s** must not occur in any premise or undischarged assumption listed on line p or prior to it. In other words, **s** does not occur in any member of Σ_p. Since p < k, the Induction Hypothesis applies to p; hence $\Sigma_p \vDash \mathbf{Z}_p$ (recall that $\mathbf{Z}_p$ is **Y**). We will prove that $\Sigma_p \vDash (\forall \mathbf{z})\mathbf{Y[z]}$. So let M be any PL model of Σ_p that is relevant to $(\forall \mathbf{z})\mathbf{Y[z]}$. We know that the name **s** occurs neither in the members of Σ_p nor in $(\forall \mathbf{z})\mathbf{Y[z]}$. So a PL interpretation that is relevant to Σ_p and $(\forall \mathbf{z})\mathbf{Y[z]}$ is under no requirement to include the name **s** in its List of Names (LN). But we want M to be relevant to **Y** and **s** occurs in **Y**. We can solve this problem by making sure that the name **s** is in LN; if it is not there, we add it to LN, and assign to it any individual in UD. This, of course, changes M slightly; but this change has no effect on the truth values of $(\forall \mathbf{z})\mathbf{Y[z]}$ and of the members of Σ_p. Our goal is to prove that $(\forall \mathbf{z})\mathbf{Y[z]}$ is true on every model of Σ_p that is relevant to $(\forall \mathbf{z})\mathbf{Y[z]}$. It should be clear, therefore, that this slight change in M will have no bearing on our proof. Thus, without further ado, we will assume that **s** is in LN of M. We will write **Y** as **Y[s]** to make it clear that the arbitrary name **s** occurs in **Y**. Since **Y[s]** is a logical consequence of Σ_p and M is a model of Σ_p that is relevant to **Y[s]**, **Y[s]** is true on M. Our proof will show that since (1) **Y[s]** is true on M, (2) **s** does not occur in any member of Σ_p, and (3) **Y[s]** is a logical consequence

---

[1] If M is relevant to **Y[t]**, it is also relevant to $(\forall \mathbf{z})\mathbf{Y}$, because **Y[t]** contains all the extra-logical vocabulary that occurs in $(\forall \mathbf{z})\mathbf{Y}$.

of Σ_P, every sentence **Y[t]** is true on M, where **Y[t]** is obtained from **Y[s]** by replacing every occurrence of the name **s** with the name **t**. So we let **t** be any name in LN of M. Moreover, we suppose that the referent that M assigns to the name **s** is σ, and that the referent that M assigns to the name **t** is τ. If σ and τ are the same individual, then the interpretations of the vocabulary in **Y[t]** are identical with the interpretations of the vocabulary in **Y[s]**. Hence M gives precisely the same interpretation of both sentences, because **Y[s]** and **Y[t]** are syntactically identical except that in **Y[s]** the name **s** occurs and in **Y[t]** the name **t** replaces **s** in all its occurrences. This entails that both sentences have the same truth value on M. Given that **Y[s]** is true on M, **Y[t]** is true on M as well. Now suppose that σ and τ are distinct individuals in the universe of discourse. This is the general case. We construct another interpretation M* that is identical with M in all aspects except that M* assigns τ as the referent of both **s** and **t**. This *might* leave σ without a name. In this case we make one additional change: we introduce a single new name in LN of M* and assign to it the referent σ. Since the name **s** does not appear in any sentence in Σ_P, and since the name **s** is the only part of the vocabulary of M (i.e., the vocabulary interpreted by M) that receives a different interpretation in M*, the truth values of the members of Σ_P are the same on M and M*. Given that M is a model of Σ_P that is relevant to **Y[s]**, M* is also a model of Σ_P that is relevant to **Y[s]**. But **Y[s]** is a logical consequence of Σ_P; hence **Y[s]** is true on M*. By construction, M* assigns the same referent to **s** and **t**—namely, τ. This entails that M* gives exactly the same interpretation of **Y[s]** and **Y[t]**. As stated above, this shows that **Y[s]** and **Y[t]** have the same truth value on M*. It follows that **Y[t]** is true on M*. Since the name **s** does not occur in **Y[t]**, M and M* agree on the interpretations of the vocabulary in **Y[t]**. Therefore **Y[t]** has the same truth value on M and M*, which entails that **Y[t]** is true on M. We established that for any name **t** in LN of M, the basic substitutional instance **Y[t]** is true on M. By the truth conditions of the universal quantifier, $(\forall z)Y$ is true on M. We conclude that $(\forall z)Y$ is true on every PL model of Σ_P that is relevant to $(\forall z)Y$. By the definition of logical consequence, $\Sigma_P \vDash (\forall z)Y[z]$. As any normal rule, UG must be applied in an open block. As explained in previous cases, this implies that Σ_P is a subset of Σ_k. Thus $\Sigma_k \vDash (\forall z)Y[z]$. Finally, given that $(\forall z)Y[z]$ is Z_k, we obtain that $\Sigma_k \vDash Z_k$.

Case 10: Z_k is the conclusion of the rule substitution (Sub). This rule has two antecedents: a sentence of the form **s = t** (or **t = s**), where **s** and **t** are any PL singular terms, and a sentence **Y[s]** in which **s** occurs. The conclusion of this rule, which is Z_k, is **Y[s, t]**, where **Y[s, t]** is obtained from **Y[s]** by replacing **s** in *one or more* of its occurrences with **t**. Since the first antecedent could be either **s = t** or **t = s**, this rule really consists of two parts. We will prove only one part. The proof of the other part is identical with the proof we will present here. The two antecedents occur on two lines that are prior to line k. Let these lines be p and q, respectively, and suppose that p < q. Thus Z_p is **s = t** and Z_q is **Y[s]**. Since p and q are less than k, the Induction Hypothesis applies to them; that is, $\Sigma_P \vDash$ **s = t** and $\Sigma_q \vDash$ **Y[s]**. As a normal rule, Sub must be applied in an open block. As explained in previous cases, this implies that Σ_P and Σ_q are subsets of Σ_k. We obtain that $\Sigma_k \vDash$ **s = t** and $\Sigma_k \vDash$ **Y[s]**. We want to show that $\Sigma_k \vDash Y[s, t]$, which is Z_k. To prove this we take M to be any model of Σ_k that is relevant to $Y[s, t]$. If **s** occurs in Σ_k or in $Y[s, t]$, then M would be relevant to $Y[s]$. If **s** occurs neither in Σ_k nor in $Y[s, t]$, then M might not be relevant to $Y[s]$. But we want M to be relevant to $Y[s]$. So in this case, we make a change in M. We add the necessary vocabulary to M so that M would be able to interpret **s**. This can always be done, and it has no effect on the truth values of $Y[s, t]$ and of the members of Σ_k on M. Since **s** = **t** and $Y[s]$ are logical consequences of Σ_k, they are true on M. By Exercise 2.5.12, $Y[s]$ and $Y[s, t]$ have the same truth value on M. Hence, $Y[s, t]$ is true on M. We conclude that $Y[s, t]$ is true on every model of Σ_k that is relevant to $Y[s, t]$. This entails that $\Sigma_k \vDash Z_k$.

The 10 cases above represent all the possible ways that Z_k could be introduced at line k. In all cases, we showed that $\Sigma_k \vDash Z_k$. Hence the Inductive Step is established. By PCI, for every positive natural number n, $\Sigma_n \vDash Z_n$. As we explained at the outset, this is sufficient for **X** to be a logical consequence of Σ_D; and the preceding fact is sufficient for **X** to be a logical consequence of Γ. This completes our proof of the Soundness Theorem for PL.

3.2 The Completeness Theorem

3.2.1 The proof of the Completeness Theorem is far more involved than the proof of the Soundness Theorem. As we stated in Chapter One, the proof presented here is due to Henkin (1949). The Completeness Theorem says that for every set Γ of PL sentences and every PL sentence **X**, if **X** is a logical consequence of Γ, then it is derivable from Γ; symbolically, if $\Gamma \vDash X$, then $\Gamma \vdash X$. However, we will not prove this statement. We will prove a statement that is equivalent to it. This statement says that if no contradiction is derivable from a set of PL sentences, then it has a model. We will prove this statement through a series of lemmas and theorems. Our first theorem is to prove that the Completeness Theorem is equivalent to this statement. However, to simplify matters we will first prove a simple lemma. Readers might remember from Chapter One that in this book we use the terms 'consistent' and 'inconsistent' to mean something different from what is usually meant by them in introductory symbolic logic books, including my book *An Introduction to Logical Theory*. 'Consistent' and 'inconsistent' in introductory symbolic logic books are typically used to refer to sets that have models and to sets that have no models, respectively. In metalogic books, including this book, a **consistent** set is a set from which no contradiction is derivable and an **inconsistent** set is a set from which a contradiction is derivable (by a 'contradiction' we mean a sentence and its negation). A set that has a model is called **satisfiable** and one that has no model is called **unsatisfiable**.

Lemma 3.2.1:

3.2.1a $\Gamma \cup \{\neg X\}$ is inconsistent iff $\Gamma \vdash X$, and $\Gamma \cup \{X\}$ is inconsistent iff $\Gamma \vdash \neg X$.

3.2.1b $\Gamma \vdash \neg(W \rightarrow Z)$ iff $\Gamma \vdash W$ and $\Gamma \vdash \neg Z$.

3.2.1c If $\Gamma \vdash Y$, the *name* **s** occurs in **Y**, the variable **z** does not occur in **Y**, and **s** does not occur in any member of Γ, then $\Gamma \vdash (\forall z)Y[z]$, where, as usual, **Y[z]** is obtained from **Y** by replacing all the occurrences of **s** with **z**.

Proof of 3.2.1a

1. We will prove only the first conjunct. The proof of the second conjunct is virtually identical with the first part. We first show that if $\Gamma \cup \{\neg X\}$ is inconsistent, then $\Gamma \vdash X$. So suppose that $\Gamma \cup \{\neg X\}$ is inconsistent.
2. From 1: there is a PL sentence **Y** such that $\Gamma \cup \{\neg X\} \vdash Y$ and $\Gamma \cup \{\neg X\} \vdash \neg Y$.
3. From 2 by the definition of derivability: there are PL derivations D_1 of **Y** and D_2 of $\neg Y$ from $\Gamma \cup \{\neg X\}$.
4. From 3: D_1 and D_2 can be combined into a single RAA derivation of **X** from Γ as follows:

```
[0    0             All the members of Γ that are invoked in D₁ and D₂
[1    1      ¬X                  RA
             ⋮      D₁
             h      Y
             ⋮      D₂
             i      ¬Y
1]    i+1    Y                   h, Reit
0]    i+2    X                   1–(i+1), RAA
```

5. From 4: $\Gamma \vdash X$.
6. From 1 through 5: if $\Gamma \cup \{\neg X\}$ is inconsistent, then $\Gamma \vdash X$.
7. Now we prove that if $\Gamma \vdash X$, then $\Gamma \cup \{\neg X\}$ is inconsistent. Thus suppose that $\Gamma \vdash X$.
8. From 7: since Γ is a subset of $\Gamma \cup \{\neg X\}$, $\Gamma \cup \{\neg X\} \vdash X$.
9. It is obvious that $\Gamma \cup \{\neg X\} \vdash \neg X$ (by the rule Reiteration).
10. From 8 and 9 by the definition of inconsistency: $\Gamma \cup \{\neg X\}$ is inconsistent.
11. From 7 through 10: if $\Gamma \vdash X$, then $\Gamma \cup \{\neg X\}$ is inconsistent.
12. From 6 and 11: $\Gamma \cup \{\neg X\}$ is inconsistent iff $\Gamma \vdash X$.

Proof of 3.2.1b

1. We first prove that if $\Gamma \vdash \neg(W \rightarrow Z)$, then $\Gamma \vdash W$ and $\Gamma \vdash \neg Z$. Suppose that $\Gamma \vdash \neg(W \rightarrow Z)$.
2. From 1: there is a PL derivation D of $\neg(W \rightarrow Z)$ from Γ.
3. The following are two PL derivations of **W** and of $\neg Z$ from Γ:

```
[0      0         All the members of Γ that are invoked in D
        ⋮         D
        i         ¬(W→Z)
[1      i+1       ¬W                  RA
[2      i+2       W                   CPA
[3      i+3       ¬Z                  RA
        i+4       ¬W                  i+1, Reit
 3]     i+5       W                   i+2, Reit
 2]     i+6       Z                   (i+3)–(i+5), RAA
        i+7       W→Z                 (i+2)–(i+6), CP
 1]     i+8       ¬(W→Z)              i, Reit
 0]     i+9       W                   (i+1)–(i+9), RAA


[0      0   All the members of Γ that are invoked in D
        ⋮   D
        i   ¬(W→Z)
[1      i+1       Z                   RA
[2      i+2       W                   CPA
 2]     i+3       Z                   i+1, Reit
        i+4       W→Z                 (i+2)–(i+3), CP
 1]     i+5       ¬(W→Z)              i, Reit
 0]     i+6       ¬Z                  (i+1)–(i+5), RAA
```

4. From 3: $\Gamma \vdash W$ and $\Gamma \vdash \neg Z$.
5. From 1 through 4: if $\Gamma \vdash \neg(W \rightarrow Z)$, then $\Gamma \vdash W$ and $\Gamma \vdash \neg Z$.
6. Now we prove that if $\Gamma \vdash W$ and $\Gamma \vdash \neg Z$, then $\Gamma \vdash \neg(W \rightarrow Z)$. Suppose that $\Gamma \vdash W$ and $\Gamma \vdash \neg Z$.
7. From 6: there is a PL derivation D_1 of **W** from Γ and there is a PL derivation D_2 of $\neg Z$ from Γ.
8. From 7: we combine these two derivations to produce a single derivation of $\neg(W \rightarrow Z)$ from Γ.

```
[0    0       All the members of Γ that are invoked in D₁ and D₂.
      ⋮       D₁
      h       W
      ⋮       D₂
      i       ¬Z
[1    i+1     W→Z              RA
      i+2     Z                h, (i+1), MP
 1]   i+3     ¬Z               i, Reit
 0]   i+4     ¬(W→Z)           (i+1)–(i+3), RAA
```

9. From 8: $\Gamma \vdash \neg(W \to Z)$.
10. From 6 through 9: if $\Gamma \vdash W$ and $\Gamma \vdash \neg Z$, then $\Gamma \vdash \neg(W \to Z)$.
11. From 5 and 10: $\Gamma \vdash \neg(W \to Z)$ iff $\Gamma \vdash W$ and $\Gamma \vdash \neg Z$.

Proof of 3.2.1c

1. Suppose that $\Gamma \vdash Y$, such that the *name* **s** occurs in **Y** but not in any member of Γ and the variable **z** does not occur in **Y**. Our goal is to show that $\Gamma \vdash (\forall z)Y[z]$ where, as usual, **Y[z]** is obtained from **Y** by replacing all the occurrences of **s** with **z**.
2. From 1: there is a PL derivation D of **Y** from Γ.
3. The following is a PL derivation of $(\forall z)Y[z]$ from Γ:

```
0]   0       All the members of Γ that are invoked in D.
     ⋮       D
     H       Y              Y is not a premise or an assumption. It is not a premise because
                             it contains s and none of the premises contains s. It is not an
                             assumption because it is the conclusion of a PL subderivation
                             and the block rules do not allow a derivation to terminate with
                             an assumption. Furthermore, all assumptions that might be
                             introduced prior to line h must be discharged by the time h is
                             reached. Otherwise, Y cannot be the conclusion of a PL
                             subderivation. Hence s is arbitrary at line h. This entails that
                             the rule Universal Generalization can be applied to line h.
0]   h+1     (∀z)Y[z]       h, UG
```

4. From 3: $\Gamma \vdash (\forall z)Y[z]$.

Now we prove that the Completeness Theorem and the statement in 3.2.1 above are equivalent.

Theorem 3.2.1: The following statement is equivalent to the Completeness Theorem: for every set Γ of PL sentences, if Γ is consistent, then Γ is satisfiable, that is, Γ has a model.

Proof

1. We first show that if the Completeness Theorem is true, then the statement above is true.

2. Assume that the Completeness Theorem is true and let Γ be any consistent set. Our goal is to show that Γ has a model.

3. Reductio Assumption: Γ has no model (i.e., Γ is unsatisfiable).

4. From 3 by the definition of logical consequence: for any PL sentence **Y**, since there is no PL interpretation that satisfies Γ, $\Gamma \vDash$ **Y** and $\Gamma \vDash \neg$**Y**. (In order for $\Gamma \vDash$ **Y** to fail, there must be a PL interpretation that satisfies Γ and makes **Y** false.)

5. From 4 by the Completeness Theorem: $\Gamma \vdash$ **Y** and $\Gamma \vdash \neg$**Y**.

6. From 5: Γ is inconsistent, which contradicts 2.

7. From 3 through 6: the Reductio Assumption is false; and hence Γ has a model.

8. Now we show that if the statement in Theorem 3.2.1 is true, then the Completeness Theorem is true.

9. Suppose that the statement in Theorem 3.2.1 is true and let Γ be any PL set and **X** be any PL sentence such that $\Gamma \vDash$ **X**. We want to show that $\Gamma \vdash$ **X**.

10. From 9 by the definition of logical consequence: there is no PL interpretation that satisfies Γ and on which $\neg$**X** is true.

11. From 10 by the definition of unsatisfiability: $\Gamma \cup \{\neg$**X**$\}$ is unsatisfiable.

12. From 9: if $\Gamma \cup \{\neg$**X**$\}$ is consistent, then $\Gamma \cup \{\neg$**X**$\}$ is satisfiable.

13. From 11 and 12: $\Gamma \cup \{\neg$**X**$\}$ is inconsistent.

14. From 13 by Lemma 3.2.1a: $\Gamma \vdash$ **X**.

Because of this equivalence, we will refer to either statement as the Completeness Theorem. We will let the context disambiguate which statement is intended. In the remainder of Section 3.2, we will be occupied with presenting a proof of the second formulation of the Completeness Theorem—namely, that every consistent PL set has a model. It is clear from the equivalence above that establishing either formulation of the Completeness Theorem establishes the other.

3.2.2 It is important to note that we are working with the economical version of PL, which, like the full version, has a countably infinite vocabulary. There are countably infinite names, variables, n-place function symbols, and n-place predicates, for every positive integer n; and there are two connectives, one quantifier, an identity predicate, and two parentheses. The total of all these symbols is countably infinite. This fact will play an important role in our proof of the Completeness Theorem. The first stage of the proof of the Completeness Theorem is to prove a theorem that is known as **Lindenbaum's Lemma**. It is historically called a "lemma" because the term "Lindenbaum's Theorem" is usually reserved for a different (though similar) statement. In order to state this theorem precisely, we need to define the notion of a maximal set. We present the following definition.

Definition 3.2.1: Let Δ be any set of PL sentences.

3.2.1a Δ is **maximal** iff for every PL sentence X, either $X \in \Delta$ or $\neg X \in \Delta$.

3.2.1b Δ is **deductively closed** iff for every PL sentence X, if $\Delta \vdash X$, then $X \in \Delta$, that is, Δ contains all its theorems.

3.2.1c Δ is **semantically closed** iff for every PL sentence X, if $\Delta \vDash X$, then $X \in \Delta$, that is, Δ contains all its logical consequences.

The set of all the PL sentences is trivially maximal, deductively closed, and semantically closed, since it contains every PL sentence and its negation and it contains all its theorems and logical consequences. But of course the set of all the PL sentences is an inconsistent set. These notions are interesting only when the sets are consistent. In fact, most logic books define a single notion of a maximal consistent set. We will have use for maximal sets later, so we decided to separate the two notions. The Soundness and Completeness Theorems entail that the notions of deductive and semantical closure are equivalent: a PL set is deductively closed iff it is semantically closed. To simplify the proof of Lindenbaum's Lemma, we prove first a simple lemma.

Lemma 3.2.2:

3.2.2a Every maximal consistent set is deductively closed.

3.2.2b A set Δ is maximal consistent iff it is consistent and for every set Δ', if $\Delta \subset \Delta'$, then Δ' is inconsistent.

Proof of 3.2.2a

1. Let Δ be any PL maximal consistent set of PL sentences.
2. Let X be any theorem of Δ, that is, $\Delta \vdash X$.
3. Reductio Assumption: $X \notin \Delta$.
4. From 1: since Δ is maximal, $X \in \Delta$ or $\neg X \in \Delta$.
5. From 3 and 4: $\neg X \in \Delta$.
6. From 5 (by Reiteration): $\Delta \vdash \neg X$.
7. From 2 and 6: $\Delta \vdash X$ and $\Delta \vdash \neg X$; hence Δ is inconsistent, which contradicts 1.
8. From 3 through 7: the Reductio Assumption is false. Therefore $X \in \Delta$.
9. From 2 through 8: for every PL sentence X, if $\Delta \vdash X$, then $X \in \Delta$.
10. From 9 by the definition of deductive closure: Δ is deductively closed.

Proof of 3.2.2b

1. First we prove the 'if' part. Let Δ be a consistent set such that for every set Δ', if $\Delta \subset \Delta'$, then Δ' is inconsistent.
2. Let **X** be any PL sentence.
3. Reductio Assumption: $X \notin \Delta$ and $\neg X \notin \Delta$.
4. From 3: $\Delta \subset \Delta \cup \{X\}$ and $\Delta \subset \Delta \cup \{\neg X\}$.
5. From 1 and 4: $\Delta \cup \{X\}$ and $\Delta \cup \{\neg X\}$ are inconsistent sets.
6. From 5 by Lemma 3.2.1a: $\Delta \vdash \neg X$ and $\Delta \vdash X$.
7. From 6: Δ is inconsistent, which contradicts 1.
8. From 3 through 7: the Reductio Assumption is false; hence $X \in \Delta$ or $\neg X \in \Delta$.
9. From 1 and 8: Δ is a maximal consistent set.
10. Now we prove the 'only if' part. Let Δ be a maximal consistent set of PL sentences.
11. Let Δ' be any PL set such that $\Delta \subset \Delta'$.
12. By definition of proper subset: there is a sentence **X** such that $X \notin \Delta$ and $X \in \Delta'$.
13. From 9 and 12 by the definition of maximal: $\neg X \in \Delta$.
14. From 11 and 13: $\neg X \in \Delta'$.
15. From 12 and 14: $\Delta' \vdash X$ and $\Delta' \vdash \neg X$.
16. From 15: Δ' is inconsistent.

We will use a new term to describe a familiar relation. If a set Σ is obtained from a set Γ by adding zero or more PL sentences to Γ, we say that Σ is an **extension of** Γ or that Γ is **extended into** Σ. In other words, for any PL sets Σ and Γ, Σ is an extension of Γ or Γ is extended into Σ just in case $\Gamma \subseteq \Sigma$. Hence, every set is an extension of itself and every set is an extension of the empty set; also, the set of all PL sentences is an extension of every PL set. Our goal now is to state and prove Lindenbaum's Lemma. This theorem is the first stage in proving the Completeness Theorem.

Lindenbaum's Lemma: Every consistent PL set can be extended into a maximal consistent PL set.

Proof

1. Let Γ be any consistent PL set. We want to establish the existence of a PL set Δ, such that $\Gamma \subseteq \Delta$ and Δ is maximal consistent.
2. Let Sent$_{PL}$ be the set that consists of all the PL sentences. Since every PL sentence is a finite string of symbols and since there are countably infinite PL symbols, the set Sent$_{PL}$ is countably infinite. Hence, there is a one-to-one correspondence between Sent$_{PL}$ and $\mathbb{N}$. This entails that the members of Sent$_{PL}$ can be listed as an infinite sequence. Let this infinite sequence be $X_0, X_1, X_2, \ldots, X_n, \ldots$.

3. Now we employ an **inductive definition** to define an increasing sequence of consistent PL sets. An inductive definition has some similarity to PMI. Using PMI, one can prove that a certain arithmetical formula E(n) is true of every natural number by proving that E(0) is true and that E(k+1) is true whenever E(k) is true. Similarly, using an inductive definition, one can define an infinite sequence of objects $q_0, q_1, q_2, \ldots, q_n, \ldots$ by defining first the 0th object q_0 and then defining the k+1st object q_{k+1} on the basis of q_k. The intuitive idea is similar to mathematical induction. If q_0 is defined, then the next object q_1 is defined; if q_1 is defined, then q_2 is defined; if q_2 is defined, then q_3 is defined; and so on. We will define the following sequence of PL sets: $\Delta_0 \subseteq \Delta_1 \subseteq \Delta_2 \subseteq \ldots \subseteq \Delta_n \subseteq \ldots$. We employ an inductive definition.

$\Delta_0 = \Gamma$
$\Delta_{k+1} = \Delta_k \cup \{\mathbf{X}_k\}$ if $\Delta_k \cup \{\mathbf{X}_k\}$ is consistent, and
$\Delta_{k+1} = \Delta_k$ if $\Delta_k \cup \{\mathbf{X}_k\}$ is inconsistent.

4. According to this definition Δ_0 is simply Γ, which is consistent. Now we add $\mathbf{X}_0$ to Δ_0. There are two cases: either the extended set $\Delta_0 \cup \{\mathbf{X}_0\}$ is consistent, in this case we let $\Delta_1 = \Delta_0 \cup \{\mathbf{X}_0\}$, or it is inconsistent, in this case we let $\Delta_1 = \Delta_0$. We repeat the process for Δ_1 and $\mathbf{X}_1$. We add $\mathbf{X}_1$ to Δ_1, and examine the extended set $\Delta_1 \cup \{\mathbf{X}_1\}$. If it is consistent, we let $\Delta_2 = \Delta_1 \cup \{\mathbf{X}_1\}$; if it is inconsistent, we let $\Delta_2 = \Delta_1$. In general, we add $\mathbf{X}_k$ to Δ_k to obtain the extended set $\Delta_k \cup \{\mathbf{X}_k\}$. If the extended set is consistent, we let $\Delta_{k+1} = \Delta_k \cup \{\mathbf{X}_k\}$; if it is inconsistent, we let $\Delta_{k+1} = \Delta_k$. This inductive process defines a set Δ_n for each natural number n. We have defined the following, possibly increasing, sequence of PL sets: $\Gamma = \Delta_0 \subseteq \Delta_1 \subseteq \Delta_2 \subseteq \ldots \subseteq \Delta_n \subseteq \ldots$.

5. Two things are clear about the sequence of PL sets we just defined. First, every set is a subset of all the sets that come after it in the sequence. This follows from the trivial fact that if $A \subseteq B \subseteq C$, then $A \subseteq C$. Second every Δ_n is consistent. The last fact is an obvious consequence of the way the sequence is "constructed." We did not permit any contradictions to be derived from the sets of this sequence. If adding a sentence to an already defined set resulted in a contradiction, we excluded this sentence and held to the original set. This is an intuitive argument for the consistency of the sets of this sequence.

6. The fact that every Δ_n is consistent can be proved rigorously by PMI. The Base Step is trivial: Δ_0 is consistent because it is identical with Γ and the latter is consistent. The Induction Hypothesis states that Δ_k is consistent for some $k \geq 0$. $\Delta_k \cup \{\mathbf{X}_k\}$ is either consistent or not. If it is consistent, then $\Delta_{k+1} = \Delta_k \cup \{\mathbf{X}_k\}$, which means that Δ_{k+1} is consistent. On the other hand, if $\Delta_k \cup \{\mathbf{X}_k\}$ is inconsistent, we let Δ_{k+1} be Δ_k, which we know, by the Induction Hypothesis, to be consistent. Thus, by PMI, Δ_n is consistent for each natural number n.

7. Let $\mathcal{F}$ be the family that consists of the PL sets defined above. In other words, $\mathcal{F} = \{\Delta_n:$ n is a natural number$\}$. Let the desired set Δ be $\cup \mathcal{F}$, that is, Δ is the union of all

the sets in $\mathcal{F}$. By the definition of union, for every natural number n, $\Delta_n \subseteq \Delta$; in particular $\Gamma = \Delta_0 \subseteq \Delta$. So the first condition is satisfied: Δ is an extension of Γ.

8. It remains to show that Δ is maximal consistent. We first need to prove that it is consistent. We make a Reductio Assumption: Δ is inconsistent. This entails that there is a sentence **Y** such that $\Delta \vdash$ **Y** and $\Delta \vdash \neg$**Y**.

9. From 8 by the definition of derivability: there is a PL derivation D of **Y** from Δ and there is a PL derivation E of $\neg$**Y** from Δ. Let Σ_D and Σ_E be the set of premises invoked in D and E, respectively. It is obvious that each of Σ_D and Σ_E is a finite subset of Δ. Let $\Sigma = \Sigma_D \cup \Sigma_E$. Since Σ_D and Σ_E are subsets of Σ, $\Sigma \vdash$ **Y** and $\Sigma \vdash \neg$**Y**.

10. Since Σ is a subset of Δ and $\Delta = \cup \mathcal{F}$, every member **Z** of Σ belongs to Δ_k for some natural number k. Of course, if **Z** belongs to Δ_k, it belongs to all Δ_j, for $j \geq k$, because every set in the sequence is a subset of the sets that come after it. We define Δ_Z to be the *first* Δ_m that contains **Z**. Take the family $\mathcal{K}$ to be the set $\{\Delta_Z : \textbf{Z} \text{ is a member } \Sigma\}$. Since Σ is finite, $\mathcal{K}$ is finite too.

10. Given the way the sequence of Δ_n's is defined, the members of $\mathcal{K}$ form a chain ordered by the relation $\subseteq$. Since $\mathcal{K}$ is finite, there is Δ^* that is the largest set in $\mathcal{K}$, in the sense that every other set in $\mathcal{K}$ is a subset of it. So Δ^* contains all the members of Σ. Therefore, $\Delta^* \vdash$ **Y** and $\Delta^* \vdash \neg$**Y**. Since Δ^* is Δ_m for some $m \geq 0$, Δ_m is inconsistent, which contradicts 6.

11. From 7 through 10: The Reductio Assumption is false, and thus Δ is consistent.

12. Finally we prove that Δ is maximal. Let Δ' be any PL set such that $\Delta \subset \Delta'$. Hence, there is a PL sentence **W** that belongs to Δ' but not to Δ. Since the list $\textbf{X}_0, \textbf{X}_1, \textbf{X}_2, \ldots, \textbf{X}_n, \ldots$ comprises *all* the PL sentences, **W** must be $\textbf{X}_p$ for some $p \geq 0$. Since $\textbf{X}_p$ is not a member of Δ, it must have been excluded from Δ at stage p, when Δ_{p+1} is formed. $\textbf{X}_p$ was not added to Δ_p to form Δ_{p+1} because $\Delta_p \cup \{\textbf{X}_p\}$ is inconsistent. Since $\Delta_p \cup \{\textbf{X}_p\}$ is a subset of Δ', Δ' is inconsistent. Hence, by Lemma 3.2.2b, Δ is maximal.

13. From 7, 11, and 12: Δ is a maximal consistent extension of Γ. This concludes our proof of Lindenbaum's Lemma.

3.2.3   The proof of the Completeness Theorem is based on two central ideas. The first idea is to take a consistent set and extend it into a maximal consistent set that has sufficiently rich resources to allow it to capture the truth conditions of all the PL sentences. Such a set is referred to as a **Henkin set**. We already proved that every consistent PL set Γ can be extended into a maximal consistent PL set Δ. Let us examine Δ to see how rich it is. First consider the conditional. If $\neg$**X** or **Y** belongs to Δ, then we would expect that $\textbf{X} \rightarrow \textbf{Y}$ belongs to Δ. Let $\neg \textbf{X} \in \Delta$. We can construct a derivation of $\textbf{X} \rightarrow \textbf{Y}$ from Δ.

```
[0  0   ¬X      P
[1  1   X       CPA
[2  2   ¬Y      RA
    3   ¬X      1, Reit
2]  4   X       2, Reit
1]  5   Y       2–4, RAA
0]  6   X→Y     1–5, CP
```

This derivation shows that $\Delta \vdash X \to Y$. Since Δ is maximal consistent, by Lemma 3.2.2 it is deductively closed. Hence $X \to Y \in \Delta$. Similarly, if $Y \in \Delta$, we can derive $X \to Y$ from Δ.

```
[0  0   Y       P
[1  1   X       CPA
1]  2   Y       0, Reit
0]  3   X→Y     1–2, CP
```

This derivation establishes that $\Delta \vdash X \to Y$, which entails, by the deductive closure of Δ, that $X \to Y \in \Delta$. Now suppose that $X \to Y \in \Delta$. We should expect that either $\neg X$ or $Y \in \Delta$. Since Δ is maximal, there are two cases: either $X \in \Delta$ or $\neg X \in \Delta$. If $\neg X \in \Delta$, then we have our desired conclusion. On the other hand, if $X \in \Delta$, then both $X \to Y \in \Delta$ and $X \in \Delta$. By Modus Ponens, there is a derivation of Y from $\{X, X \to Y\}$. It follows that $\Delta \vdash Y$. By the deductive closure of Δ, $Y \in \Delta$. So we conclude that for all PL sentences X and Y, $X \to Y \in \Delta$ iff either $\neg X \in \Delta$ or $Y \in \Delta$. This membership relation mirrors the truth conditions of the conditional: $X \to Y$ is true iff $\neg X$ is true or Y is true.

We continue our examination of the maximal consistent set Δ to see whether it is rich enough to be a Henkin set. We have, so far, demonstrated that it captures the truth conditions of the conditional. Let us consider the negation. We would expect that $X \in \Delta$ iff $\neg X \notin \Delta$—it is clear that both X and $\neg X$ cannot be members of Δ; otherwise, $\Delta \vdash X$ and $\Delta \vdash \neg X$, which means that Δ is inconsistent. This contradicts our previous conclusion that Δ is consistent. Thus by the consistency of Δ, if $X \in \Delta$, $\neg X \notin \Delta$. On the other hand, if $\neg X \notin \Delta$, $X \in \Delta$, since by the maximality of Δ either $X \in \Delta$ or $\neg X \in \Delta$. It follows that $X \in \Delta$ iff $\neg X \notin \Delta$. This membership relation mirrors the truth conditions of the negation: X is true iff $\neg X$ false.

It remains to discuss the case of the universal quantifier. Unfortunately the maximal consistent set Δ is not rich enough to capture the truth conditions of the universal quantifier. These truth conditions state that a universally quantified sentence $(\forall z)Y$ is true iff all its basic substitutional instances are true. Hence we should expect that $(\forall z)Y \in \Delta$ iff $Y[t] \in \Delta$ for every PL name t. The 'only-if' part is true: if $(\forall z)Y \in \Delta$, then every $Y[t] \in \Delta$. The proof is trivial. Let $(\forall z)Y \in \Delta$. Using the rule Universal Instantiation, we can derive from Δ every substitutional instance of $(\forall z)Y$; in particular, for every PL name t, $\Delta \vdash Y[t]$. By the deductive closure of Δ, $Y[t] \in \Delta$ for every PL name t. The 'if' part however is not always true. To see the point, consider a specific case. Consider a 1-place predicate A and let Ψ be the set $\{As: s \text{ is a PL name}\}$. So Ψ consists of all the PL sentences that result from appending a PL name s to A. Let J be the PL interpretation whose UD is $\{-1, 0, 1, 2, 3, \ldots\}$ and whose LN consists of all the PL names with the addition of a single name α (a PL interpretation may bring its own names). J makes the following semantical assignments: it assigns –1 to α; it assigns to every PL name s a unique natural number; and it assigns to the predicate A the property of being a natural number. It is clear that on J every sentence of the form As is true, where s is any PL name (recall that every PL name is in LN, but LN contains an additional name). $(\forall x)Ax$, however, is false on J. Hence $\Psi \not\models (\forall x)Ax$. By the Soundness Theorem, $\Psi \not\vdash (\forall x)Ax$. By Lemma 3.2.1a, $\Psi \cup \{\neg(\forall x)Ax\}$ is consistent. By Lindenbaum's Lemma $\Psi \cup \{\neg(\forall x)Ax\}$ can be extended into a maximal consistent set. This set contains every PL sentence At where t is any PL name, but it contains $\neg(\forall x)Ax$ instead of $(\forall x)Ax$.

This example shows that if Δ contains every basic substitutional instance of $(\forall z)\mathbf{Y}$, there is no guarantee that it would contain $(\forall z)\mathbf{Y}$. So what is missing from Δ? Recall that in order to derive $(\forall z)\mathbf{Y}$ from any set Λ, an *arbitrary basic substitutional instance* of $(\forall z)\mathbf{Y}$ must be derivable from Λ. Since PL derivations are finite sequences, this is equivalent to saying that there must be a finite $\Sigma \subseteq \Lambda$ such that $\Sigma \vdash \mathbf{Y}[\mathbf{c}]$ and $\mathbf{c}$ is a name that does not occur in any member of Σ. Informally speaking, we need an *arbitrary name* $\mathbf{c}$ for every universally quantified sentence $(\forall z)\mathbf{Y}$, so that if $\mathbf{Y}[\mathbf{c}]$ is derivable, then $(\forall z)\mathbf{Y}$ is derivable. The problem we face is that the consistent set Γ we started with might have exhausted all the PL names. We need more names to serve as arbitrary names for the universally quantified sentences of PL. There is only one option: we need to add new names to PL. Since there are countably infinitely many universally quantified sentences of PL, we need to add a countably infinite sequence of names. Let these names be $\alpha_0, \alpha_1, \alpha_2, \alpha_3, ..., \alpha_n, ...$ We will refer to these names as the α names. We call the Predicate Logic whose vocabulary is identical with the vocabulary of PL except for the addition of the new α names PL⁺. Of course, every PL expression is a PL⁺ expression, but the converse is not true. For instance, $e_1 = \alpha_3$, $A\alpha_1$, and $(\forall x)(\neg Ax \rightarrow Bx\alpha_5)$ are sentences of PL⁺, which contain PL vocabulary; but they are not PL sentences. Our original set Γ consists of PL sentences only. Since every PL expression is also a PL⁺ expression, we can think of Γ as consisting of PL⁺ sentences.

We now collect all the universally quantified sentences of PL⁺. Since there are countably infinitely many such sentences, we can list all of them in an infinite sequence: $(\forall z_0)\mathbf{W}_0, (\forall z_1)\mathbf{W}_1, (\forall z_2)\mathbf{W}_2, (\forall z_3)\mathbf{W}_3, ..., (\forall z_n)\mathbf{W}_n, ...$. Given that every PL⁺ expression is a finite string of symbols, any finite collection of PL⁺ strings can only use finitely many of the new names. So for any such collection, there are infinitely many α names that are not used. This allows us to construct an infinite sequence of names that meets a certain condition. Let $\mathbf{c}_0, \mathbf{c}_1, \mathbf{c}_2, \mathbf{c}_3, ..., \mathbf{c}_n, ...$ be an infinite sequence of α names such that

$\mathbf{c}_0$ is the first α name that does not occur in $(\forall z_0)\mathbf{W}_0$
$\mathbf{c}_1$ is the first α name that does not occur in $(\forall z_0)\mathbf{W}_0$ or $(\forall z_1)\mathbf{W}_1$
$\mathbf{c}_2$ is the first α name that does not occur in $(\forall z_0)\mathbf{W}_0, (\forall z_1)\mathbf{W}_1,$ or $(\forall z_2)\mathbf{W}_2$; and in general
$\mathbf{c}_n$ is the first α name that does not occur in $(\forall z_0)\mathbf{W}_0, (\forall z_1)\mathbf{W}_1, (\forall z_2)\mathbf{W}_2, (\forall z_3)\mathbf{W}_3, ...,$ or $(\forall z_n)\mathbf{W}_n$

Notice that $\mathbf{c}_n$, for every natural number n, is boldfaced, which means that it is a metalinguistic name that stands for some α name. Precisely speaking, the $\mathbf{c}$ names are not

PL⁺ names. They are metalinguistic expressions that stand for α names, which are PL⁺ names. Also their numeric subscripts do not have to match the numeric subscripts of the α names. For instance c_3 does not have to be α_3. It is most likely not α_3. For instance if $(\forall z_0)\mathbf{W}_0$ is the PL⁺ sentence $(\forall x)(\forall y)((Px\alpha_3 \to \neg Ry\alpha_1) \to \neg By\alpha_6)$, then c_0 is α_2 since α_2 is the first α name that does not occur in $(\forall z_0)\mathbf{W}_0$; also if

$(\forall z_1)\mathbf{W}_1$ is $(\forall w)\neg(\forall x)(\neg(Gh^2w\alpha_2 \to R\alpha_4\alpha_7) \to \neg Kx\alpha_5)$

then c_1 is α_8 because α_8 is the first α name that does not occur in $(\forall z_0)\mathbf{W}_0$ or $(\forall z_1)\mathbf{W}_1$.

Now for every $(\forall z_n)\mathbf{W}_n$, we define a conditional θ_n that we call a "W-conditional":

θ_0: $\quad \mathbf{W}_0[c_0] \to (\forall z_0)\mathbf{W}_0$
θ_1: $\quad \mathbf{W}_1[c_1] \to (\forall z_1)\mathbf{W}_1$
θ_2: $\quad \mathbf{W}_2[c_2] \to (\forall z_2)\mathbf{W}_2$
θ_3: $\quad \mathbf{W}_3[c_3] \to (\forall z_3)\mathbf{W}_3$; and in general
θ_n: $\quad \mathbf{W}_n[c_n] \to (\forall z_n)\mathbf{W}_n$

As usual, $\mathbf{W}_n[c_n]$ is obtained from $\mathbf{W}_n$ by replacing every occurrence of z_n in $\mathbf{W}_n$ with c_n. As we will see later, the purpose of these definitions is to construct an arbitrary basic substitutional instance of every universally quantified PL⁺ sentence, that is, a basic substitutional instance that contains an arbitrary name. Let Θ be the set of all the W-conditionals; symbolically, $\Theta = \{\theta_0, \theta_1, \theta_2, \theta_3, ..., \theta_n, ...\}$. We have the following theorem. (Recall that our original set Γ is a consistent set of PL sentences.)

Theorem 3.2.2: $\Gamma \cup \Theta$ is consistent.

Proof

1. Reductio Assumption: $\Gamma \cup \Theta$ is inconsistent, where Γ is a consistent set of PL sentences and Θ is the set of all the W-conditionals.

2. From 1: there is a PL⁺ sentence $\mathbf{Y}$ such that $\Gamma \cup \Theta \vdash \mathbf{Y}$ and $\Gamma \cup \Theta \vdash \neg\mathbf{Y}$.

3. From 2: there is PL⁺ derivation D of $\mathbf{Y}$ from $\Gamma \cup \Theta$ and a PL⁺ derivation E of $\neg\mathbf{Y}$ from $\Gamma \cup \Theta$.

4. From 3: since derivations are finite sequences, there are two finite sets Σ_D and Σ_E that are subsets of $\Gamma \cup \Theta$ and such that $\Sigma_D \vdash \mathbf{Y}$ and $\Sigma_E \vdash \neg\mathbf{Y}$. ($\Sigma_D$ is the set of the premises invoked in D and Σ_E is the set of the premises invoked in E.)

5. Let $\Sigma = \Sigma_D \cup \Sigma_E$. So $\Sigma \subseteq \Gamma \cup \Theta$.

6. From 4 and 5: $\Sigma \vdash \mathbf{Y}$ and $\Sigma \vdash \neg\mathbf{Y}$. Hence Σ is inconsistent.

7. From 1 and 6: $\Sigma \not\subseteq \Gamma$ (otherwise, Γ would be inconsistent).

8. From 5 and 7: $\Sigma \cap \Theta \neq \emptyset$. In other words, Σ contains some W-conditionals. Call the nonempty set of the W-conditionals that Σ contains Φ; hence $\Sigma \cap \Theta = \Phi$. Since $\Sigma \subseteq \Gamma \cup \Theta$, Σ might also contain some members of Γ. Let $\Sigma \cap \Gamma = \Psi$. We have the following: (1) $\Psi \cup \Phi = \Sigma$; (2) Ψ is a finite subset of Γ; (3) Φ is a nonempty finite subset of Θ; and $\Psi \cup \Phi$ is inconsistent (because Σ is inconsistent). It follows that $\Gamma \cup \Phi$ is inconsistent. In other words, there are finitely many W-conditionals such that their

union with Γ forms an inconsistent set. Given this fact, we can now define Λ to be the set of W-conditionals $\theta_0, \theta_1, \theta_2, \ldots, \theta_{q-1}$, and θ_q such that q is the *least* number for which $\Gamma \cup \Lambda$ is inconsistent. In other words, $\Gamma \cup \{\theta_0, \theta_1, \theta_2, \ldots, \theta_q\}$ is inconsistent, while $\Gamma \cup \{\theta_0, \theta_1, \theta_2, \ldots, \theta_{q-1}\}$ is *consistent*. We know for certain that there is such a nonempty set Λ, because we know that the union of Γ and the nonempty finite set Φ, which consists of W-conditionals, is inconsistent.

9. From 8 by Lemma 3.2.1a: $\Gamma \cup \{\theta_0, \theta_1, \theta_2, \ldots, \theta_{q-1}\} \vdash \neg \theta_q$ (because $\Gamma \cup \{\theta_0, \theta_1, \theta_2, \ldots, \theta_{q-1}\} \cup \{\theta_q\}$ is inconsistent).

10. From 9 by the definition of θ_q: $\Gamma \cup \{\theta_0, \theta_1, \theta_2, \ldots, \theta_{q-1}\} \vdash \neg(\mathbf{W}_q[\mathbf{c}_q] \to (\forall \mathbf{z}_q)\mathbf{W}_q)$.

11. From 10 by Lemma 3.2.1b: $\Gamma \cup \{\theta_0, \theta_1, \theta_2, \ldots, \theta_{q-1}\} \vdash \mathbf{W}_q[\mathbf{c}_q]$ and $\Gamma \cup \{\theta_0, \theta_1, \theta_2, \ldots, \theta_{q-1}\} \vdash \neg(\forall \mathbf{z}_q)\mathbf{W}_q$.

12. From 11: since $\mathbf{c}_q$ is an α name, it does not occur in any member of Γ (Γ is a set of PL sentences), and since, by definition, $\mathbf{c}_q$ does not occur in $\theta_0, \theta_1, \theta_2, \ldots$, or θ_{q-1}, $\mathbf{c}_q$ does not occur in any member of $\Gamma \cup \{\theta_0, \theta_1, \theta_2, \ldots, \theta_{q-1}\}$. Hence relative to $\Gamma \cup \{\theta_0, \theta_1, \theta_2, \ldots, \theta_{q-1}\}$, $\mathbf{c}_q$ occurs arbitrarily in $\mathbf{W}_q[\mathbf{c}_q]$.

13. From 12 by Lemma 3.2.1c: $\Gamma \cup \{\theta_0, \theta_1, \theta_2, \ldots, \theta_{q-1}\} \vdash (\forall \mathbf{z}_q)\mathbf{W}_q[\mathbf{z}_q]$, where $\mathbf{W}_q[\mathbf{z}_q]$ is obtained from $\mathbf{W}_q[\mathbf{c}_q]$ by replacing all the occurrences of $\mathbf{c}_q$ with $\mathbf{z}_q$. Since, by definition, $\mathbf{c}_q$ does not occur in $(\forall \mathbf{z}_q)\mathbf{W}_q$, and since $\mathbf{W}_q[\mathbf{c}_q]$ is obtained from $(\forall \mathbf{z}_q)\mathbf{W}_q$ by substituting $\mathbf{c}_q$ for $\mathbf{z}_q$ in all its occurrences, $(\forall \mathbf{z}_q)\mathbf{W}_q[\mathbf{z}_q]$ is simply $(\forall \mathbf{z}_q)\mathbf{W}_q$. Thus $\Gamma \cup \{\theta_0, \theta_1, \theta_2, \ldots, \theta_{q-1}\} \vdash (\forall \mathbf{z}_q)\mathbf{W}_q$.

14. From 11 and 13: $\Gamma \cup \{\theta_0, \theta_1, \theta_2, \ldots, \theta_{q-1}\} \vdash \neg(\forall \mathbf{z}_q)\mathbf{W}_q$ and $\Gamma \cup \{\theta_0, \theta_1, \theta_2, \ldots, \theta_{q-1}\} \vdash (\forall \mathbf{z}_q)\mathbf{W}_q$. This contradicts 8—namely, that $\Gamma \cup \{\theta_0, \theta_1, \theta_2, \ldots, \theta_{q-1}\}$ is consistent.

15. From 1 through 14: the Reductio Assumption is false; and hence $\Gamma \cup \Theta$ is consistent.

By Lidenbaum's Lemma $\Gamma \cup \Theta$ can be extended into a maximal consistent set Π. (Π is a set of PL⁺ sentences.) This set is a Henkin set: it is maximal consistent and its membership relations mirror the truth conditions of all the PL⁺ Sentences. We argued previously that $\mathbf{X} \to \mathbf{Z} \in \Pi$ iff $\neg \mathbf{X} \in \Pi$ or $\mathbf{Z} \in \Pi$, and that $\mathbf{X} \in \Pi$ iff $\neg \mathbf{X} \notin \Pi$. We also argued that if $(\forall \mathbf{y})\mathbf{W} \in \Pi$, then $\mathbf{W}[\mathbf{s}] \in \Pi$, for every PL⁺ name $\mathbf{s}$. Now we will show that if $\mathbf{W}[\mathbf{s}] \in \Pi$ for every PL⁺ name $\mathbf{s}$, then $(\forall \mathbf{y})\mathbf{W} \in \Pi$. So assume that $\mathbf{W}[\mathbf{s}] \in \Pi$ for every PL⁺ name $\mathbf{s}$. $(\forall \mathbf{y})\mathbf{W}$ must be $(\forall \mathbf{z}_k)\mathbf{W}_k$ for some natural number k, since the list above exhausts all the universally quantified sentences of PL⁺. Given that $\Theta \subseteq \Pi$, the W-conditional $\theta_k \in \Pi$. But θ_k is $\mathbf{W}_k[\mathbf{c}_k] \to (\forall \mathbf{z}_k)\mathbf{W}_k$. It follows that $\{\mathbf{W}_k[\mathbf{c}_k], \mathbf{W}_k[\mathbf{c}_k] \to (\forall \mathbf{z}_k)\mathbf{W}_k\} \subseteq \Pi$. By Modus Ponens, there is a PL⁺ derivation of $(\forall \mathbf{z}_k)\mathbf{W}_k$ from Π. Symbolically, $\Pi \vdash (\forall \mathbf{z}_k)\mathbf{W}_k$, which is $(\forall \mathbf{y})\mathbf{W}$. By the deductive closure of Π (Lemma 3.2.2a), $(\forall \mathbf{y})\mathbf{W} \in \Pi$. Thus Π mirrors the truth conditions of the universal quantifier: $(\forall \mathbf{y})\mathbf{W} \in \Pi$ iff every substitutional instance $\mathbf{W}[\mathbf{s}] \in \Pi$, where $\mathbf{s}$ is any PL⁺ name. Π, therefore, is a Henkin set.

The substitutional interpretation of the quantifiers we adopted in this book requires us to show a slightly stronger claim about Henkin sets. In the following Subsection, we will need to invoke this claim. We proved above that if Π is a Henkin set, then $(\forall \mathbf{y})\mathbf{W} \in \Pi$ iff

W[s] ∈ Π for every PL⁺ name **s**. In fact, Π satisfies a slightly stronger condition: if Π is a Henkin set, then (∀**y**)**W** ∈ Π iff **W[t]** ∈ Π for every PL⁺ *singular term* **t**. We first show the 'if' part. So assume that Π is a Henkin set and that **W[t]** ∈ Π for every PL⁺ *singular term* **t**. Since, by definition, every PL⁺ name is a PL⁺ singular term, **W[s]** ∈ Π for every PL⁺ name **s**. Given that Π is a Henkin set, it follows that (∀**y**)**W** ∈ Π (where the variable **y** does not occur in **W[s]**). Now we prove the 'only if' part. Assume that (∀**y**)**W** ∈ Π. The rule Universal Instantiation allows us to derive every substitutional instance **W[t]** from (∀**y**)**W**, whether **t** is a simple PL⁺ name or a complex PL⁺ singular term. Hence for every PL⁺ singular term **t**, there is a PL⁺ derivation of **W[t]** from (∀**y**)**W**. This entails that for every PL⁺ singular term **t**, Π ⊢ **W[t]** (recall that (∀**y**)**W** ∈ Π). Since Π is a Henkin set, it is maximal consistent, which implies that it is deductively closed. It follows that **W[t]** ∈ Π for every PL⁺ singular term **t**. An immediate corollary of this fact is that any Henkin set that contains **W[s]** for every PL⁺ name **s** also contains **W[t]** for every PL⁺ singular term **t**; for if all basic substitutional instances **W[s]**, where **s** is any PL⁺ name, belong to a Henkin set, then a universally quantified sentence (∀**y**)**W** also belongs to the set, and this ensures that all substitutional instances **W[t]**, where **t** is any PL⁺ singular term, belong to the set as well.

3.2.4   We began with a consistent set Γ of PL sentences. We extended PL into PL⁺ by adding to PL infinitely many new names. Then we constructed a W-conditional for every universally quantified PL⁺ sentence. We added those W-conditionals to Γ, and we proved that the resulting set is consistent. We used Lindenbaum's Lemma to extended Γ∪Θ, where Θ is the set of the W-conditionals, into a maximal consistent set Π. We argued that Π is a Henkin set. Our goal now is to construct a model H^Π of Π. H^Π, of course, is a model of every subset of Π; and since Γ is a subset of Π, H^Π is a model of Γ. Thus we will establish our desired conclusion—namely, that Γ has a model—and with that, the Completeness Theorem.

The model H^Π is referred to as the **Henkin Model** of Π. Constructing a Henkin Model is the second central idea of Henkin's proof of the Completeness theorem. The first central idea was the "construction" of a Henkin set. There is a very intimate relationship between a Henkin set and its Henkin *Interpretation*: for any PL⁺ sentence **X**, **X** is a member of a Henkin set iff **X** is true on its Henkin Interpretation. This statement is referred to as the **Truth-Membership Theorem**. An immediate corollary of this theorem is that the Henkin Interpretation of a Henkin set is a model of it. Our goal in this section is to construct the Henkin Interpretation of Π and to prove the Truth-Membership Theorem. To simplify matters, we will assume that PL⁺ does not contain an identity predicate. The case of the identity predicate will be considered later. We will not give PL⁺ a different name. We only need to keep in mind that we are working with a version of Predicate Logic that has no identity predicate. Observe that a Henkin Interpretation is defined for every set of PL⁺ sentences. The set need not be a Henkin set; however, if it is, then its Henkin Interpretation is a model of it. If Σ is a set of PL⁺ sentences, Voc(Σ) is the set that consists of the extra-logical vocabulary that occurs in the members of Σ and of the logical vocabulary of PL (which is the same as the logical vocabulary of PL⁺).

The Henkin Interpretation H^Σ of a Set Σ of PL⁺ Sentences

UD: The set of all the PL⁺ singular terms

LN: All the members of UD. H^Σ is free to treat PL⁺ singular terms as names since a PL⁺ interpretation may bring with it its own collection of names. Thus for a Henkin Interpretation, LN = UD.

Semantical Assignments

For every name **s** in LN, H^Σ(**s**): **s**. That is, H^Σ assigns the singular term **s** as the referent of **s** itself.

For every 1-place predicate **P** in Voc(Σ), H^Σ(**P**): {**s**: **Ps** ∈ Σ}. In other words, the extension of **P** on H^Σ consists of all the singular terms **s** such that the atomic sentence **Ps** belongs to Σ.

For every n-place predicate **R** in Voc(Σ), H^Σ(**R**): {⟨t_1, t_2, t_3, ..., t_n⟩: **R**$t_1t_2t_3$...t_n ∈ Σ}. Informally speaking, we collect in the extension of **R** every n-tuple of singular terms that are part of any atomic sentence of the form **R**$t_1t_2t_3$...t_n that belongs to Σ.

For every n-place function symbol **g** in Voc(Σ), H^Σ(**g**): {⟨t_1, t_2, t_3, ..., t_n, **g**$t_1t_2t_3$...t_n⟩: t_1, t_2, t_3, ..., and t_n are PL⁺ singular terms}. So for instance, if g^3 and h^2 are function symbols in Voc(Σ), and *a*, *b*, *n*, *p*, and *q* are PL⁺ names, then ⟨*a*, *b*, *n*, g^3abn⟩, ⟨h^2pq, *n*, *a*, g^3h^2pqna⟩, and ⟨*a*, g^3nbb, h^2pq, $g^3ag^3nbbh^2pq$⟩ ∈ H^Σ(g^3), and ⟨*n*, *p*, h^2np⟩, ⟨*n*, h^2aa, h^2nh^2aa⟩, and ⟨g^3nbb, h^2pq, $h^2g^3nbbh^2pq$⟩ ∈ H^Σ(h^2). (Of course, H^Σ(g^3) and H^Σ(h^2) contain infinitely more ordered tuples.)

To make our definition concrete, let us consider an example of a set Σ and its Henkin Interpretation.

Σ = {*Ade*, Bkg^2res, Bh^1rg^2diq, Ag^2iiq, *Lq*, Lh^1c, Lg^2ei, (∀*x*)(Ag^2xxq→Lg^2ex), ¬(*Mc*→*Axe*)}

Voc(Σ) = {*d*, *e*, *r*, *s*, *i*, *q*, *c*, h^1, g^2, *A*, *B*, *L*, *M*, ¬, →, ∀, *x*, *y*, *z*, ..., (,)}

The Henkin Interpretation H^Σ

UD: The set of all PL⁺ singular terms

LN: All the members of UD

Semantical Assignments

    H^Σ(**t**): **t**, for every **t** in LN (every singular term names itself)

    H^Σ(*A*): {⟨*d*, *e*⟩, ⟨g^2ii, *q*⟩} because the atomic sentences *Ade* and Ag^2iiq belong to Σ.

    H^Σ(*B*): {⟨*k*, g^2re, *s*⟩, ⟨h^1r, g^2di, *q*⟩} because the atomic sentences Bkg^2res and Bh^1rg^2diq belong to Σ.

    H^Σ(*L*): {*q*, h^1c, g^2ei} because the atomic sentences *Lq*, Lh^1c, and Lg^2ei belong to Σ.

    H^Σ(*M*): ∅ because no atomic sentence of the form *Ms*, where **s** is a PL⁺ singular term, belongs to Σ.

$H\Sigma(h^1)$: {⟨**s**, h^1**s**⟩: **s** is in LN}: {⟨e, h^1e⟩, ⟨d, h^1d⟩, ⟨g^2qq, h^1g^2qq⟩, ⟨h^1r, h^1h^1r⟩, ...}

$H\Sigma(g^2)$: {⟨**t**, **s**, g^2**st**⟩: **t** and **s** are in LN}: {⟨e, i, g^2ei⟩, ⟨e, d, g^2ed⟩, ⟨q, q, g^2qq⟩, ⟨$h^1r, g^2ek, g^2h^1rg^2ek$⟩, ...}

By construction, all the atomic sentences in Σ are true on $H\Sigma$. The compound sentences are a different story. Some of them might be true on $H\Sigma$ and others might be false on $H\Sigma$. For instance, $(\forall x)(Ag^2xxq \rightarrow Lg^2ex)$ is true on $H\Sigma$ and $\neg(Mc \rightarrow Ade)$ is false on $H\Sigma$.

In the remainder of this section, we will state and prove the Truth-Membership Theorem. Recall that Π is a Henkin set, which is an extension of Γ, and that H^Π is the Henkin Interpretation of Π. The reader will notice that the proof could be made simpler if we avail ourselves of certain intermediate conclusions. However, we chose a longer route because we want to prove every clause independently of the other clauses.

The Truth-Membership Theorem: For every PL⁺ sentence **X**, **X** is true on H^Π iff **X** $\in \Pi$.

Proof

1. We will use mathematical induction (PCI) on the complexity of **X**, where the complexity of a PL⁺ sentence is the number of connectives and quantifiers that occur in the sentence.

2. **The Base Step**: Let the complexity of **X** be 0. Hence **X** is an atomic sentence. **X** is of the form $P^n t_1 t_2 t_3 ... t_n$, where P^n is an n-place predicate and $t_1, t_2, t_3, ...,$ and t_n are any singular terms. (Recall this version of PL⁺ does not contain the identity predicate, =.)

3. From 1 by the truth conditions of the atomic sentences: $P^n t_1 t_2 t_3 ... t_n$ is true on H^Π iff ⟨$H^\Pi(t_1), H^\Pi(t_2), H^\Pi(t_3), ..., H^\Pi(t_n)$⟩ $\in H^\Pi(P^n)$.

4. By the construction of H^Π: ⟨$H^\Pi(t_1), H^\Pi(t_2), H^\Pi(t_3), ..., H^\Pi(t_n)$⟩ $\in H^\Pi(P^n)$ iff ⟨$t_1, t_2, t_3, ..., t_n$⟩ $\in H^\Pi(P^n)$; and ⟨$t_1, t_2, t_3, ..., t_n$⟩ $\in H^\Pi(P^n)$ iff $P^n t_1 t_2 t_3 ... t_n \in \Pi$.

5. From 3 and 4: $P^n t_1 t_2 t_3 ... t_n$ is true on H^Π iff $P^n t_1 t_2 t_3 ... t_n \in \Pi$. This establishes the Base Step.

6. **The Inductive Step**: Suppose that the Truth-Membership Theorem is true for every **X** of complexity m, where $0 \leq m < k$. This is the Induction Hypothesis. Our goal is to prove that the theorem holds for sentences of complexity k. Since k > 0, **X** is a compound sentence. Hence it is either a negation, a conditional, or a universally quantified sentence. We consider each case in turn.

7. Suppose that **X** is ¬**Y**, for some PL⁺ sentence **Y**. Since **Y** has a complexity less than k, the Induction Hypothesis applies to it. Therefore, **Y** is true on H^Π iff **Y** $\in \Pi$.

8. Since Π is a Henkin set, **Y** $\in \Pi$ iff ¬**Y** $\notin \Pi$ (see Subsection 3.2.3).

9. From 7 and 8: **Y** is false on H^Π iff **Y** $\notin \Pi$, and **Y** $\notin \Pi$ iff ¬**Y** $\in \Pi$.

10. From 9 by the truth conditions of the negation: ¬**Y** is true on H^Π iff ¬**Y** $\in \Pi$; that is, **X** is true on H^Π iff **X** $\in \Pi$.

11. Now we consider the case of the conditional. So let **X** be **Y→Z**. Since the complexities of **Y** and **Z** are less than k, the Induction Hypothesis applies to them. Thus **Y** is true on H$^\Pi$ iff **Y** ∈ Π, and **Z** is true on H$^\Pi$ iff **Z** ∈ Π.

12. Suppose that **Y→Z** is true on H$^\Pi$.

13. From 12 by the truth conditions of the conditional: ¬**Y** is true on H$^\Pi$ or **Z** is true on H$^\Pi$.

14. Assume first that ¬**Y** is true on H$^\Pi$; that is, **Y** is false on H$^\Pi$.

15. From 11 and 14: **Y** ∉ Π.

16. From 15: since Π is maximal consistent: ¬**Y** ∈ Π.

17. Since Π is a Henkin set: **Y→Z** ∈ Π iff ¬**Y** ∈ Π or **Z** ∈ Π (see Subsection 3.2.3).

18. From 16 and 17: **Y→Z** ∈ Π.

19. Now assume that **Z** is true on H$^\Pi$.

20. From 11 and 19: **Z** ∈ Π.

21. From 17 and 20: **Y→Z** ∈ Π.

22. From 12 through 21: if **Y→Z** is true on H$^\Pi$, then **Y→Z** ∈ Π.

23. Now suppose that **Y→Z** ∈ Π.

24. From 23: since Π is a Henkin set, ¬**Y** ∈ Π or **Z** ∈ Π.

25. Assume first that ¬**Y** ∈ Π.

26. From 8 and 25: **Y** ∉ Π.

27. From 11 and 26: **Y** is false on H$^\Pi$.

28. From 27 by the truth conditions of the conditional: **Y→Z** is true on H$^\Pi$.

29. Assume now that **Z** ∈ Π.

30. From 11 and 29: **Z** is true on H$^\Pi$.

31. From 30 by the truth conditions of the conditional: **Y→Z** is true on H$^\Pi$.

32. From 23 through 31: if **Y→Z** ∈ Π, then **Y→Z** is true on H$^\Pi$.

33. From 22 and 32: **Y→Z** is true on H$^\Pi$ iff **Y→Z** ∈ Π; that is, **X** is true on H$^\Pi$ iff **X** ∈ Π.

34. Finally we consider the case of the universal quantifier. So let **X** be (∀z)**W**.

35. For every name **t** in LN, the substitutional instance **W[t]** has complexity less than k. Hence the Induction Hypothesis applies to it: **W[t]** is true on H$^\Pi$ iff **W[t]** ∈ Π.

36. Suppose that (∀z)**W** is true on H$^\Pi$.

37. From 36 by the truth conditions of the universal quantifier: for every name **t** in LN, **W[t]** is true on H$^\Pi$.

38. From 35 and 37: for every name **t** in LN, **W[t]** ∈ Π.

39. Since Π is a Henkin set: (∀z)**W** ∈ Π iff **W[t]** ∈ Π, for every PL$^+$ singular term **t** (see Subsection 3.2.3).

40. From 38 and 39: given that the names in LN are precisely the PL⁺ singular terms, it follows that $(\forall z)W \in \Pi$.
41. From 36 through 40: if $(\forall z)W$ is true on H$^\Pi$, then $(\forall z)W \in \Pi$.
42. Now assume that $(\forall z)W \in \Pi$.
43. From 42: since Π is a Henkin set, every substitutional instance $W[t] \in \Pi$, where t is a PL⁺ singular term.
44. From 35 and 43: for every name t in LN, $W[t]$ is true on H$^\Pi$ (since the names of LN are the PL⁺ singular terms).
45. From 44 by the truth conditions of the universal quantifier: $(\forall z)W$ is true on H$^\Pi$.
46. From 42 through 45: if $(\forall z)W \in \Pi$, then $(\forall z)W$ is true on H$^\Pi$.
47. From 41 and 46: $(\forall z)W$ is true on H$^\Pi$ iff $(\forall z)W \in \Pi$; that is, X is true on H$^\Pi$ iff $X \in \Pi$.
48. From 10, 33, 47: if X has a complexity k, then X is true on H$^\Pi$ iff $X \in \Pi$. This establishes the Inductive Step.
49. From 5 and 48 by PCI: for every PL⁺ sentence X of any complexity, X is true on H$^\Pi$ iff $X \in \Pi$.

An immediate consequence of the Truth-Membership Theorem is that the Henkin Interpretation of a Henkin set is a model of it. Since we proved in Subsection 3.2.3 that every consistent set could be extended into a Henkin set, every consistent set has a model, i.e., is satisfiable. This concludes our proof of the Completeness Theorem for PL without the identity predicate.

3.2.5 The proof we gave of the Completeness Theorem pertains to a version of PL that contains no identity predicate. Our goal in this Subsection is to show how to carry out the previous proof for the standard version of PL, i.e., Predicate Logic with the identity predicate. The problem arises when we consider identity sentences. Let us consider again our set Π, which is a Henkin set. It is clear that Π contains every sentence of the form $t = t$, since these sentences are logical theorems (they are derivable from the empty set) and Π is maximal consistent. In general, for every logical theorem X, $\Pi \vdash X$, and, given that Π is deductively closed (see Lemma 3.2.2), $X \in \Pi$. But what about sentences of the form $t = s$, where t and s are distinct singular terms? None of these sentences are logical theorems and neither are their negations. Since Π is maximal, for every sentence $t = s$, either it or its negation belongs to Π. Hence Π might contain many identity sentences. Suppose for the sake of argument that the sentence $a = b \in \Pi$, where a and b are two specific PL names. The Henkin Interpretation we constructed for Π treats a and b as two distinct individuals in UD and assigns to them the names a and b. Thus H$^\Pi(a)$ is a and H$^\Pi(b)$ is b. Since a and b are two distinct individuals in UD, the sentence $a = b$ is false on H$^\Pi$. Of course, this is a problem, because our purpose of constructing the Henkin Interpretation for Π is to have a model of Π. We want H$^\Pi$ to make every member of Π, including $a = b$, true. Thus something needs to be done about the UD of H$^\Pi$ and the semantical assignments of referents.

The most natural solution is to *partition* the set of all the PL⁺ singular terms. In Subsection 2.3.1 we defined what we meant by a **partition**. Let us revisit that definition. If A is a nonempty set and $\mathcal{F}$ is a family of mutually exclusive and collectively exhaustive subsets of A, then F is a partition of A. Mutually exclusive sets are sets that do not share any common elements, and sets are collectively exhaustive of another set if their union is identical with that set. Here is an example. Let A be the set {n: n is a natural number such that $0 \leq n < 100$}. Let B_0 be the set of the natural numbers that are greater than or equal to 0 and less than 10; that is, $B_0 = \{0, 1, 2, \ldots, 9\}$. Let B_1 be the set of the natural numbers that are greater than or equal to 10 and less than 20; so $B_1 = \{10, 11, 12, \ldots, 19\}$. In a similar fashion, we define $B_2, B_3, \ldots$, and B_9. We take $\mathcal{F}$ to be the family $\{B_k: 0 \leq k \leq 9\}$. It is clear that the members of this family are mutually exclusive, since $B_i \cap B_j = \emptyset$, whenever i and j are distinct. They are also collectively exhaustive of A, since $\cup \mathcal{F} = A$. Hence $\mathcal{F}$ is a partition of A.

We are going to partition UD into mutually exclusive and collectively exhaustive subsets. Let **t** be any member of UD of H^Π. By construction, **t** is a PL⁺ singular term. We define a set E[**t**], which we call the **equivalence class of t**, as follows: for every **s**, $s \in E[t]$ iff **s** is a PL⁺ singular term such that the sentence **t = s** appears in Π. More formally, E[**t**] = {**s**: the sentence **t = s** belongs to Π}. Since UD is the set of all the PL⁺ singular terms and E[**t**] is a collection of PL⁺ singular terms, E[**t**] $\subseteq$ UD. The idea behind the equivalence class E[**t**] is sufficiently intuitive. Let us say that "the Henkin set Π declares the singular terms **t** and **s** identical" just in case the sentence **t = s** belongs to Π. We begin by selecting any member **t** of UD (i.e., a PL⁺ singular term); we search Π for any sentence of the form **t = s**; if we find such a sentence, we place the singular term **s** inside the equivalence class of **t**; we continue this process until we collect all the singular terms that Π declares to be identical with **t**. We denote this collection "E[**t**]." Now we select another member **s** of UD, and we repeat the same process. We obtain the equivalence class E[**s**]. We do the same for every member of UD; that is, for every PL⁺ singular term **q**, we construct its equivalence class E[**q**].

Once we construct an equivalence class for every PL⁺ singular term, we collect all these equivalence classes in a single family $\mathcal{U}$. In other words, $\mathcal{U}$ is the family {E[**q**]: **q** is a PL⁺ singular term}. Now we modify the Henkin Interpretation H^Π for Π as follows:

(1) Replace UD with the family $\mathcal{U}$—that is, the individuals of the modified H^Π are no longer the PL⁺ singular terms but the equivalence classes of these terms.

(2) Do not make any changes to LN, so that LN still consists of all the PL⁺ singular terms.

(3) To every name in LN assign the equivalence class to which it belongs (we will prove later that every singular term belongs to exactly one equivalence class). In other words, if the singular term **q** belongs to the equivalence class E[**t**], then the interpretation assigns E[**t**] as the referent of the name **q**.

(4) For every n-place predicate P^n, define the extension of P^n to be the set $\{\langle E[t_1], E[t_2], E[t_3], \ldots, E[t_n]\rangle: P^n t_1 t_2 t_3 \ldots t_n \in \Pi\}$. Said differently, the n-tuple of the equivalence

classes of the singular terms t_1, t_2, t_3, ..., and $t_n \in$ the extension of $\mathbf{P}^n$ iff the sentence $\mathbf{P}^n t_1 t_2 t_3 ... t_n \in \Pi$.

(5) For every n-place function symbol $\mathbf{g}^n$, define the extension of $\mathbf{g}^n$ to be the set $\{\langle E[t_1], E[t_2], E[t_3], ..., E[t_n], E[\mathbf{g}^n t_1 t_2 t_3 ... t_n]\rangle: t_1, t_2, t_3, ...,$ and t_n are PL$^+$ singular terms$\}$; in this case rather than taking the extension of $\mathbf{g}^n$ to be the set $\{\langle t_1, t_2, t_3, ..., t_n, \mathbf{g} t_1 t_2 t_3 ... t_n\rangle: t_1, t_2, t_3, ...,$ and t_n are PL$^+$ singular terms$\}$, we take it to be the set of the n+1-tupels of the equivalence classes of all these terms.

The resulting Henkin Interpretation, which we will denote also as H$^\Pi$, is a model of the set Π, whose language contains the identity predicate.

Before we modify the Truth-Membership Theorem for H$^\Pi$, we need to prove that the family $\mathcal{U}$ is indeed a partition of the old UD. As stated above, it is clear that all these equivalence classes are subsets of UD. Also none of them is empty, since for every PL$^+$ singular term $\mathbf{t}$, the sentence $\mathbf{t} = \mathbf{t}$ belongs to Π (all logical theorems belong to Π); and hence $\mathbf{t} \in E[\mathbf{t}]$. We must show now that they are mutually exclusive. Let $E[\mathbf{t}]$ and $E[\mathbf{s}]$ be two equivalence classes. We want to show that either they are disjoint or they are precisely the same class. So we will prove that if they are not disjoint then they are identical. We assume that they are not disjoint. It follows that there must be a singular term $\mathbf{p}$ that belongs to both classes. Let $x \in E[\mathbf{t}]$. By the definition of $E[\mathbf{t}]$, the sentence $\mathbf{t} = x$ belongs to the set Π. Since $\mathbf{p} \in E[\mathbf{t}] \cap E[\mathbf{s}]$, the sentences $\mathbf{t} = \mathbf{p}$ and $\mathbf{s} = \mathbf{p}$ also belong to Π. Thus Π contains $\mathbf{t} = x$, $\mathbf{t} = \mathbf{p}$, and $\mathbf{s} = \mathbf{p}$. By repeated application of the rule Substitution, we can construct a PL derivation of $\mathbf{s} = x$ from Π. Since Π is deductively closed, the sentence $\mathbf{s} = x$ belongs to Π. By the definition of $E[\mathbf{s}]$, $x \in E[\mathbf{s}]$. This proves that $E[\mathbf{t}] \subseteq E[\mathbf{s}]$. By almost identical reasoning, we can show that $E[\mathbf{s}] \subseteq E[\mathbf{t}]$. By the Principle of Extensionality, $E[\mathbf{t}]$ and $E[\mathbf{s}]$ are identical sets. Hence if two equivalence classes share any element, they are identical, which means that distinct equivalence classes are disjoint. This proves that $\mathcal{U}$ is a pairwise disjoint family of nonempty subsets of our old UD. It remains to show that $\cup\mathcal{U}$ and UD are identical. Since every equivalence class in $\mathcal{U}$ is a subset of UD, it is obvious that $\cup\mathcal{U} \subseteq$ UD. We only need to show that UD $\subseteq \cup\mathcal{U}$. So let $\mathbf{t}$ be any member of UD. By the definition of UD, $\mathbf{t}$ is a PL$^+$ singular term. We proved above that $\mathbf{t} \in E[\mathbf{t}]$. This entails that $\mathbf{t}$ belongs to the union $\cup\mathcal{U}$. Hence UD $\subseteq \mathcal{U}$, which is the desired conclusion. Therefore, the new universe of discourse $\mathcal{U}$ is indeed a partition of the old universe of discourse UD. We can express the fact that $\mathcal{U}$ is a partition of UD by saying that $\mathcal{U}$ is a family of subsets of UD such that every member of UD belongs to exactly one class in $\mathcal{U}$.

Now we prove the Truth-Membership Theorem for the new Henkin Interpretation H$^\Pi$. There are no changes to the Inductive Step. However the Base Step requires a new proof. We begin by considering the case of identity sentences. We need to prove that for any PL$^+$ singular terms $\mathbf{t}$ and $\mathbf{s}$, the sentence $\mathbf{t} = \mathbf{s}$ is true on the modified Henkin Interpretation H$^\Pi$ iff it belongs to Π. We first prove the 'only-if' part. So let $\mathbf{t} = \mathbf{s}$ be true on H$^\Pi$. We want to show that $\mathbf{t} = \mathbf{s}$ belongs to Π. By the truth conditions of the identity sentences, H$^\Pi$ assigns the same referent to $\mathbf{t}$ and $\mathbf{s}$. By definition, $\mathbf{t}$ belongs to its referent and $\mathbf{s}$ belongs to its referent.

But these referents are identical. Thus **t** and **s** belong to the same equivalence class E[**q**]. By the definition of E[**q**], the sentences **q** = **t** and **q** = **s** belong to the set Π. By Substitution, Π ⊢ **t** = **s**. Since Π is a Henkin set, it is deductively closed. It follows that **t** = **s** belongs to Π. Second, we prove the 'if' part. So let the sentence **t** = **s** be a member of Π. We want to show that **t** = **s** is true on H^Π. By the definition of equivalence class, **t** and **s** ∈ E[**t**] (recall that **t** = **t** also belongs to Π). Therefore H^Π assigns E[**t**] as the referent of both **t** and **s**. Hence, **t** = **s** is true on H^Π. We conclude that for any PL⁺ singular terms **t** and **s**, the sentence **t** = **s** is true on H^Π iff it belongs to Π. Now consider the case of the atomic sentence $P^n t_1 t_2 t_3 \ldots t_n$. Suppose that $P^n t_1 t_2 t_3 \ldots t_n$ is true on H^Π. By the truth conditions of the subject-predicate sentences, the n-tuple ⟨E[t_1], E[t_2], E[t_3], ..., E[t_n]⟩ ∈ the extension of P^n on H^Π. By the definition of the extension of P^n on H^Π, $P^n t_1 t_2 t_3 \ldots t_n$ ∈ Π. Now suppose that $P^n t_1 t_2 t_3 \ldots t_n$ ∈ Π. Again by the definition of the extension of P^n on H^Π, ⟨E[t_1], E[t_2], E[t_3], ..., E[t_n]⟩ ∈ the extension of P^n. But each E[t_k] is the referent that H^Π assigns to t_k; hence by the truth conditions of $P^n t_1 t_2 t_3 \ldots t_n$, this sentence is true on H^Π. It follows that for every PL⁺ sentence $P^n t_1 t_2 t_3 \ldots t_n$, it is true on H^Π iff it belongs to Π. Therefore, the Truth-Membership Theorem is true of the modified Henkin Interpretation H^Π, which entails that H^Π is a model of Π. This means that the Completeness Theorem also holds for PL with identity.

3.3 The Compactness Theorem

3.3.1 An important corollary of the Soundness and Completeness Theorems is the Compactness Theorem. It is stated below.

The Compactness Theorem: For every PL set Γ and every PL sentence **X**, if **X** is a logical consequence of Γ, then **X** is a logical consequence of a **finite subset** of Γ.

We will give a generic proof of the Compactness Theorem that works for any logical system **L** that has a sound and complete proof theory whose derivations are finite sequences. It is understood that **L** has a well-defined notion of logical consequence (if it did not, its Soundness and Completeness Theorems would not make much sense). We denote the relations of logical consequence and derivability in **L** as ⊨_L and ⊢_L, respectively.

Proof of the Compactness Theorem for **L**

1. Suppose that Γ is any set of **L** sentences and **X** is any **L** sentence such that Γ ⊨_L **X**.
2. From 1 by the Completeness Theorem for **L**: Γ ⊢_L **X**.
3. From 2 by the definition of derivability: there is an **L** derivation D of **X** from Γ.
4. Let Σ be the subset of Γ whose members are all the sentences in Γ that D invokes.
5. From 4: since D is a finite sequence of **L** sentences, Σ must be a finite subset of Γ.
6. From 3 and 4 by the definition of derivation: Σ ⊢_L **X**.
7. From 6 by the Soundness Theorem for **L**: Σ ⊨_L **X**.

8. From 5 and 7: **X** is a logical consequence of a finite subset of Γ.

The Compactness Theorem has an interesting philosophical implication about the expressive powers of PL. Suppose that Δ/Z is a natural language argument whose set of premises, Δ, is infinite. Suppose further that all the premises are needed for the validity of the argument, that is, if a single premise is removed from Δ, the resulting argument would be invalid.[2] The Compactness Theorem implies that such an argument cannot be translated faithfully into PL (or into any logical system for which the Compactness Theorem holds). There are two possibilities: either the PL translation of Δ/Z is valid or it is invalid. If it is invalid, then the PL translation fails to capture the logical relation between the premises and the conclusion of the original argument; and hence the PL translation fails to be faithful. On the other hand, if it is valid, then by the Compactness Theorem there is a finite collection of the PL premises of which the PL translation of Z is a logical consequence. Hence the PL translation fails to capture the fact that the conclusion of the original argument is not a logical consequence of any finite collection of the premises (we assumed that all the members of Δ are needed for the argument to be valid); therefore the PL argument fails to be a faithful translation of the original argument Δ/Z.

3.3.2   Sometimes the Compactness Theorem is given a different formulation. We will give this version here and show that it is equivalent to the original formulation. We call this second formulation the **Finite-Satisfiability Theorem**. If every finite subset of some set Γ of PL sentences has a model, we say that this set is **finitely satisfiable**.

The Finite-Satisfiability Theorem for PL: Every PL set that is finitely satisfiable is satisfiable (or, if every finite subset of a PL set has a model, then the set also has a model).

We give two proofs. The first shows that the Compactness Theorem entails the Finite-Satisfiability Theorem, and the second shows that the Finite-Satisfiability Theorem entails the Compactness Theorem. These two proofs establish that the two theorems are equivalent.

---

2   There are many examples of such arguments. Here is one of them. The following infinite list of sentences offers an English translation of the Arabic word 'thalathah':

Thalathah is a positive integer.
Thalathah is not one.
Thalathah is not two.
Thalathah is not four.
Thalathah is not five.
  ⋮

Thalathah can only be three if the list above contains, for every positive integer n that is not three, a sentence asserting that *thalathah* is not n. This list may be considered the set of premises of an argument whose conclusion is '*Thalathah* is three'. This argument seems to be a perfectly good argument. Someone who does not know what *thalathah* means in English can learn its English translation on the basis of this argument.

Proof that the Compactness Theorem entails the Finite-Satisfiability Theorem
1. Let Γ be a finitely satisfiable set of PL sentences and **X** be any PL sentence.
2. Reductio Assumption: Γ is unsatisfiable.
3. From 2 by the definition of unsatisfiability: there is no PL interpretation that satisfies Γ.
4. From 3 by the definition of logical consequence: $\Gamma \vDash$ **X** and $\Gamma \vDash \neg$**X**, since in order for **X** or $\neg$**X** not to be a logical consequence of Γ there must be a PL interpretation that satisfies Γ and makes **X** or $\neg$**X** false.
5. From 4 by the Compactness Theorem: there are two finite subsets Σ_1 and Σ_2 of Γ such that $\Sigma_1 \vDash$ **X** and $\Sigma_2 \vDash \neg$**X**. Let $\Sigma = \Sigma_1 \cup \Sigma_2$. It is clear that Σ is a finite subset of Γ, and that $\Sigma \vDash$ **X** and $\Sigma \vDash \neg$**X**.
6. From 1 and 5 by the definition of finite satisfiability: there is a PL interpretation I that is a model of Σ.
7. If I is not relevant to **X**, we expand I into I* such that I* interprets all the additional vocabulary in **X** without changing any of the semantical assignments made by I. Since I is a model of Σ, I* is also a model of Σ.
8. From 5 and 7 by the definition of logical consequence: **X** and $\neg$**X** are true on I*, which is impossible.
9. From 2 through 8: the Reductio Assumption is false; and hence Γ is satisfiable.

Proof that the Finite-Satisfiability Theorem entails the Compactness Theorem
1. Let Γ be any PL set and **X** any PL sentence such that $\Gamma \vDash$ **X**.
2. Reductio Assumption: **X** is not a logical consequence of *any* finite subset of Γ.
3. From 2 by the definition of logical consequence: for any finite subset Σ of Γ, there is a PL interpretation that satisfies Σ and makes **X** false.
4. From 3: for every finite subset Σ of Γ, there is a PL interpretation that satisfies Σ and makes $\neg$**X** true.
5. From 4 by the definition of satisfiability: for every finite subset Σ of Γ, Σ is satisfiable and $\Sigma \cup \{\neg$**X**$\}$ is satisfiable.
6. From 5: every finite subset of $\Gamma \cup \{\neg$**X**$\}$ is satisfiable (a finite subset of $\Gamma \cup \{\neg$**X**$\}$ is either a finite subset of Γ alone or a finite subset of Γ with the addition of $\neg$**X**).
7. From 6 by the definition of finite satisfiability: the set $\Gamma \cup \{\neg$**X**$\}$ is finitely satisfiable.
8. From 7 by the Finite-Satisfiability Theorem: $\Gamma \cup \{\neg$**X**$\}$ is satisfiable.
9. From 8 by the definition of satisfiability: there is a PL interpretation that satisfies $\Gamma \cup \{\neg$**X**$\}$, that is, there is a PL interpretation that satisfies Γ and makes **X** false.
10. From 9 by the definition of logical consequence: **X** is not a logical consequence of Γ, which contradicts 1.

11. From 2 through 10: the Reductio Assumption is false; and hence **X** is a logical consequence of some finite subset of Γ.

3.4 Elementary Equivalence and Isomorphism

In this section we define two relationships between PL interpretations: elementary equivalence and isomorphism. So let Σ be any set of PL sentences, and let I and J be two interpretations for Σ. As usual, Voc(Σ)—the vocabulary of Σ—is the set of all the extra-logical vocabulary that occurs in the members of Σ together with the PL logical vocabulary. We have the following definition. To simplify our notation, we will denote Voc(Σ) as V.

Elementary Equivalence: I and J are elementarily equivalent with respect to V just in case for every PL sentence **X** that is composed of V, **X** is true on I iff **X** is true on J. We denote this relationship as $I \equiv_V J$.

We now define the relationship of isomorphism between PL interpretations. Let Σ, I, J, and V be as defined at the outset of this section. Furthermore, let UD_I be the universe of discourse of I and UD_J be the universe of discourse of J.

Isomorphism: A function h is an isomorphism between I and J just in case h is a one-to-one correspondence between UD_I and UD_J such that

(1) For every name **c** in V, h(I(**c**)) = J(**c**).
(2) For each 1-place predicate $\mathbf{P^1}$ in V, and for each individual β in UD_I, β ∈ I($\mathbf{P^1}$) iff h(β) ∈ J($\mathbf{P^1}$).
(3) For every n-place predicate $\mathbf{P^n}$ in V, where n > 1, and for each n-tuple ⟨$β_1, β_2, β_3, ..., β_n$⟩ of individuals in UD_I, ⟨$β_1, β_2, β_3, ..., β_n$⟩ ∈ I($\mathbf{P^n}$) iff ⟨h($β_1$), h($β_2$), h($β_3$), ..., h($β_n$)⟩ ∈ J($\mathbf{P^n}$).
(4) For every n-place function symbol $\mathbf{g^n}$ in V, and for each n-tuple ⟨$β_1, β_2, β_3, ..., β_n$⟩ of individuals in UD_I, h(I($\mathbf{g^n}$)($β_1, β_2, β_3, ..., β_n$)) = J($\mathbf{g^n}$)(h($β_1$), h($β_2$), h($β_3$), ..., h($β_n$)).

If there is an isomorphism between I and J, we say that I and J are **isomorphic** with respect to V, and we denote this relation as $I \sim_V J$. If $I \sim_V J$, the constituents of I and the constituents of J that interpret V are **structurally identical**, that is, they exhibit similar relations and properties. I and J are not only elementarily equivalent with respect to V—their corresponding structures also mirror each other. As we will see later, elementarily equivalent interpretations are not necessarily isomorphic.

To illustrate this definition we will consider an example of isomorphic PL interpretations. Let Σ be the set of the flowing sentences:

S1 $(\forall y)(Dsg^1y \rightarrow \neg Kg^1y)$
S2 $(\forall x)(Dxs \rightarrow Exg^1a)$
S3 $(\forall y)(\neg Ky \rightarrow Eya)$

Voc(Σ) = V = {a, s, g^1, D, E, K, ¬, ∀, →, =, x, y, z, ..., (,)}

PL Interpretation I
UD_I: {Jamal, Camilla, Helga, Sophia, Michael, Marco}
LN: a, b, t, s, m, e

Semantical Assignments

I(a): Jamal; I(b): Camilla; I(t): Helga; I(s): Sophia; I(m): Michael; I(e): Marco
I(Dxy): "x is in love with y": {⟨Sophia, Michael⟩, ⟨Marco, Sophia⟩}
I(Kz): "z is older than forty": {Michael, Helga}
I(Ezw): "z is a brother of w": {⟨Marco, Michael⟩, ⟨Jamal, Camilla⟩, ⟨Marco, Helga⟩, ⟨Michael, Helga⟩}
I(Lz): "z took a class at Stanford": {Sophia, Michael, Helga}
I(g^1x): "The instructor of x": {⟨Jamal, Michael⟩, ⟨Camilla, Michael⟩, ⟨Marco, Sophia⟩, ⟨Sophia, Helga⟩, ⟨Michael, Helga⟩, ⟨Helga, Sophia⟩}

PL Interpretation J
UD_J: {2, 3, 5, 9, 11, 13}
LN: a, c, r, s, b, o

Semantical Assignments

J(a): 2; J(c): 3; J(r): 5; J(s): 9; J(b): 11; J(o): 13
J(Dxy): {⟨9, 11⟩, ⟨13, 9⟩}
J(Kz): {11, 5}
J(Ezw): {⟨13, 11⟩, ⟨2, 3⟩, ⟨13, 5⟩, ⟨11, 5⟩}
J(Tz): {2, 5, 11, 13}
J(g^1x): {⟨2, 11⟩, ⟨3, 11⟩, ⟨13, 9⟩, ⟨9, 5⟩, ⟨11, 5⟩, ⟨5, 9⟩}

Note that the vocabularies that I and J interpret must include the vocabulary of the set Σ, but they may also include symbols that do not occur in any member of Σ. If we denote the vocabularies that I and J interpret together with the logical vocabulary of PL as Voc(I) and Voc(J), respectively, we may state the preceding requirement as follows: V ⊆ Voc(I) and V ⊆ Voc(J), and Voc(I) and Voc(J) need not be identical. In fact, according to the examples above, Voc(I) ≠ Voc(J), since for instance L belongs to Voc(I) but not Voc(J), and T belongs to Voc(J) but not Voc(I).

    The function h that assigns the values 2, 3, 5, 9, 11, and 13 to the arguments Jamal, Camilla, Helga, Sophia, Michael, and Marco, respectively, is an isomorphism between I and J with respect to V. To see this, observe (1) that for every name **c** in V, h(I(**c**)) = J(**c**), since h(I(a)) = h(Jamal) = 2 = J(a) and h(I(s)) = h(Sophia) = 9 = J(s); (2) that the ordered pairs ⟨h(Sophia), h(Michael)⟩ and ⟨h(Marco), h(Sophia)⟩ (which are ⟨9, 11⟩, ⟨13, 9⟩) are the members of J(Dxy); (3) that the numbers h(Michael) and h(Helga) (which are 11 and 5) are the members of J(Kz); (4) that the ordered pairs ⟨h(Marco), h(Michael)⟩, ⟨h(Jamal), h(Camilla)⟩, ⟨h(Marco), h(Helga)⟩, and ⟨h(Michael), h(Helga)⟩ (which are ⟨13, 11⟩, ⟨2, 3⟩, ⟨13, 5⟩, and ⟨11, 5⟩) are the members of J(Ezw); and (5) that the ordered pairs ⟨h(Jamal), h(Michael)⟩, ⟨h(Camilla), h(Michael)⟩, ⟨h(Marco), h(Sophia)⟩, ⟨h(Sophia), h(Helga)⟩,

⟨h(Michael), h(Helga)⟩, and ⟨h(Helga), h(Sophia)⟩ (which are ⟨2, 11⟩, ⟨3, 11⟩, ⟨13, 9⟩, ⟨9, 5⟩, ⟨11, 5⟩, and ⟨5, 9⟩) are the members of the function $J(g^1x)$.

It is clear that the constituents of I and J that interpret V are structurally identical. It does not follow that I as a whole is structurally identical with J. I contains an extension for the predicate L, which J does not contain, and J contains an extension for the predicate T, which I does not contain. These two extensions are associated with different predicates, and they are structurally different from each other (e.g., they have different cardinalities). The sentence S1 is false on I and J, S2 is true on I and J, and S3 is false on I and J. In fact, every sentence that is composed of the vocabulary V has the same truth value on I and J. This implies that I and J are elementarily equivalent with respect to V. We state the following theorem.

Theorem 3.4.1: Let I and J be any two PL interpretations for the PL set Σ, whose vocabulary is V. If I ∼v J, then I ≡v J.

The truth of this theorem is intuitive enough. We will not present a proof of it here. It can be proved using mathematical induction on the complexity of sentences. We will see later that there are PL interpretations that are elementarily equivalent but not isomorphic with respect to the same vocabulary. Isomorphism is a much stronger relationship than elementary equivalence.

3.5   Properties of PL Sets

3.5.1   In this subsection we will deal with an important type of PL sets. But we first redefine some familiar terms. We previously said that a set Σ is **deductively closed** iff for every PL sentence **X**, Σ ⊢ **X** only if **X** ∈ Σ, and that Σ is **semantically closed** iff for every PL sentence **X**, Σ ⊨ **X** only if **X** ∈ Σ. Given the Soundness and Completeness Theorems, deductive closure and semantical closure are equivalent notions. A set is **maximal** iff for every PL sentence, either it or its negation belongs to the set. We introduce below a new term.

Complete PL Set: A PL set Σ is **complete** iff for every PL sentence **X** that is composed of Voc(Σ), either Σ ⊢ **X** or Σ ⊢ ¬**X**.

This notion of completeness is different from the completeness of a proof theory, which we discussed in Section 3.2.

Much of the discussion in the rest of this book concerns a specific type of PL sets. We call these sets "PL theories." Here is the formal definition.

PL Theory: A set Σ of PL sentences is a **theory** iff for every PL sentence **X** that is composed of Voc(Σ), if Σ ⊨ **X**, then **X** ∈ Σ.

The defining condition of PL theory is similar to semantical closure, except a semantically closed PL set must contain *all* its logical consequences regardless of whether these consequences are expressed in the vocabulary of the set or not. For instance, a semantically closed PL set must contain every PL sentence that is valid, since a valid sentence is a logical consequence of every set; a PL theory, on the other hand, need not contain every valid PL sentence, but only those that are expressed in its vocabulary. We have a few corollaries that follow from the definitions above and the Soundness and Completeness Theorems.

Corollary 3.5.1:

3.5.1a For every PL set Σ, Σ is a theory iff for every PL sentence **X** that is composed of Voc(Σ), if $\Sigma \vdash \mathbf{X}$, then $\mathbf{X} \in \Sigma$.

3.5.1b For every PL set Σ, Σ is complete iff for every sentence **X** that is composed of Voc(Σ), $\Sigma \vDash \mathbf{X}$ or $\Sigma \vDash \neg \mathbf{X}$.

3.5.1c For every PL set Σ, the set of all the logical consequences of Σ that are composed of Voc(Σ) is a theory. It is denoted as "Th(Σ)."

3.5.1d For every PL set Σ, Th(Σ) = {**X**: **X** is composed of Voc(Σ) and **X** is a theorem of Σ}.

3.5.1e If V is a PL vocabulary that consists of the logical vocabulary of PL together with some (but not necessarily all) of the extra-logical vocabulary interpreted by a PL interpretation J, then Th$_V$(J) is a consistent complete PL theory, where Th$_V$(J) consists of all the sentences composed of V that are true on J. If V is Voc(J), that is, the logical vocabulary of PL together with *all* the extra-logical vocabulary interpreted by J, then we denote Th$_V$(J) simply as Th(J). "Th$_V$(J)" is referred to as the V theory of J, and "Th(J)" as the PL theory of J.

3.5.1f Every deductively or semantically closed PL set is a PL theory. (The converse is not true.)

To simplify our proofs, we will denote Voc(Σ) as V, and if **X** is a PL sentence that is composed of Voc(Σ), we will simply say that **X** is a **V sentence**.

Proof of 3.5.1a

1. By the definition of a theory: Σ is a theory iff for every V sentence **X**, $\Sigma \vDash \mathbf{X}$ only if $\mathbf{X} \in \Sigma$.
2. By the Soundness and Completeness Theorems: $\Sigma \vDash \mathbf{X}$ iff $\Sigma \vdash \mathbf{X}$.
3. From 1 and 2: Σ is a theory iff for every V sentence **X**, $\Sigma \vdash \mathbf{X}$ only if $\mathbf{X} \in \Sigma$.

Proof of 3.5.1b

1. By the definition of completeness: Σ is complete iff for every V sentence **X**, $\Sigma \vdash \mathbf{X}$ or $\Sigma \vdash \neg \mathbf{X}$.

2. By the Soundness and Completeness Theorem: $\Sigma \vDash X$ iff $\Sigma \vdash X$, and $\Sigma \vDash \neg X$ iff $\Sigma \vdash \neg X$.
3. From 1 and 2: Σ is complete iff for every V sentence X, $\Sigma \vDash X$ or $\Sigma \vDash \neg X$.

Proof of 3.5.1c

1. Let Σ be any set of PL sentences. Define $Th(\Sigma) = \{X: X \text{ is a V sentence and } \Sigma \vDash X\}$. We wish to prove that $Th(\Sigma)$ is a theory.
2. Take X to be any V sentence that is a logical consequence of $Th(\Sigma)$. We want to show $X \in Th(\Sigma)$.
3. If Σ has no models (i.e., Σ is unsatisfiable), then every V sentence is a logical consequence of Σ. In this case, every V sentence is a member of $Th(\Sigma)$; and hence $X \in Th(\Sigma)$.
4. On the other hand, if Σ is satisfiable, let M be any model of Σ. (Since X is a V sentence, any interpretation for Σ is relevant to X.)
5. From 1 and 4 by the definition of logical consequence: every member of $Th(\Sigma)$ is true on M.
6. From 5: M is a model of $Th(\Sigma)$.
7. From 2 and 6: X is true on M.
8. From 4 and 7: since M is arbitrary, X is true on every model of Σ.
9. From 8 by the definition of logical consequence: $\Sigma \vDash X$.
10. From 1 and 9: since X is a V sentence, $X \in Th(\Sigma)$.
11. From 2 and 10: every V sentence that is a logical consequence of $Th(\Sigma)$ belongs to $Th(\Sigma)$.
12. From 11 by the definition of PL theory: $Th(\Sigma)$ is a theory.

Proof of 3.5.1d

1. By the definition of $Th(\Sigma)$: for every V sentence X, $X \in Th(\Sigma)$ iff $\Sigma \vDash X$.
2. By the Soundness and Completeness Theorems: $\Sigma \vDash X$ iff $\Sigma \vdash X$.
3. From 1 and 2: for every V sentence X, $X \in Th(\Sigma)$ iff $\Sigma \vdash X$, that is, $Th(\Sigma) = \{X: X \text{ is a V sentence and } X \text{ is a theorem of } \Sigma\}$.

Proof of 3.5.1e

1. Let J be any PL interpretation (for any PL set) and V be a PL vocabulary that consists of the PL logical vocabulary together with some or all of the extra-logical vocabulary interpreted by J. Define $Th_V(J) = \{X: X \text{ is a V sentence that is true on J}\}$.
2. We first prove that $Th_V(J)$ is a PL theory. Then we prove that it is consistent and complete. So take X to be any V sentence that is a logical consequence of $Th_V(J)$.
3. From 1: J is a model of $Th_V(J)$.

4. From 2 and 3: **X** is true on J.
5. From 1 and 4: **X** ∈ Thv(J).
6. From 2 and 5: every V sentence that is a logical consequence of Thv(J) belongs to Thv(J).
7. From 6 by the definition of PL theory: Thv(J) is a PL theory whose vocabulary is V.
8. Now we prove that Thv(J) is complete. Let **X** be any V sentence.
9. From 8: either **X** is true on J or ¬**X** is true on J.
10. From 1 and 9: either **X** ∈ Thv(J) or ¬**X** ∈ Thv(J).
11. From 10: either Thv(J) ⊢ **X** or Thv(J) ⊢ ¬**X**. (Every sentence is a theorem of any set to which it belongs by the inference rule Reiteration.)
12. From 11 by the definition of completeness: Thv(J) is complete.
13. From 3: Thv(J) is satisfiable, and hence by the Soundness Theorem, it is consistent. (See Exercise 3.6.4.)
14. From 7, 12, and 13: Thv(J) is a consistent complete PL theory.

Proof of 3.5.1f

A semantically closed PL set contains all its logical consequences, including the ones expressed in its vocabulary. Hence it is a PL theory. The same reasoning applies to the case of deductive closure by invoking Corollary 3.5.1a.

Due to Corollary 3.5.1d, when working in PL, we use "Th(Σ)" to stand either for the set of the logical consequences of Σ or the set of the theorems of Σ that are composed of Voc(Σ); both sets are identical.

3.5.2   In the mathematical sciences, practitioners are interested in **axiomatic systems**. An axiomatic system is typically a PL theory with some fixed vocabulary and a set of axioms of which all the members of the theory are logical consequences. Since PL has a complete and sound proof theory, we can restate our definition of an axiomatic system in terms of derivability. We can say that an axiomatic system is a PL theory all of whose members are derivable from a certain set of axioms. It is a trivial fact that all the members of a PL theory Σ are logical consequences of Σ itself. In this sense Σ could be considered its own set of axioms. However, this type of axioms is neither helpful nor interesting. It is neither helpful nor interesting, for instance, to tell us that the set of all the true sentences of Euclidean geometry is the set of Euclidean axioms. This set is "unmanageable." We want a "reasonable" set of axioms, from which we can derive all the true sentences of Euclidean geometry. Hence we need to place some restriction on what counts as a set of axioms for a given theory. As a first reaction, one might insist that an acceptable set of axioms must be finite. This, however, is too severe a restriction. Many important axiomatic systems, such as number theory and set theory, have infinitely many axioms. On a second reflection, we can see that what matters for an acceptable set of axioms is that we should be able to determine for any given sentence **X** whether it is an axiom or not. This is so because our interest is in proving theorems from theses axioms, and in order to have a correct proof we need to know what the axioms that we can invoke in these proofs are. In other words, we need an *effective decision procedure* that would allow us to determine whether any given sentence is an axiom or not, so that we may know whether we can invoke it in the proof or not.

In Chapter One, we defined what we meant by an effective decision procedure. Let us revisit the issue here briefly. An **effective procedure** is a procedure that can be followed mechanically (i.e., without any creative steps) and after a finite sequence of deterministic steps produces a certain result (output). An **effective decision procedure** is an effective procedure that can fully decide a Yes-No question. In other words, if we ask whether something is the case, an effective decision procedure is an effective procedure that would answer this question affirmatively if that is the correct answer and it would answer it negatively if that is the correct answer. Not every effective procedure is a decision procedure. There are effective **Yes-procedures** that would answer a question affirmatively when and only when the correct answer is Yes, but might not generate any answer when the correct answer is No. Similarly, there are effective **No-procedures**, which would answer a question negatively when and only when this is the correct answer, but might produce no answer when the correct answer is Yes. An effective decision procedure is an effective Yes-No procedure. Any concept whose applicability to any object can be decided by an effective decision procedure is said to be a **decidable concept**. A concept for which there is only an effective Yes-procedure is described as a **semidecidable concept**. If there is an effective decision procedure for determining membership in a set, we call it a **decidable set**. If there is only an effective Yes-procedure for determining membership in a set, we call it a **semidecidable set** or an **effectively enumerable set**. The members of a semidecidable set can be effectively listed in some order; this explains why we call semidecidable sets effectively enumerable. We will return to these issues in greater detail in the next chapter. Here we only need the notion of a decidable set. The reasonable restriction we should place on a set of axioms for a given theory is that it must be a decidable set such that all the members of the theory are logical consequences of it, or equivalently, derivable from it. We have the following definitions.

Definition 3.5.1:

3.5.1a For any PL theory Σ whose vocabulary is V, Φ is a **set of axioms** for Σ iff Φ is a decidable set of V sentences and $\Sigma = \text{Th}(\Phi)$, where $\text{Th}(\Phi) = \{X: X \text{ is a V sentence and } \Phi \vDash X\} = \{X: X \text{ is a V sentence and } \Phi \vdash X\}$.

3.5.1b A PL theory is **axiomatizable** iff it has a set of axioms.

3.5.1c A PL theory is **finitely axiomatizable** iff it has a finite set of axioms.

It is obvious that if a theory has a set of axioms, then this set is a subset of the theory.

If we consider the entire PL vocabulary, the theory whose set of axioms is the empty set is the set of all the valid PL sentences, or equivalently, the set of all the logical theorems of PL (recall that a logical theorem is a PL sentence that is derivable from the empty set). As we said in Chapter One, in books of elementary logic, including my book *An Introduction to Logical Theory*, a valid sentence is called "logically true" and a contradictory sentence is called "logically false." The terms 'valid' and 'contradictory' are standard in books of metalogic. There is no standard symbol for the set of all valid PL sentences. We will refer to it simply as "Th($\emptyset$)." A very important theorem of the metatheory of PL is that Th($\emptyset$) is undecidable. This is Church's Undecidability Theorem, which we will prove near the end of this book. Th($\emptyset$) is, in fact, semidecidable: there is only an effective Yes-procedure for it. At the other end of the spectrum, if the set of axioms Φ is inconsistent and we allow V to be the entire vocabulary of PL, then Th(Φ) is the largest of all theories: it is the set of *all* the PL sentences, since every PL sentence is derivable from an inconsistent set. This feature of classical (and some systems of non-classical) logic is referred to as **Explosion**.

In Chapter Five, we will introduce an effective procedure for encoding PL terms, formulas, sentences, and sequences of PL sentences into natural numbers. It will become clear then that this effective procedure *in principle* allows us to produce mechanically (i.e., effectively) an infinite list whose terms are all the PL derivations from a decidable PL set of premises. In other words, if we are given a decidable PL set Ψ, we would be able in principle to generate mechanically an infinite sequence ρ_0, ρ_1, ρ_2, ρ_3, ..., ρ_n, ... such that every ρ_n is a PL derivation from Ψ and every PL derivation from Ψ is ρ_k for some natural number k. Now we can prove on the basis of this fact a very important theorem about PL theories. We denote Voc(Σ) as V. We always assume, throughout this book, that the set of all the sentences that are composed of V is decidable, that is, there is an effective decision procedure for determining whether any string of PL symbols is a V sentence or not.

Theorem 3.5.1: Every complete axiomatizable PL theory is decidable.

Proof

1. Let Σ be a complete axiomatizable PL theory. If Σ is inconsistent, then it consists of all the V sentences, and this set is decidable. So we assume now that Σ is consistent.

2. From 1 by the definition of theory: Σ contains all the V sentences that are logical consequences of it. By Corollary 3.5.1a, for every V sentence **X**, $\Sigma \vdash \mathbf{X}$ only if $\mathbf{X} \in \Sigma$.

3. From 1 and 2: since Σ is consistent, if $\Sigma \vdash \neg \mathbf{X}$, then $\mathbf{X} \notin \Sigma$ (for otherwise, **X** and $\neg \mathbf{X}$ would belong to Σ).

4. From 1: since Σ is complete, for every V sentence **X**, $\Sigma \vdash \mathbf{X}$ or $\Sigma \vdash \neg \mathbf{X}$.

5. From 1 by the definition of axiomatizability: Σ has a set of axioms, that is, there is a decidable set Φ of V sentences such that $\Sigma = \text{Th}(\Phi) = \{\mathbf{X}: \mathbf{X}$ is a V sentence and $\Phi \vdash \mathbf{X}\}$.

6. From 5: since Φ is decidable, there is an effective procedure for listing all the PL derivations from Φ. Let this infinite list be ρ_0, ρ_1, ρ_2, ρ_3, ..., ρ_n, ...

7. From 2, 3, 4, and 6: we now describe an effective decision procedure for determining membership in the set Σ. We take any PL sentence **X**. First we

determine whether it is a sentence composed of V or not. We assumed that we could make such a determination effectively, since we assumed that the set of the V sentences was decidable. If **X** is not a V sentence, we determine that it does not belong to Σ. If it is a V sentence, we examine the last sentence of the first derivation ρ_0 in the list above. If this sentence is **X**, we determine that **X** belongs to Σ; if this sentence is ¬**X**, we determine that **X** does not belong to Σ. If it is neither, we proceed to the next derivation in the list, ρ_1. We repeat the same process. Since Σ is complete, we must eventually reach a derivation of **X** or a derivation of ¬**X**. Finding a derivation of **X** means that **X** belongs to Σ, and finding a derivation of ¬**X** means that **X** does not belong to Σ.

8. From 7: since there is an effective decision procedure for determining membership in Σ, it is decidable.

Theorem 3.5.1 demonstrates a very important fact about consistent PL theories: the triad of (1) completeness, (2) axiomatizability, and (3) undecidability is inconsistent. Hence if a consistent PL theory is complete and undecidable, it cannot be axiomatizable; and if it is axiomatizable and undecidable, it cannot be complete.[3] In Chapter Five, we will see examples of these types of theories.

3.5.3   We conclude this section by indicating a connection between complete PL sets and the relationships of elementary equivalence and isomorphism between their models. We have the following theorem.

Theorem 3.5.2: For any PL set Σ whose vocabulary is V, Σ is complete iff all its models are elementarily equivalent with respect to V.

Proof

1. First, we prove the 'only-if' part. So let Σ be any complete PL set, and let I and J be any models of Σ.[4] We want to show that I ≡v J.
2. Take **X** to be any V sentence that is true on I.
3. Since Σ is complete, by Corollary 3.5.1b, either **X** or ¬**X** is a logical consequence of Σ.
4. From 1: if ¬**X** is a logical consequence of Σ, then it is true on all the models of Σ, including I, which contradicts the fact that **X** is true on I.
5. From 3 and 4: **X** is a logical consequence of Σ.

---

3   I am indebted to an anonymous reviewer of Broadview Press for formulating the import of Theorem 3.5.1 in these terms.
4   If Σ is inconsistent (and hence has no models), then it is trivially true that Σ is complete (all sentences are derivable from an inconsistent set), and that all the models of Σ are elementarily equivalent since Σ has no models. In order for the statement that all the models of Σ are elementarily equivalent to be false, there must *exist* two models of Σ that are not elementarily equivalent; but since Σ has no models, this statement can never be false.

6. From 1 and 5: **X** is true on all Σ's models; hence it is true on J.
7. From 2 through 6: if **X** is true on I, it is true on J.
8. By the same reasoning, if **X** is true on J, it is true on I.
9. From 2, 7, and 8: For every V sentence **X**, **X** is true on I iff it is true on J.
10. From 9 by the definition of elementary equivalence: I ≡v J.
11. Now we prove the 'if' part. So assume that all the models of Σ are elementarily equivalent with respect to V.
12. Take **X** to be any V sentence.
13. For any given model M of Σ, **X** is either true on M or ¬**X** is true on M.
14. From 11: if **X** is true on M, then it is true on all the models of Σ since all these models are elementarily equivalent to M.
15. From 14: if **X** is true on M, then it is a logical consequence of Σ.
16. Similarly, if ¬**X** is true on M, then it is true on all the models of Σ, since all these models are elementarily equivalent to M.
17. From 16: if ¬**X** is true on M, then it is a logical consequence of Σ.
18. From 13, 15, and 17: either **X** is a logical consequence of Σ or ¬**X** is a logical consequence of Σ.
19. From 18 by Corollary 3.5.1b: Σ is complete.

Theorem 3.5.2 demonstrates that the notions of the completeness of a PL set and the elementary equivalence of all its models are equivalent. Since isomorphic models are elementarily equivalent, a PL set whose models are isomorphic with respect to its vocabulary is complete. But isomorphism is a much stronger notion than elementary equivalence. Thus a PL set all of whose models are isomorphic with respect to its vocabulary is a much "stronger" set than simply a complete one. We have a term for such a set—**categorical**. A satisfiable categorical PL set is a set that "captures" the structure of its models, since the constituents of these models that interpret the vocabulary of this set have the same structure. It can be claimed with justice that a satisfiable categorical PL set "defines" the structure of its models relative to its vocabulary. As we will see in Chapter Five, no PL theory that is "sufficiently strong" is categorical. We state the following definition.

Categoricity: A PL set Σ is said to be **categorical** iff all of its models are isomorphic with respect to Voc(Σ). It is called $\aleph_0$-**categorical** iff all its countably infinite models (i.e., the models of Σ whose cardinality is $\aleph_0$) are isomorphic with respect to Voc(Σ). In general, for any cardinality κ (finite or infinite), Σ is said to be **κ-categorical** iff all its models whose cardinality is κ are isomorphic with respect to Voc(Σ).

A trivial consequence of the definition above is that every inconsistent set is categorical, because for a set not to be categorical it must have at least two models that are not isomorphic. Given that an inconsistent set has no models, it cannot fail to be categorical.

3.6 The Löwenheim-Skolem Theorem

3.6.1   We restate what we have said previously. The PL language with which we are working is countably infinite. Everything about it is countably infinite: its sets of names, predicates, function symbols, variables, terms, formulas, and sentences. Only its connectives, quantifiers, and punctuation marks are finite. Hence the cardinality of our PL language is $\aleph_0$. In Henkin's proof of the Completeness Theorem, when we extended a consistent PL set Γ into a Henkin set Π, we added to the language of PL a countably infinite set of new names. We designated the Predicate Logic with the expanded language as PL⁺. It is clear that the expanded language of PL⁺ is also countably infinite. We first considered the case of PL⁺ without the identity predicate, and we constructed the Henkin model H^Π of Π. The universe of discourse, UD, of H^Π consists of all the singular terms of the language of PL⁺. Since the language of PL⁺ is countably infinite, the set of all the PL⁺ singular terms is countably infinite as well. Thus the size of H^Π is $\aleph_0$. Second, we considered the case of PL⁺ with the identity predicate. The universe of discourse $\mathcal{U}$ of the Henkin model H^Π of the Henkin set Π consists of all the equivalence classes of the PL⁺ singular terms. As we explained previously, $\mathcal{U}$ is a partition of UD. As any partition, card($\mathcal{U}$) ≤ card(UD). We will prove below this fact in a general setting. But for now, we want to continue our chain of reasoning. In either case, whether we include the identity predicate or not, so long as the language is countably infinite, the cardinality of the Henkin model of a PL set is countable, that is, its cardinality is less than or equal to $\aleph_0$. Since PL with the identity predicate is the general case, we will only consider Henkin models of PL sets whose vocabularies include identity. Henkin's proof of the Completeness Theorem entails that every consistent PL set has a countable model. This is the celebrated Löwenheim-Skolem Theorem, which is due to the German mathematician Leopold Löwenheim and the Norwegian mathematician Thoralf Skolem

The Löwenheim-Skolem Theorem: For every satisfiable PL set Σ, if the language of PL is countably infinite, then Σ has a countable model.

Our preceding discussion supplies a **proof of this theorem**. If Σ is satisfiable, then by the Soundness Theorem, it is consistent. By Henkin's proof of the Completeness Theorem, Σ can be extended into a Henkin set Π. If the language of PL is countably infinite, Π has a countable model—namely, the Henkin model H^Π. It follows that Σ has a countable model.

We conclude this subsection by proving the claim made above about the cardinality of a partition. Let A be any nonempty set and $\mathcal{F}$ any partition of it. By the definition of a partition, $\mathcal{F}$ is a pairwise disjoint family of nonempty subsets of A that are collectively exhaustive of A. Take any member of $\mathcal{F}$. Call this member B. Since B is nonempty, we chose a *single* member of B to be the *representative* of B. Since the members of $\mathcal{F}$ are mutually exclusive, no distinct members of $\mathcal{F}$ have the same representative. We denote the selected representative of B as "ρ(B)." Now we define the function G from $\mathcal{F}$ into A as follows: for every X in $\mathcal{F}$, G(X) = ρ(X). In other words, G assigns to every set in $\mathcal{F}$ its representative. It is clear that G is a one-to-one function from $\mathcal{F}$ into A, and that the range of G is a subset of A. Therefore, by definition 2.3.4a, card($\mathcal{F}$) ≤ card(A).[5]

3.6.2   There is a very interesting and philosophically significant implication of the Löwenheim-Skolem Theorem. In exercise 1.5.6, we listed two satisfiable PL sets whose models are all infinite. We reproduce one of these sets: $\{(\forall x)(\forall y)(\forall z)((Kxy \wedge Kyz) \to Kxz)$, $(\forall x)(\exists y)Kxy$, $(\forall x)(\forall y)(Kxy \to \neg Kyx)\}$. The fact that there is a PL set whose models are all infinite is not remarkable in its own right. There are many such sets. Perhaps the most obvious and least interesting one is the following set {**a** ≠ **b**: **a** and **b** are any distinct PL names}. In other words, this set consists of *all* the sentences of the form **a** ≠ **b** where **a** and **b** are distinct PL names. This is an infinite set, and it is satisfiable. Any interpretation that has a countably infinite UD and that assigns to every name occurring in this set a unique individual in UD is a model of it. It is clear that no finite interpretation can satisfy this set. Thus saying that PL can capture the concept of the infinite is not philosophically significant if by 'capture the concept of the infinite' it is meant that there is a satisfiable PL set whose models are all infinite. What is remarkable and philosophically significant is that PL can capture the concept of the infinite through a finite set of sentences each of which can be true in a finite interpretation. This fact implies that the concept of actual infinity is an emergent concept that may be based on three finite concepts. It is said in this strong sense that PL can capture the concept of the infinite. Observe that since this set is finite, the conjunction of all its members is a single PL sentence whose models are all infinite.

The Löwenheim-Skolem Theorem entails that the same thing cannot be said about the concept of the uncountable. There is no PL sentence whose models are all uncountable; and hence there is no finite PL set whose models all have cardinalities greater than $\aleph_0$. In fact, the situation is even worse: there is no satisfiable PL set, whether finite or infinite, whose models are all uncountable, if the language of PL is countably infinite. The Löwenheim-Skolem Theorem states that if a PL set has a model, then it has a countable model. Thus there can be no satisfiable PL set that fails to have a countable model, if the language of PL is countable. But now we have a puzzle. There are sets of PL sentences that entail the existence of uncountable sets. The most obvious of these PL sets is the set of the set-theoretic axioms. On the basis of these axioms, we can prove the existence of uncountable sets, such as the set $\mathbb{R}$ of all the real numbers and the set $\mathbb{C}$ of all the complex numbers. If there is a model of the standard axioms of set theory, then by the Löwenheim-Skolem Theorem they must have a countable model M (the language of standard set theory is countable). But these axioms logically imply that there are uncountably many objects;

---

5   In order for this argument to work for every partition of any nonempty set, the Axiom of Choice is needed. In exercise 2.5.2f, we defined a choice set. The Axiom of Choice says that every nonempty pairwise disjoint family of nonempty sets has a choice set.

hence M must contain uncountably many objects, which contradicts the fact that M is countable. This puzzle is known as **Skolem's Paradox**. The paradox can be dissolved by observing that there are two perspectives involved in this puzzle. There is the perspective from within the model M itself and there is the perspective from outside M. From outside M, there is a one-to-one correspondence between the universe of discourse of M and the set of natural numbers ℕ; thus from this external perspective M is definitely a countable model. However, from the perspective within M, there is no one-to-one correspondence between UD (or a certain subset of UD) and ℕ (or whatever set in M that represents ℕ) because all the functions that are one-to-one from ℕ onto UD are not represented in M—they are not accessible to M. So from within M, there is a subset of UD that *cannot be enumerated*, that is, it cannot be put into a one-to-one correspondence with ℕ. Any function that reveals the true nature of M—namely, that it is countable—is not represented in M. Metaphorically speaking, for those folks who inhabit M, their universe *appears* uncountable to them because they have no way of enumerating the objects of their universe—all the functions that can enumerate the objects of their universe are not available to them; so based on what is available to them, their universe appears to them uncountable, when "in reality" it is countable.

PL sets, such as the set of the set-theoretic axioms, that entail the existence of uncountably many objects have models that are "really" uncountable, that is, they are uncountable from a within-model perspective and from an outside-model perspective. Let Γ be the set of the standard axioms of set theory and let J be one of those "really" uncountable models of Γ. For simplicity, we denote the vocabulary of Γ as V. As usual, Thv(J) is the set {X: X is a PL sentence composed of V and is true on J}. By Corollary 3.5.1e, Thv(J) is a satisfiable complete theory whose vocabulary is V. The language of set theory is countable; hence V is countable. By the Löwenheim-Skolem Theorem Thv(J) has a countable model M. J and M, of course, are not isomorphic, since there can be no one-to-one correspondence between their UD's (the UD of J is uncountable and the UD of M is countable). However, by Theorem 3.5.2, since Thv(J) is complete, all of its models are elementarily equivalent with respect to V. Therefore, J and M are elementarily equivalent with respect to V. This conclusion establishes that there are PL interpretations that are elementarily equivalent but not isomorphic. In the last exercise of this book, we will give another example of elementarily equivalent models that are not isomorphic.

3.7 Exercises

Note: All answers must be adequately justified. Unless stated otherwise, 'PL' in the following exercises refers to the full version of PL.

3.7.1 Extend the Inductive Step of the proof of the Soundness Theorem to show that the rules Modus Tollens, Disjunctive Syllogism, and Explosion are sound (see Subsection 1.4.5a for descriptions of these rules).

3.7.2 Extend the Inductive Step of the proof of the Soundness Theorem to show that the rules Existential Generalization and Existential Instantiation are sound (see Subsections 1.4.5a and 1.4.5b for descriptions of these rules).

3.7.3* Give a partial demonstration of the soundness of the replacement rule Contraposition by proving that for every PL sentence **X**, if **X** contains a formula of the form **Y**→**Z**, and if **W** is the PL sentence that is obtained from **X** by replacing the formula **Y**→**Z** with the formula ¬**Z**→¬**Y**, then **X** and **W** are logically equivalent. Restrict your reasoning to the economical version of PL.

3.7.4 Prove that the Soundness Theorem is equivalent to the following statement: every satisfiable set of PL sentences is consistent.

3.7.5 Prove that there is a maximal consistent set that contains the sentence $(\exists x)Px$ and all the sentences of the form ¬P**a**, where **a** is any PL name. Is this maximal consistent set a Henkin set?

3.7.6 Construct the Henkin Interpretations for the following sets, and evaluate the truth values of their members on these interpretations. Describe Voc(Σ) for each problem.

3.7.6a* Σ = {$Kadg^1c$, Rdb, Kag^1bb, Rg^1ad, Rg^1bc, Kcg^1ag^1d, $(\forall x)(\forall y)(Rxy \rightarrow (\exists w)(\exists z)Kwxz)$, $(\exists x)(\exists y)Kxg^1yg^1y$}.

3.7.6b Σ = {Eab, Dh^2staa, Dh^2bbqp, Lq, Lp, Lh^2rr, Est, Epr, Eqr, $(\forall x)(Lx \rightarrow ((x = p \vee x = q) \vee (\exists y)x = h^2yy))$, $(\exists x)(Dh^2xxqp \wedge Exb)$, $(\exists x)Dh^2stxx \leftrightarrow \neg(\exists y)Dh^2yyqp$}.

3.7.7 Let Φ and Ψ be nonempty satisfiable sets of PL sentences. Assume that for any I, if I is an interpretation for both Φ and Ψ, then I cannot be a model of Φ and a model of Ψ. Prove that there is a PL sentence **X**∧**Y** such that **X**∧**Y** is contradictory, every model of Φ is a model of **X**, and every model of Ψ is a model of **Y**.

3.7.8* Construct PL sentences $\chi_2, \chi_3, \chi_4, \ldots, \chi_n, \ldots$ from the existential quantifier, identity predicate, and PL variables such that each χ_n, where $n \geq 2$, is considered to express the

English sentence 'There are at least n individuals'. Let Ω be the set that consists of all these sentences. Prove that Ω is consistent.

3.7.9   Let Ω be as defined in Exercise 3.7.8. Prove that it is impossible for Ω to have a logical consequence that is considered to express the English sentence 'There are infinitely many individuals'.

3.7.10   Prove that if Σ is a PL set that has arbitrarily large finite models,[6] then it has an infinite model.

3.7.11   The conclusion of this exercise is meant to show that the English sentence 'Most individuals are P', where P is a 1-place PL predicate, is not expressible by any single PL sentence.

    3.7.11a   Construct, for each natural number $n \geq 2$, a PL sentence π_n that is considered to express the English sentence 'There are at least n individuals that are *not* P.

    3.7.11b   Invoke the sentences constructed above to prove that there is no PL sentence θ such that for every interpretation I for θ, I is a model of θ iff I assigns an extension to P whose cardinality is strictly greater than the cardinality of its complement in UD, that is, $card(I(P)) > card(UD-I(P))$.

3.7.12*   Consider the logical system NL ("Number Logic") that is obtained from PL by making the following modifications to PL syntax and semantics: (1) the names of NL include infinitely many new names $\beta_0, \beta_1, \beta_2, \beta_3, \ldots, \beta_n, \ldots$; (2) the vocabulary of NL contains an additional 1-place predicate $\tilde{N}$; (3) the list of names of every NL interpretation includes the names $\beta_0, \beta_1, \beta_2, \beta_3, \ldots, \beta_n, \ldots$; (4) the universe of discourse of every NL interpretation contains every natural number, that is, $\mathbb{N} \subseteq UD$; (5) every NL interpretation assigns the natural number n as the referent of the name β_n; and (6) every NL interpretation assigns $\mathbb{N}$ as the extension of the predicate $\tilde{N}$. All the logical concepts in NL are defined exactly as in PL. Prove the following.

    3.7.12a   The sentences $(\forall x)(\forall y) x = y$ and $(\exists x)(\exists y)(x \neq y \wedge (\forall z)(z = x \vee z = y))$ are contradictory in NL and every sentence of the form $\tilde{N}\beta_n$ is valid in NL.

    3.7.12b   If the proof theory of PL is introduced as a proof theory for NL, then NL proof theory is sound but incomplete.

    3.7.12c   The sentence $\neg \tilde{N}a$, where a is an NL (PL) name, is a logical consequence of the infinite NL set $\{a \neq \beta_0, a \neq \beta_1, a \neq \beta_2, a \neq \beta_3, \ldots, a \neq \beta_n, \ldots\} = \{a \neq \beta_n : n \in \mathbb{N}\}$.

    3.7.12d   If NL derivations are finite sequences of NL sentences, then NL cannot have a sound and complete proof theory.

---

6   A set has arbitrarily large finite models iff for every natural number n, the set has a finite model whose size is $\geq n$.

3.7.13   Suppose that a PL set Σ has a model of size $\aleph_0$. Demonstrate that for every infinite cardinality κ, Σ has a model whose size $\geq \kappa$. Assume that the Finite-Satisfiability Theorem applies to sets of any cardinalities, and not only to countable sets.[7]

3.7.14*   Let V be Voc($(\exists x)Px$). Show that all finite models of $(\exists x)Px$ that are elementarily equivalent with respect to V are isomorphic with respect to V.

3.7.15   Let M be any finite model of $(\exists x)Px$ and let V be Voc($(\exists x)Px$). Prove that Thv(M) is decidable, where Thv(M) = {**X**: **X** is a V sentence that is true on M}.

3.7.16*   Demonstrate that for every PL set Φ, if Th(Φ) is finitely axiomatizable, then there is a PL set Ψ such that Ψ is a finite subset of Φ and Th(Φ) = Th(Ψ).

3.7.17   Let Φ and Ψ be two PL sets that have the same vocabulary. Establish that if $\Phi \subseteq \Psi$, Φ is a complete theory, and Ψ is satisfiable, then $\Phi = \Psi$.

3.7.18   Prove that every effectively enumerable PL theory is axiomatizable.

Solutions to the Starred Exercises

SOLUTION TO 3.7.3

First, we define the complexity of a PL sentence **X**[**Y**→**Z**] in which a formula of the form **Y**→**Z** occurs to be the number of its quantifiers and connectives other than the → in **Y**→**Z**. We will prove using PCI (see Subsection 2.2.2) that the following statement E(n) is true for all natural numbers.

E(n)   For every natural number n and every PL sentence **X**[**Y**→**Z**] in which a formula of the form **Y**→**Z** occurs, if n is the complexity of **X**[**Y**→**Z**], then the PL sentence **X**[¬**Z**→¬**Y**] that is obtained from **X**[**Y**→**Z**] by replacing **Y**→**Z** with ¬**Z**→¬**Y** is logically equivalent to **X**[**Y**→**Z**].

---

[7] We need to make this assumption explicit because the proof we gave of the Finite-Satisfiability Theorem is based on the Compactness Theorem whose proof invokes the Completeness Theorem. The reader would recall that our proof of the Completeness Theorem presupposes that the language of PL is countably infinite; hence the vocabularies of the PL sets we considered are countable as well. In this problem we will have to consider a set whose vocabulary is uncountable.

Proof by PCI

1. **The Base Step**: Let n = 0. Thus X[Y→Z] is simply Y→Z. A simple truth table shows that this sentence is logically equivalent to ¬Z→¬Y.

2. **The Inductive Step**: Assume that for some k > 0, E(m) is true for every m such that 0 ≤ m < k. This is the Induction Hypothesis. We want to prove that E(k) is true. Let X[Y→Z] be of complexity k. Since k > 0, X[Y→Z] contains a quantifier or a connective in addition to the → of Y→Z. Since we are working with the economical version of PL, this gives us three cases: ¬, →, and ∀.

   2.1. Assume that X[Y→Z] is ¬W[Y→Z]. The complexity of W[Y→Z] is less than k. Hence the Induction Hypothesis applies to W[Y→Z]. This implies that W[Y→Z] and W[¬Z→¬Y] are logically equivalent. Therefore, ¬W[Y→Z] and ¬W[¬Z→¬Y] are logically equivalent. But ¬W[¬Z→¬Y] is simply X[¬Z→¬Y]. Thus X[Y→Z] and X[¬Z→¬Y] are logically equivalent.

   2.2. Suppose that X[Y→Z] is a conditional. The formula Y→Z could be a component of the antecedent, the consequent, or both. Without loss of generality, we assume that it is a component of both the antecedent and consequent. Hence X[Y→Z] is W[Y→Z]→V[Y→Z]. Each of W[Y→Z] and V[Y→Z] has a complexity less than k; thus the Induction Hypothesis applies to both of them. We have that W[Y→Z] and W[¬Z→¬Y] are logically equivalent, and V[Y→Z] and V[¬Z→¬Y] are logically equivalent. It follows that W[Y→Z]→V[Y→Z] and W[¬Z→¬Y]→V[¬Z→¬Y] are logically equivalent. But W[¬Z→¬Y]→V[¬Z→¬Y] is X[¬Z→¬Y]. Therefore X[Y→Z] and X[¬Z→¬Y] are logically equivalent.

   2.3. Assume that X[Y→Z] is (∀u)W[Y→Z]. Let I be any PL interpretation for (∀u)W[Y→Z]. Since the complexity of W[Y→Z] is less than k, the complexity of every basic substitutional instance W[Y→Z][s] of (∀u)W[Y→Z] is less than k. Hence the Induction Hypothesis applies to all those basic substitutional instances. It follows that for every name s in LN of I, the basic substitutional instances W[Y→Z][s] and W[¬Z→¬Y][s] are logically equivalent, which implies that they have the same truth value on I. In other words, every basic substitutional instance of (∀u)W[Y→Z] has the same truth value as the corresponding basic substitutional instance of (∀u)W[¬Z→¬Y]. By the truth conditions of the universal quantifier, (∀u)W[Y→Z] and (∀u)W[¬Z→¬Y] have the same truth value on I. Since I is arbitrary, they have identical truth values on every interpretation that is relevant to them. Therefore X[Y→Z] and X[¬Z→¬Y] are logically equivalent.

3. From 1 and 2 by PCI: E(n) is true for every natural number n.

Solution to 3.7.6

3.7.6a $\Sigma = \{Kadg^1c, Rdb, Kag^1bb, Rg^1ad, Rg^1bc, Kcg^1ag^1d, (\forall x)(\forall y)(Rxy \rightarrow (\exists w)(\exists z)Kwxz), (\exists x)(\exists y)Kxg^1yg^1y\}$

$Voc(\Sigma) = \{a, b, c, d, R^2, K^3, g^1, \neg, \wedge, \vee, \rightarrow, \leftrightarrow, \forall, \exists, (,), u, v, w, x, y, z, u_1, v_1, w_1, x_1, y_1, z_1, \ldots\}$

The Henkin Interpretation $H\Sigma$

UD: The set of all PL singular terms (since Σ is a set of PL sentences, we work with PL instead of PL⁺)
LN: All the members of UD, including the names a, b, c, and d
Semantical Assignments

$H\Sigma(a)$: a; $H\Sigma(b)$: b; $H\Sigma(c)$: c; $H\Sigma(d)$: d; in general for every **t** in LN, $H\Sigma(\mathbf{t})$: **t**
$H\Sigma(Rxy)$: $\{\langle d, b \rangle, \langle g^1a, d \rangle, \langle g^1b, c \rangle\}$
$H\Sigma(Kxyz)$: $\{\langle a, d, g^1c \rangle, \langle a, g^1b, b \rangle, \langle c, g^1a, g^1d \rangle\}$
$I(g^1x)$: $\{\langle a, g^1a \rangle, \langle b, g^1b \rangle, \langle c, g^1c \rangle, \langle d, g^1d \rangle, \langle g^1a, g^1g^1a \rangle, \ldots\}$: $\{\langle \mathbf{t}, g^1\mathbf{t} \rangle$: **t** is a singular term in UD$\}$

By construction, the atomic sentences in Σ are all true on $H\Sigma$.
The sentence $(\forall x)(\forall y)(Rxy \rightarrow (\exists w)(\exists z)Kwxz)$ is true on $H\Sigma$ because the first coordinate of any ordered pair in the extension of R is the middle coordinate of some ordered triple in the extension of K.
The sentence $(\exists x)(\exists y)Kxg^1yg^1y$ is false on $H\Sigma$ because there is no ordered triple in the extension of K of the form $\langle \mathbf{s}, g^1\mathbf{t}, g^1\mathbf{t} \rangle$.

Solution to 3.7.8

Define
χ_2 to be $(\exists x_1)(\exists x_2)x_1 \neq x_2$
χ_3 to be $(\exists x_1)(\exists x_2)(\exists x_3)(x_1 \neq x_2 \wedge x_1 \neq x_3 \wedge x_2 \neq x_3)$
χ_4 to be $(\exists x_1)(\exists x_2)(\exists x_3)(\exists x_4)(x_1 \neq x_2 \wedge x_1 \neq x_3 \wedge x_1 \neq x_4 \wedge x_2 \neq x_3 \wedge x_2 \neq x_4 \wedge x_3 \neq x_4)$
In general, for each natural number $n \geq 2$, define
χ_n to be $(\exists x_1)\ldots(\exists x_n)(x_1 \neq x_2 \wedge \ldots \wedge x_1 \neq x_n \wedge x_2 \neq x_3 \wedge \ldots \wedge x_2 \neq x_n \wedge x_3 \neq x_4 \wedge \ldots \wedge x_3 \neq x_n \wedge \ldots \wedge x_{n-1} \neq x_n)$ which can be expressed compactly as $(\exists x_1)\ldots(\exists x_n) \bigwedge_{1 \leq i < j \leq n} x_i \neq x_j$, where the last notation stands for a repeated conjunction (for example, $\bigwedge_{1 \leq i \leq 4} X_i$ stands for the conjunction $X_1 \wedge X_2 \wedge X_3 \wedge X_4$).

It is clear that for any interpretation I, I is a model of χ_n iff the size of I is equal to or greater than n. Thus it is natural to consider χ_n as expressing the English sentence 'There are at least n individuals'.

Let Ω be the set $\{\chi_n$: n is a natural number $\geq 2\}$. Every finite subset Σ of Ω contains a sentence χ_m such that m is the greatest index of all the members of Σ, that is, $m \geq k$ for each χ_k in Σ. Hence any PL interpretation whose size is equal to or greater than m is a model of Σ. For instance, if $\Sigma = \{\chi_3, \chi_5, \chi_8, \chi_{12}, \chi_{22}\}$, then the greatest index is 22. Any interpretation of size ≥ 22

RESOURCES OF THE METATHEORY

is a model of Σ. It follows that Ω is finitely satisfiable. By the Finite-Satisfiability Theorem, Ω is satisfiable. By the Soundness Theorem, it is consistent.

SOLUTIONS TO 3.7.12

3.7.12a The sentences $(\forall x)(\forall y)x = y$ and $(\exists x)(\exists y)(x \neq y \land (\forall z)(z = x \lor z = y))$ are contradictory in NL since every NL interpretation is infinite and the former sentence can only be true on an interpretation of size 1 and the latter on an interpretation of size 2. Hence these sentences are false on all NL interpretations. Every sentence of the form $\tilde{N}\beta_n$ is valid in NL because by stipulation every NL interpretation assigns natural numbers to the β-names, and the extension of the predicate $\tilde{N}$ on every NL interpretation consists of all the natural numbers. Hence the referent of β_n on any NL interpretation always belongs to the extension of $\tilde{N}$. This shows that every sentence of the form $\tilde{N}\beta_n$ is true on all NL interpretations.

3.7.12b Suppose that we introduce NDS, as described in Subsection 1.4.5, as the deduction system for NL. First observe that every NL interpretation is a PL interpretation, but the converse is not true: there are many PL interpretations that are not NL interpretations. In other words, the class of all NL interpretations is a proper subclass of the class of all PL interpretations. With some reflection, one can see that this fact entails that if **X** is a logical consequence of Γ in PL, then **X** is also a logical consequence of Γ in NL. Using the language of validity, this says that any argument that is valid in PL is also valid in NL. The converse, however, is not true: there are many arguments that are valid in NL but invalid in PL. (Recall that the sense in which an argument is "valid" is different from the sense in which a sentence is "valid"; see Chapter One.) Keeping this in mind, we can see why NDS is sound but incomplete in NL. Let **X** be a sentence that is derivable from some set Γ by means of NDS. Since NDS is sound in PL, **X** is a logical consequence of Γ in PL. By the preceding observation, **X** is a logical consequence of Γ in NL. This shows that NDS is sound in NL. From the previous solution, we know that $(\forall x)(\forall y)x = y$ is contradictory in NL. Equivalently, $\neg(\forall x)(\forall y)x = y$ is valid in NL. But this means that $\neg(\forall x)(\forall y)x = y$ is a logical consequence of the empty set in NL. If NDS were complete in NL, $\neg(\forall x)(\forall y)x = y$ would have to be derivable from the empty set using the rules of NDS. But this is false: the rules of NDS do not allow us to derive $\neg(\forall x)(\forall y)x = y$ from the empty set (otherwise, $\neg(\forall x)(\forall y)x = y$ would be a logical theorem in PL, which is not the case). It follows that $\neg(\forall x)(\forall y)x = y$ is a logical consequence of the empty set in NL, but it is not a theorem of the empty set. Thus NDS is an incomplete deduction system for NL. Of course, this does not show that NL cannot have a sound and complete proof theory. This only shows that the proof theory of PL is a sound but incomplete proof theory for NL. Below we will show that, under a reasonable assumption, NL indeed cannot have a sound and complete proof theory.

3.7.12c The sentence $\neg \tilde{N}a$ is a logical consequence of the infinite NL set $\{a \neq \beta_n: n \in \mathbb{N}\}$. To see this let M be any NL interpretation for the set $\{a \neq \beta_n: n \in \mathbb{N}\}$ and for $\neg \tilde{N}a$ that is also a model of the set $\{a \neq \beta_n: n \in \mathbb{N}\}$. Because M is a model of this set, every sentence of the form $a \neq \beta_n$ is true on M. But the referent of each β_n is the natural number n. It follows that the referent M(a) is not a natural number. Since the extension of $\tilde{N}$ on M is the set $\mathbb{N}$ of all the natural numbers, the referent M(a) does not belong to the extension of $\tilde{N}$. Therefore $\neg \tilde{N}a$ is true on M. So $\neg \tilde{N}a$ is true on every model of the set $\{a \neq \beta_n: n \in \mathbb{N}\}$ that is also relevant to $\neg \tilde{N}a$. Hence $\neg \tilde{N}a$ is a logical consequence of $\{a \neq \beta_n: n \in \mathbb{N}\}$.

3.7.12d
1. From Exercise 3.7.12c: $\neg \tilde{N}a$ is a logical consequence of $\{a \neq \beta_n: n \in \mathbb{N}\}$.
2. Let Σ be any nonempty finite subset of $\{a \neq \beta_n: n \in \mathbb{N}\}$.
3. From 2: Σ consists of finitely many sentences each of which is of the form $a \neq \beta_n$ for some natural number n.
4. From 3: There is a sentence $a \neq \beta_m$ in Σ such that m is the greatest index of all the members of Σ.
5. Let I be the NL interpretation whose UD is $\mathbb{N}$ and whose LN consists of a and all the β-names. Further let I assign the number m+1 as the referent of a. As stipulated, the referent of every β_n is the natural number n and the extension of $\tilde{N}$ is $\mathbb{N}$.
6. From 4 and 5: I is a model of Σ since the members of Σ prevent I(a) from being at most 0, 1, 2, …, or m, and I(a) is none of these numbers—it is m+1. Given that I(a) belongs to the extension of $\tilde{N}$ on I, it follows that $\neg \tilde{N}a$ is false on I.
7. From 6: There is a model of Σ on which $\neg \tilde{N}a$ is false. Hence $\neg \tilde{N}a$ is not a logical consequence of Σ.
8. From 2 and 7: since Σ is arbitrary, $\neg \tilde{N}a$ is not a logical consequence of any nonempty finite subset of $\{a \neq \beta_n: n \in \mathbb{N}\}$.
9. From 6: Since $\neg \tilde{N}a$ is false on I, it is not a valid sentence, and hence it is not a logical consequence of the empty set.
10. From 1, 8, and 9: $\neg \tilde{N}a$ is a logical consequence of the set $\{a \neq \beta_n: n \in \mathbb{N}\}$, but it is not a logical consequence of any finite subset of $\{a \neq \beta_n: n \in \mathbb{N}\}$.
11. From 10: the Compactness Theorem does not hold for NL.
12. From the proof of the Compactness Theorem in Subsection 3.3.1: The Compactness Theorem holds for any logical system that has a complete and

RESOURCES OF THE METATHEORY

sound proof theory and whose formal derivations are finite sequences of sentences that meet the standard conditions of formal derivations.

13. From 11 and 12: NL cannot have a sound and complete proof theory, so long as an NL formal derivation is defined as a finite sequence of sentences that meet the standard conditions of formal derivations.

SOLUTION TO 3.7.14

1. V is $Voc((\exists x)Px)$. Let J and M be any two finite models of $(\exists x)Px$, and assume that $J \equiv_V M$. We want to prove that $J \sim_V M$. Since the extension of P on J, $J(P)$, is non-empty, $J(P)$ is $\{\alpha_1, \alpha_2, \alpha_3, ..., \alpha_m\}$, where $m \geq 1$. Since J is finite, its UD_J consists of n individuals, where $n \geq m$. We arrange the n members of UD_J such that the first m individuals of UD_J are the members of $J(P)$. Thus UD_J is $\{\alpha_1, \alpha_2, \alpha_3, ..., \alpha_m, ..., \alpha_n\}$.

2. Define the V sentence δ_J as follows:

$(\exists x_1)...(\exists x_n)((x_1 \neq x_2 \wedge ... \wedge x_1 \neq x_n \wedge x_2 \neq x_3 \wedge ... \wedge x_2 \neq x_n \wedge x_3 \neq x_4 \wedge ... \wedge x_3 \neq x_n \wedge ...$
$\wedge x_{n-1} \neq x_n) \wedge (\forall y)(y = x_1 \vee y = x_2 \vee ... \vee y = x_n) \wedge (Px_1 \wedge Px_2 \wedge ... \wedge Px_m) \wedge$
$(\forall u)(Pu \rightarrow (u = x_1 \vee u = x_2 \vee ... \vee u = x_m)))$.

We can express δ_J compactly as follows:

$(\exists x_1)...(\exists x_n)((\bigwedge_{1 \leq i < j \leq n} x_i \neq x_j) \wedge (\forall y)(\bigvee_{1 \leq j \leq n} y = x_j)) \wedge ((\bigwedge_{1 \leq j \leq m} Px_j)$
$\wedge (\forall u)(Pu \rightarrow (\bigvee_{1 \leq j \leq m} u = x_j))$

As explained in the Solution to 3.7.8, the symbols $\bigwedge_\vartheta X_j$ and $\bigvee_\vartheta X_j$ represent, respectively, repeated conjunction and repeated disjunction, whose conjuncts and disjuncts are indexed according to the formula ϑ. For example if ϑ is "$0 \leq j \leq 3$," then $\bigwedge_{0 \leq j \leq 3} X_j$ is the conjunction $X_0 \wedge X_1 \wedge X_2 \wedge X_3$, and $\bigvee_{0 \leq j \leq 3} X_j$ is the disjunction $X_0 \vee X_1 \vee X_2 \vee X_3$. δ_J describes J completely with respect to V. Informally speaking, δ_J asserts that there are exactly n individuals in UD_J, and $J(P)$ consists of exactly m individuals. It is clear that δ_J is true on J.

3. From 1 and 2: since $J \equiv_V M$ and since δ_J is composed of V and true on J, δ_J must be true on M as well. The truth of δ_J entails that UD_M of M has exactly n individuals and that $M(P)$ has exactly m individuals, where $1 \leq m \leq n$. This means that M has a structure identical to J relative to V. We arrange the individuals in UD_M and $M(P)$ in a fashion similar to the way in which the members of UD_J and $J(P)$ are arranged. Hence UD_M is $\{\varepsilon_1, \varepsilon_2, \varepsilon_3, ..., \varepsilon_m, ..., \varepsilon_n\}$ and $M(P)$ is $\{\varepsilon_1, \varepsilon_2, \varepsilon_3, ..., \varepsilon_m\}$ (i.e., the members of $M(P)$ are the first m individuals of UD_M).

4. From 1 and 3: define the one-to-one correspondence h between UD_J and UD_M such that for each $k = 1, 2, ..., n$, $h(\alpha_k) = \varepsilon_k$. It is clear that h is an isomorphism, since $h(\alpha_k) = \varepsilon_k$ for each $k = 1, 2, ..., m$, which entails that for every α_k in UD_J, $\alpha_k \in J(P)$ iff $h(\alpha_k) \in M(P)$ (V contains no names). Therefore J and M are isomorphic with respect to V. Symbolically, $J \sim_V M$. (Observe that the assumption that M is finite was not invoked in the proof. It is a consequence of the fact that the sentence δ_J is true on M.)

Solution to 3.7.16

1. Assume $Th(\Phi)$ is finitely axiomatizable, where Φ is some PL set. We wish to prove that there is a PL set Ψ such that Ψ is a finite subset of Φ and $Th(\Phi) = Th(\Psi)$. Let V be $Voc(\Phi)$ (it should be clear that $Voc(Th(\Phi)) = Voc(\Phi)$).

2. From 1: since $Th(\Phi)$ is finitely axiomatizable, $Th(\Phi) = Th(\Delta)$ for some finite PL set Δ whose vocabulary is V.

3. From 2: since Δ is finite, we let σ be the conjunction of the V sentences in Δ. It is obvious that $\Delta \vDash \sigma$ and that σ is a V sentence. Hence $\sigma \in Th(\Delta)$.

4. From 2 and 3: $\sigma \in Th(\Phi)$.

5. From 4: $\Phi \vDash \sigma$.

6. From 5 by the Compactness Theorem: there is a finite subset Ψ of Φ such that $\Psi \vDash \sigma$. We will prove now that $Th(\Phi) = Th(\Psi)$.

7. Let $X \in Th(\Phi)$.

8. From 2 and 7: $X \in Th(\Delta)$.

9. From 8: $\Delta \vDash X$.

10. From 3 and 9 by the construction of σ: $\{\sigma\} \vDash X$ (recall that σ is the conjunction of the sentences in Δ).

11. From 6 and 10: given that $\Psi \vDash \sigma$ and $\{\sigma\} \vDash X$, it follows that $\Psi \vDash X$.

12. From 11 and the fact that X is a V sentence: $X \in Th(\Psi)$.

13. From 7 through 12: for every X, if $X \in Th(\Phi)$, then $X \in Th(\Psi)$.

14. Now let $X \in Th(\Psi)$.

15. From 14: X is a V sentence and $\Psi \vDash X$.

16. From 6 and 15: since $\Psi \subseteq \Phi$, $\Phi \vDash X$.

17. From 16: $X \in Th(\Phi)$.

18. From 14 through 17: for every X, if $X \in Th(\Psi)$, then $X \in Th(\Phi)$.

19. From 13 and 18: for every X, $X \in Th(\Phi)$ iff $X \in Th(\Psi)$. Therefore $Th(\Phi) = Th(\Psi)$ by the Principle of Extensionality.

20. From 6 and 19: there is a PL set Ψ such that Ψ is a finite subset of Φ and $Th(\Phi) = Th(\Psi)$.

Chapter Four

Computability

4.1 Effective Procedures and Computable Functions

In previous chapters we discussed effective procedures, effective decision procedures, decidable and semidecidable concepts, and decidable and semidecidable sets. The central notion in all of these definitions is that of an effective procedure. For instance, a decidable set may be defined as a set such that there is an effective procedure determining whether an object is a member of the set and an effective procedure determining whether an object is not a member of the set. This is equivalent to saying that a decidable set is a set such that there is an effective *decision* procedure for determining membership in the set, since an effective decision procedure is an effective Yes-No procedure. In this chapter we will study two different ways of formalizing the notion of an effective procedure. Although we will not prove this fact here, these two different formalizations are equivalent: every procedure that is deemed effective by one of the formalizations is also deemed effective by the other formalization.

According to these formalizations, a procedure is a function, and an **effective procedure** is a **computable function**. It is intuitive to think of procedures as functions. After all, a procedure transforms a certain input into some output, and a function takes an argument (an input) and returns a value (an output). However, 'computable' in the *narrow* sense suggests that those functions are numerical: their arguments and values are numbers. Indeed, both of these formalizations attempt to characterize the class of all computable numerical functions. More precisely, they concern functions whose arguments are n-tuples of natural numbers and whose values are single natural numbers. In other words, these are *partial* or *total* functions from $\mathbb{N}^n$ into $\mathbb{N}$ where, as usual, $\mathbb{N}^n$ is the set of all the n-tuples whose coordinates are natural numbers. As explained in Chapter Two, a total function from A into B is a function that assigns to *every* member of A *exactly one* value in B. A partial function from A into B is a function that assigns to *every* member of A *at most one* value in B. Since 'at most one' means "one or none," it is clear that every total function is a partial function but the converse is not true. For instance, the addition function is a total function from $\mathbb{N}^2$ *onto* $\mathbb{N}$, since for every ordered pair $\langle n, m \rangle$ in $\mathbb{N}^2$, there is exactly one natural number, n+m, that is the sum of n and m. On the other hand, the subtraction function is only a partial function, since there are many ordered pairs of natural numbers whose subtractions are not natural numbers, e.g., 2–7 and 5–14. Contrary to our practice in Chapter Two, in this chapter we use the term 'function' to mean 'partial function'. It is important to keep in mind that total functions are a special type of partial function.

Intuitively speaking, by 'computable function' we understand a numerical function such that when applied to an n-tuple of natural numbers it generates at most one natural number *via a mechanical computational procedure*. A mechanical computational procedure is a numerical process that consists of finitely many deterministic steps, which can be followed without any creativity. It is the sort of numerical procedure that could be, in principle, executed by a computing machine, such as a calculator, if there were no limits on time, memory, and hardware material. This is an ideal notion of a mechanical computational procedure; but it is necessary, since any stipulated limits on time, memory, or hardware material seem to be arbitrary and not part of the concept of a mechanical computational procedure. The two approaches we will study in this chapter aim at formalizing the informal notion of a computable function, which replaces the equally informal notion of an effective procedure. There seems to be a conceptual problem here. Not all effective procedures are arithmetical. For instance, the procedure of constructing truth tables for PL sentences that contain only sentential connectives is an effective procedure for determining the truth values of the compound sentences on all possible distributions of truth values to their sentential components. This is not an arithmetical procedure. In the *wide* sense of 'computational', it is standardly described as a computational procedure. However, we will reserve the terms 'computation' and 'computational' for arithmetical procedures. If constructing truth tables is not an arithmetical procedure, yet it is an effective procedure, what justifies our narrow focus on computable numerical functions?

A fully convincing answer requires a good deal of tedious mathematical constructions that would show that many of what we think of as effective procedures can be "arithmetized," that is, they can be transformed into computable numerical functions. We will carry out such an arithmetization in the next chapter. But for now we will give a simple example of an effective procedure that can be transformed into a computable numerical function.

Let us consider the logic that is based on the connectives $\neg$ and $\rightarrow$, and contains no quantifiers. We will refer to this logic as Sentence Logic (SL). SL contains, in addition to the sentential connectives, infinitely many sentence letters A_1, A_2, A_3, ..., A_n, ... We define Sents$_{SL}$ to be the set of all the SL sentences. This set is constructed as follows: (1) every sentence letter belongs to Sents$_{SL}$; (2) if **X** and **Y** belong to Sents$_{SL}$, then so do $\neg$**X** and (**X**$\rightarrow$**Y**); (3) Sents$_{SL}$ contains no members that cannot be generated by the first two clauses. Sents$_{SL}$ is a decidable set: any given finite string of SL symbols can be examined to see whether it is a sentence letter or is built from sentence letters by means of (possibly, repeated applications of) the two formation rules described in clause (2). We will show that the effective decision procedure for determining membership in Sents$_{SL}$ can be turned into a computable numerical function.

We begin by assigning **basic numerical codes** to the basic vocabulary of SL.

Symbol	¬	→	(	)	A_1	A_2	A_3	...	A_n	...
Basic Numerical Code	2	3	5	7	11	13	17	...	nth prime number after 7	...

We will generate a numerical code for every SL sentence. If **X** is an SL sentence, we denote its **numerical code** as [X]. If **X** is a sentence letter, [X] = its basic numerical code. The numerical codes of the compound sentences (i.e., the SL sentences that are not sentence letters) are assigned as follows (the dot '.' represents multiplication)

[¬**X**] = $2^2 \cdot 3^{[X]}$
[(**X**→**Y**)] = $2^5 \cdot 3^{[X]} \cdot 5^3 \cdot 7^{[Y]} \cdot 11^7$

We are using the prime numbers to hold the places of the symbols, and the powers of these prime numbers are the numerical codes of the SL expressions that occur at these places. For example, the numerical code of ¬**X** is the product of 2 (the first prime number) raised to the basic numerical code of the symbol ¬ and 3 (the second prime number) raised to the numerical code of the sentence **X**. The use of 2 indicates that the symbol whose numerical code is the power of 2 occurs first in the string ¬**X**, and the use of 3 indicates that the sentence whose numerical code is the power of 3 occurs immediately after the negation sign. An important observation to make is that the numerical codes of the basic vocabulary are prime numbers (in fact, all the prime numbers), while the numerical codes of the compound sentences are composite numbers.

This encoding procedure is itself mechanical (i.e., effective). Let us consider an example. We will generate the numerical code of the SL sentence ¬(A_1→¬A_2). We follow the following encoding procedure.

[¬(A_1→¬A_2)] = $2^2 \cdot 3^{[(A_1 \to \neg A_2)]}$
[(A_1→¬A_2)] = $2^5 \cdot 3^{11} \cdot 5^3 \cdot 7^{[\neg A_2]} \cdot 11^7$
[¬A_2] = $2^2 \cdot 3^{13}$

The decoding of these numbers is also a mechanical procedure. For instance, if we are presented with the number $2^2 \cdot 3^{[(A_1 \to \neg A_2)]}$, we first prime-factor it into $2^2 \cdot 3^k$. Prime-factorizations are unique: every number has exactly one prime-factorization. So this number can only be prime-factored into $2^2 \cdot 3^k$. Since the power of 2 is a prime number, we know that this is the numerical code of a basic vocabulary item. According to the table above this is the symbol ¬. The power k of 3 is a composite number, so we know it cannot be the numerical code of a basic vocabulary item. It could be the numerical code of an SL sentence or some ungrammatical expression. In order to find out, we need to prime-factor k; we discover that it is $2^5 \cdot 3^{11} \cdot 5^3 \cdot 7^m \cdot 11^7$. Again, since there are prime powers, we decode these into their corresponding symbols. We obtain $(A_1 \rightarrow 7^m)$. m is a composite number, so we prime-factor it into $2^2.3^{13}$, which have only prime powers. This gives us $\neg A_2$. Finally we put these symbols in the order of their prime-factorizations; we get the sentence $\neg(A_1 \rightarrow \neg A_2)$.

We can articulate an arithmetical procedure that would determine which number is a numerical code of an SL sentence and which is not. Let this procedure be C. We want the procedure C to generate the number 1 when and only when it is applied to a natural number that is the numerical code of an SL sentence, and to generate the number 0 when and only when it is applied to a number that is not the numerical code of an SL sentence. Now we collect into a single set SENT$_{SL}$ all the natural numbers such that C generates 1 when applied to them. SENT$_{SL}$ is the set of all the numerical codes of the SL sentences. It is the numerical counterpart of Sent$_{SL}$, which is the set of all the linguistics expressions that form grammatical SL sentences. As we will see, the procedure C is a mechanical computational procedure. Furthermore, it is an effective decision procedure, since for any number n it determines whether n is a member of the set SENT$_{SL}$ or not. This shows that SENT$_{SL}$ is a decidable set. Since the encoding procedure is itself an effective procedure, the set Sent$_{SL}$ is also a decidable set. To determine whether an SL expression is a sentence of SL or not, we first use the encoding procedure to generate its numerical code n; and then we apply the arithmetical procedure C to determine whether n belongs to SENT$_{SL}$ or not. If it does, then the original expression is an SL sentence, and if it does not, then the original expression is not an SL sentence.

To complete the arithmetization of SL "sentencehood," it remains to describe the arithmetical procedure C. As stated previously, C will assign 1 to n iff n is the numerical code of an SL sentence and will assign 0 to n iff n is not the numerical code of an SL sentence. So let n be any natural number. We follow the steps below.

1. Prime-factor n.
2. If this prime-factorization is not of the form $2^2.3^k$ or $2^5.3^j.5^3.7^m.11^7$, where k, j, and m > 0, then assign 0 and halt.
3. If this prime-factorization is of the form $2^2.3^k$, where k > 0, then execute Case 1; and if it is of the form $2^5.3^j.5^3.7^m.11^7$, where j and m > 0, then execute Case 2.
4. Case 1 (n is $2^2.3^k$): if k is a prime number ≤ 7, assign 0 and halt; if it is a prime number > 7, assign 1 and halt. Otherwise, if k is a composite number, prime-factor k and go back to Step 2.
5. Case 2 (n is $2^5.3^j.5^3.7^m.11^7$): if either j or m is a prime number ≤ 7, assign 0 and halt; if both are prime numbers > 7, assign 1 and halt. Otherwise, if j is a composite number, prime-factor j and go back to Step 2, and if m is a composite number, prime-factor m and go back to Step 2.

The existence of the arithmetical procedure C demonstrates that the function F from $\mathbb{N}$ into $\mathbb{N}$ that assigns 1 to every numerical code of an SL sentence and 0 to everything else is a computable function. Assuming that every effective procedure can be arithmetized, the task of giving a formal definition of an effective procedure reduces to the task of giving a formal definition of a computable (numerical) function. The first approach is due to Alan Turing (1936) and the second is due to Kurt Gödel (1931).

4.2 Turing Computability

4.2.1 We begin by describing a theoretical computing device that is called a **Turing machine**. A Turing machine is a simple machine that takes natural numbers or n-tuples of natural numbers as inputs and produces single natural numbers as outputs. The inputs and outputs are represented as tallies. So the number 3, for example, is represented as a sequence of three 1's, that is, as '111', number 5 as '11111', and so on. A Turing machine consists of four components: (1) an infinite tape that is divided into identical squares, each of which contains the numeral '0' or the numeral '1'; (2) a pointer that points at one square at a time, and which is capable of reading the numeral on the square, erasing what is on the square, writing instead 0 or 1, moving one square to the left, and moving one square to the right; (3) a register that keeps track of the internal states of the machine; and (4) a set of instructions that represents the *program* of the machine. Because there are no limits on the size of inputs and outputs, on the length of the tape, on the time required to carry out the computation, or on the size of the instruction set, Turing machines, in spite of their theoretical simplicity, are extremely powerful computing devices. Every known computable function is **Turing-computable**, that is, it can be computed by a Turing machine. It is no wonder that almost all discussions of artificial intelligence, functionalism, or any computational theory of the mind invoke Turing machines as the paradigm of a computing device.

Each Turing machine is uniquely defined by a **set of instructions** (or, an **instruction set**). A set of instructions consists of an *even number* of lines: $\{l_1, l_2, l_3, l_4, l_5,..., l_m\}$. So m must be an even number ≥ 2. The lines in an instruction set are referred to as **instruction lines**. Every instruction line l_s consists of four symbols: q_iXYq_k, where X is either 0 or 1 and Y is either 0, 1, R, or L. For example, l_3 might be q_20Lq_3, q_21Rq_4, q_301q_2, q_210q_4, or any other combination. If l_s is q_iXYq_k, we say that q_i is the **initial state** of the line l_s, q_k is the **terminal state** of the line l_s, X is the **input** of the line l_s, and Y is the **output** of the line l_s. If T is the Turing machine whose set of instructions is $\{l_1, l_2, l_3, l_4, l_5,..., l_m\}$, the initial state of T is q_0, which is the initial state of the first instruction line, namely l_1. The terminal state of the Turing machine T is denoted as q_e. No instruction line begins with the state q_e. In other words, the terminal state of T is not the initial state of any instruction line. q_e must be the terminal state of at least one line of the instruction set of T. If $\{q_0, q_1, q_2, q_3, q_4,..., q_e\}$ is the set of *all* the initial and terminal states of the instruction lines $l_1, l_2, l_3, l_4, l_5,...,$ and l_m, then this set is called the **set of the internal states** of the Turing machine T. No two distinct instruction lines share the same first two symbols. Thus if l_s is q_iXYq_k and l_t is $q_iXY'q_n$, then $l_s = l_t$, that is, Y and Y' are identical and q_k and q_n are identical. For every internal state q_i that is not the terminal state q_e, there are exactly two instruction lines l_s and l_t that begin with q_i, such that l_s is $q_i 0 Y q_k$ and l_t is $q_i 1 Y' q_n$.

A Turing machine T starts at the internal state q_0, with its pointer positioned at the leftmost square of a non-zero input. If the input is 0, its pointer is positioned at a 0-square (that is, a square that contains 0). If T halts, it halts at the internal state q_e, with its pointer positioned at the leftmost square of a non-zero output, or at a 0-square if the output is 0. If T does not halt, then its output is undefined. T may fail to halt by continuing to move to the left or to the right, oscillating between two or more squares, stopping at a 0-square while there are 1-squares (that is, squares that contain 1), stopping at a square while its internal state is not q_e, or any other erratic behavior. It is important to note that a Turing machine can execute only one instruction line at a time.

4.2.2   Let S be the function from $\mathbb{N}$ into $\mathbb{N}$ such that $S(n) = n+1$. S is called the **successor function.** Here is an example of a Turing machine T_S that computes the successor function. We will discuss the behavior of T when it receives the input 2 and produces the output 3. At this stage we are only interested in the behavior of this Turing machine when the input is 2. Later we will give the instruction set for T_S.

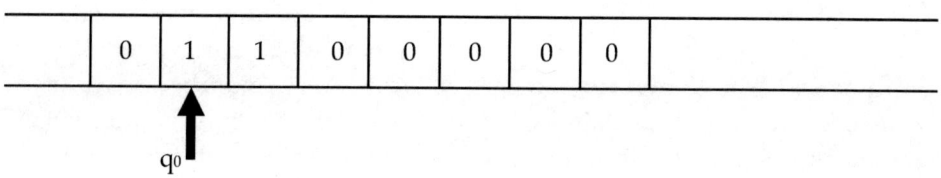

The pointer of T_S is positioned at the leftmost square of the input, which is 2. T_S is at the initial (internal) state q_0. The pointer reads 1 and moves one square to the right; T_S remains at state q_0. We get the following diagram.

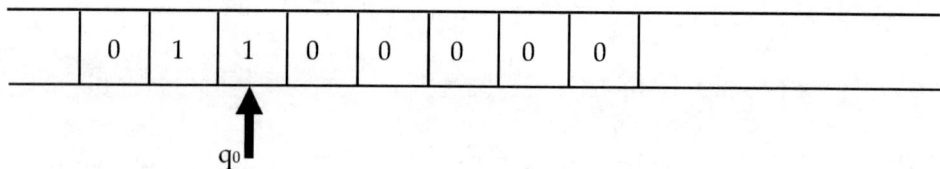

The pointer repeats the same operation as before: it reads 1 and moves one square to the right; T_S remains at state q_0. Here is the situation now.

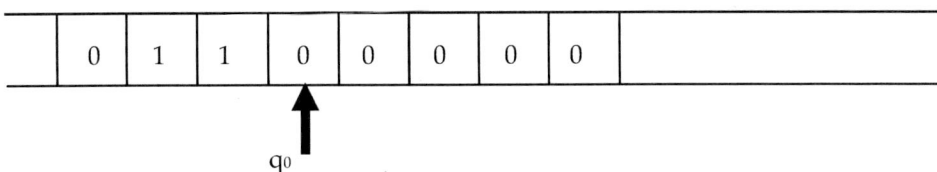

The pointer reads 0. It "realizes" that the input is complete. Now it adds 1 to generate the output 3. Thus the pointer writes 1 instead of 0 and remains at its position. The internal state of Ts changes to q_1. So we have the following situation.

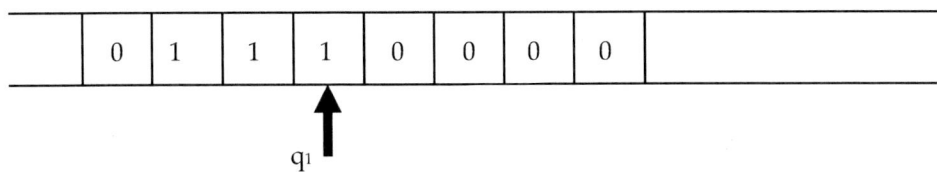

Now the pointer reads 1 and moves one square to the left and Ts remains at state q_1. Then it reads 1 again and moves one more square to the left and Ts remains at state q_1. Again the pointer reads 1 and moves one more square to the left and Ts remains at state q_1. These are three operations, which are the outcome of Ts executing the same instruction line three times. After these operations we have the following situation.

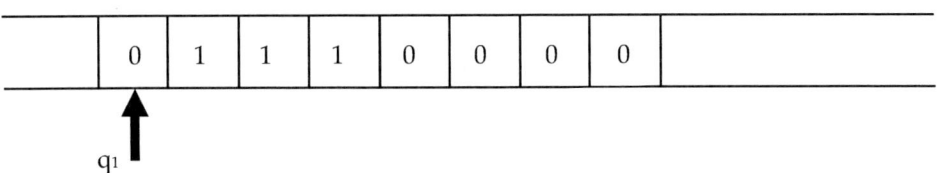

The pointer now reads 0. Hence Ts "realizes" that the output is complete. It moves one step to the right and halts. Recall that the machine halts at the leftmost position of the output. Ts enters the terminal state q_e. This is the halting position.

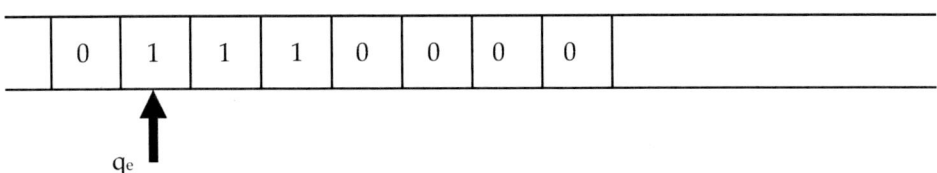

4.2.3 As stated above, the instruction set of a Turing machine consists of a finite sequence of instruction lines. Each instruction line is one of four types. Each type instructs the machine to behave in a certain deterministic way. Here are the four types.

$q_i X 0 q_k$: If T is at state q_i and the pointer reads X (X being 1 or 0), then the pointer writes 0 and T enters state q_k (k could be the same as i or a different number).

q_iX1q_k: If T is at state q_i and the pointer reads X, then the pointer writes 1 and T enters state q_k.

q_iXRq_k: If T is at state q_i and the pointer reads X, then the pointer moves one square to the right and T enters state q_k.

q_iXLq_k: If T is at state q_i and the pointer reads X, then the pointer moves one square to the left and T enters state q_k.

If an instruction line instructs the machine to enter the terminal state q_e, the machine halts and executes no further instructions.

A Turing machine can be represented by a diagram. A diagram of a Turing machine consists of cells connected by arrows. A **diagram cell**, except the terminal cell (the q_e-cell), represents two instruction lines that begin with the same initial state. Each diagram cell that is not the terminal cell is a triangle with two "existing" arrows (the terminal cell is a triangle that has no "existing" arrows). The inside of the triangle represents the initial state of the two instruction lines. Each "existing" arrow represents the input, output, and the terminal state of one of the two instruction lines. Recall that for any internal state q_i that is not the terminal state q_e, there are exactly two instruction lines, each of which begins with the state q_i. A cell that represents the internal state q_i is referred to as the q_i-cell.

Here is an example of a diagram cell. Let l_s be the instruction line q_i0Lq_k and l_t be the instruction line $q_i1 0 q_n$.

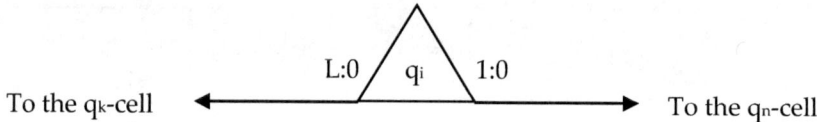

This cell represents the following commands to be executed by the Turing machine.

Right-hand arrow: if the internal state is q_i and the pointer reads 1, write 0 in place of 1 and enter state q_n.

Left-hand arrow: if the internal state is q_i and the pointer reads 0, move the pointer one square to the left and enter state q_k.

It is important to note that the instruction of the left-hand arrow is arranged in a reverse order of the instruction of the line it represents. The left-hand arrow above is arranged as follows: q_k L:0 q_i. This means that if the initial state is q_i and the pointer reads 0, then the pointer must move one square to the left and the internal state must be changed to q_k. Thus this left-hand arrow represents the instruction line: q_i0Lq_k. Right-hand arrows present no problem, since they are arranged in the same order as the instruction lines they represent. But left-hand arrows could be confusing unless one keeps in mind that a left-hand arrow reverses the order of the information given by the instruction line that the arrow represents. This unfortunate annoyance is an outcome of the way a diagram cell represents the two relevant instruction lines. But with a little care the reader will have no difficulty in converting diagram cells into instruction lines and conversely.

The arrows connect to other cells: the right-hand arrow connects to the cell that represents the two instruction lines whose initial state is q_n (the q_n-cell), and the left-hand arrow connects to the cell that represents the two instruction lines whose initial state is q_k (the q_k-cell).

4.2.4   In this subsection we discuss a few examples of Turing machines. First, we discuss the Turing machine T_Z that computes the **zero function** Z. The zero function assigns 0 to every natural number (i.e., $Z(n) = 0$, for each n). When T_Z is given a certain number as input, it replaces all its 1's with 0's. Once T_Z replaces the last 1 with 0, the pointer moves one square to the right and T_Z halts. Below is the instruction set of T_Z.

l_1: $q_0 1 0 q_0$

l_2: $q_0 0 R q_1$

l_3: $q_1 1 1 q_0$

l_4: $q_1 0 0 q_e$

Here is the diagram of T_Z.

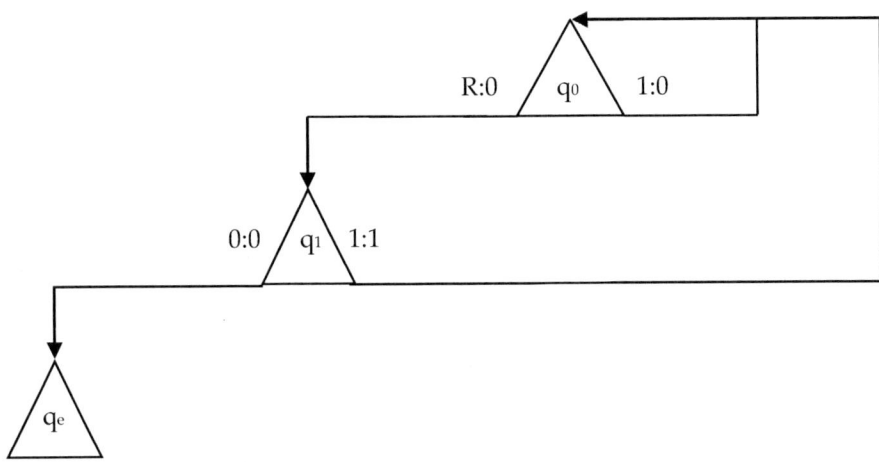

It is usually easier to work with diagrams than with instruction sets. In order to understand how to interpret this diagram, let us see how T_Z behaves when it is given the input 3. We will describe its behavior in steps.

1. There is a sequence of three squares, each of which contains 1. The rest of the squares have 0's in them. The pointer is positioned at the leftmost square that contains 1. The machine is at state q_0.

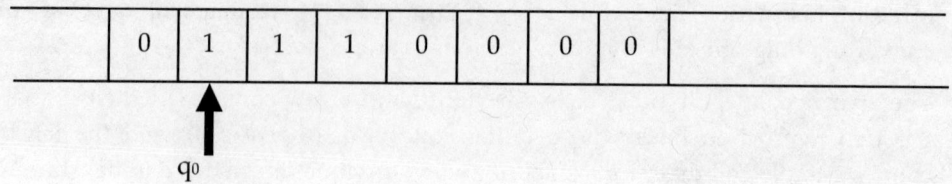

Since the state is q_0 and the input is 1, the right-hand arrow of the first cell (the q_0-cell) indicates the following command: replace 1 with 0 and stay at state q_0.

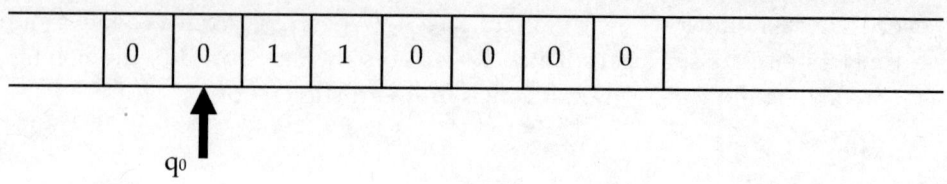

2. Since the state is q_0 and the input is 0, the left-hand arrow of the q_0-cell indicates the following command: move one square to the right and enter state q_1.

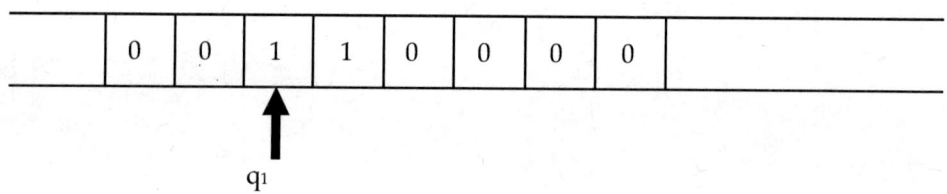

3. Since the state now is q_1 and the input is 1, the relevant arrow is the right-hand arrow of the q_1-cell. It indicates the following command: leave 1 as is but change the state again to q_0.

4. Tz is back to the q_0-cell. Step 1 is repeated. Since the input is 1, the instruction of the right-hand arrow directs the pointer to replace 1 with 0 and it leaves the state unchanged (q_0).

5. Step 2 is repeated. Since the input is 0, the left-hand arrow of the q_0-cell directs the pointer to move one square to the right and it changes the state to q_1. So we have the following situation.

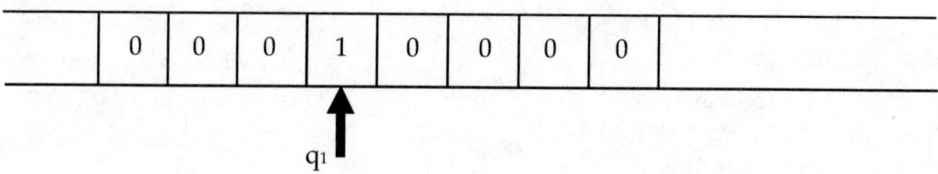

COMPUTABILITY 199

6. T_Z is back to the q_1-cell. The situation is identical to the one described in step 3. So we repeat step 3. The outcome is that the pointer is still positioned at the same square but the state is changed to q_0.

7. Since T_Z is again at the q_0-cell and since the input is 1, we repeat step 1. Thus the pointer replaces 1 with 0 and the internal state is unchanged. Here is how the tape looks now.

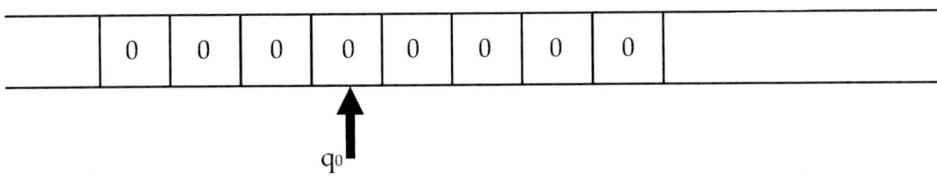

8. The situation now is similar to that of step 2. The instruction of the left-hand arrow of the q_0-cell directs the pointer to move one square to the right and it changes the state to q_1.

9. T_Z is at the q_1-cell. It is a new situation. The state is q_1 and the input is 0. The instruction of the left-hand arrow of the q_1-cell is the applicable instruction now. It directs the pointer to remain in its place and to leave 0 as is and it changes the state to the terminal state of the machine q_e. Once the machine enters the terminal state, it halts. Notice that the q_e-cell has no "existing" arrows. The q_e-cell always looks like this. Here is how the machine looks in its halting position.

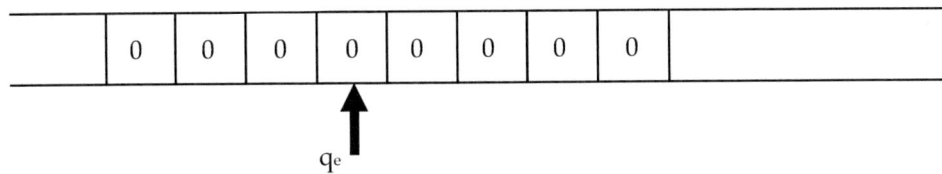

The second example we consider is the Turing machine T_S that computes the **successor function** S. We already encountered this function in 4.2.2. The instruction set of T_S consists of the following four lines.

l_1: $q_0 1 R q_0$
l_2: $q_0 0 1 q_1$
l_3: $q_1 1 L q_1$
l_4: $q_1 0 R q_e$

Here is the diagram of T$_S$.

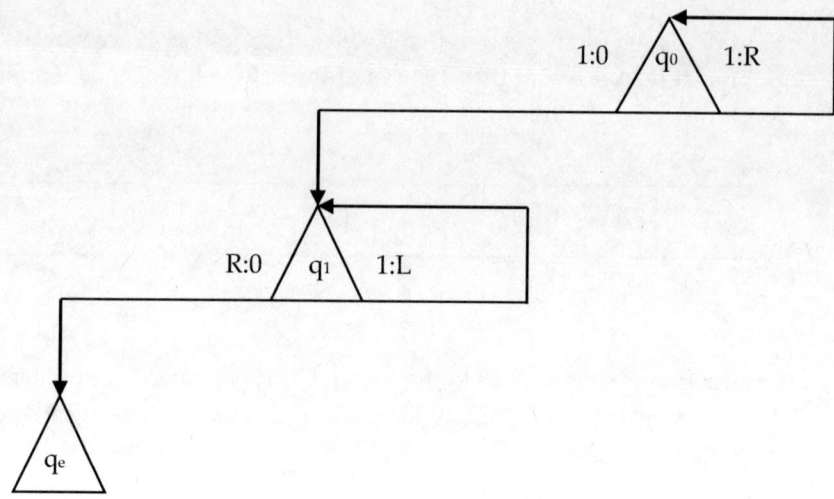

The steps described in 4.2.2 show the behavior of T$_S$ when the input is 2.

The last example we discuss is the Turing machine T$_A$ that computes the **addition function** A. A is a function from $\mathbb{N}^2$ into $\mathbb{N}$ such that A(n, m) = n+m. Hence this machine takes two numbers as input and produces their sum as output. An input for T$_A$ consists of two sequences of 1's that are separated by a square containing 0. The pointer is initially positioned at the leftmost square that contains 1. Here is how the initial position looks when the input is 2 and 3.

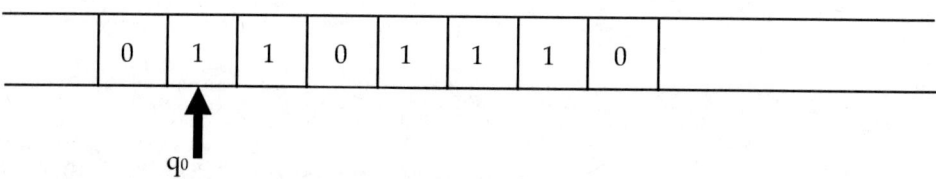

We are going to program this machine to follow this algorithm: (1) it moves to the right until it finds the 0-square between the two input numbers, (2) it replaces 0 with 1 in that square (now we have six 1's), (3) it moves to the left until it reaches the 0-square immediately to the left of the first 1, (4) it moves to the right and replaces the 1 with 0 (now we have the right number of 1's, namely five 1's), (5) it moves one square to the right and halts (now the pointer is positioned at the leftmost square of a sequence of five 1's). We want T$_A$ to follow this algorithm for *any* two-number input (not just 2 and 3). The basic idea is simple: replace the 0 between the two numbers with 1 and erase the first 1. So we add 1 and then we remove 1. The instruction set of T$_A$ consists of the following six lines.

l_1: $q_0 1 R q_0$

l_2: $q_0 0 1 q_1$

l_3: $q_1 1 L q_1$

l_4: $q_1 0 R q_2$

l_5: $q_2 1 0 q_2$

l_6: $q_2 0 R q_e$

Below is the diagram of T_A.

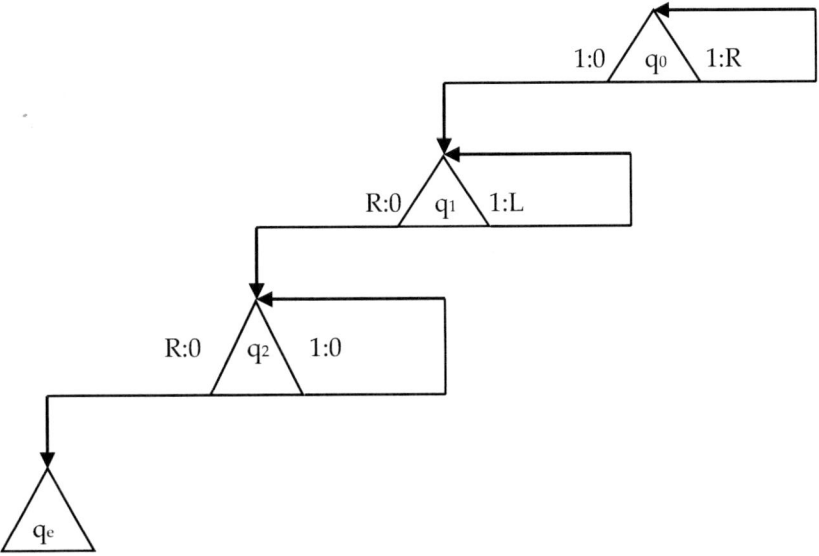

We describe the behavior of this machine when its inputs are 2 and 3. We follow the diagram above in a series of steps to generate the output 5.

1. T_A is at the initial state q_0 with its pointer positioned at the leftmost 1. In the diagram, T_A is at the q_0-cell. The pointer reads 1, so T_A follows the instruction of the right-hand arrow: move one square to the right and remain at state q_0. Since the pointer still reads 1, the same instruction is executed again.

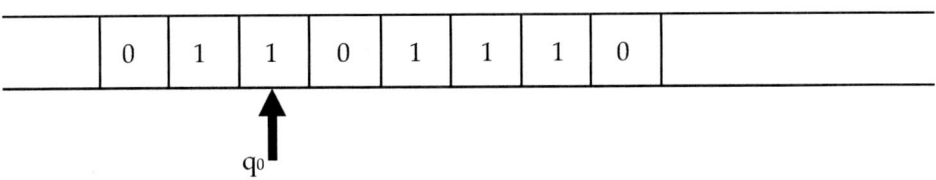

2. T_A is still at the q_0-cell, but this time the pointer reads 0, as shown in the figure below.

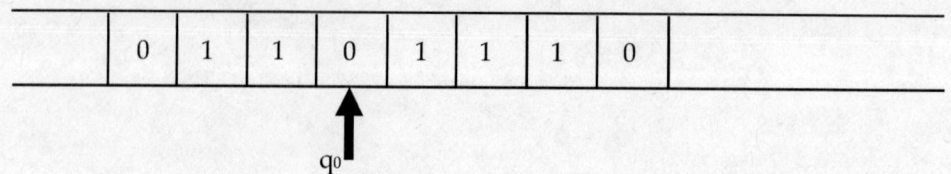

The left-hand arrow of the q_0-cell gives the following instruction: replace 0 with 1 and change the state to q_1. We have the following situation.

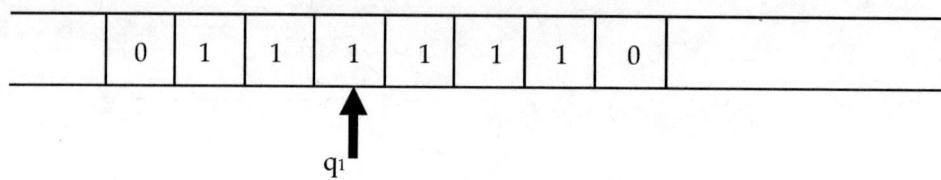

3. T_A is at the q_1-cell and its pointer is reading 1. The right-hand arrow instructs the machine to move the pointer one square to the left and remain at state q_1. This instruction is executed two more times until the pointer reaches the 0-square that is immediately to the left of the first 1. The figure below describes the current situation. (Notice that we've moved the tape to the left so the pointer's current environment can be seen.)

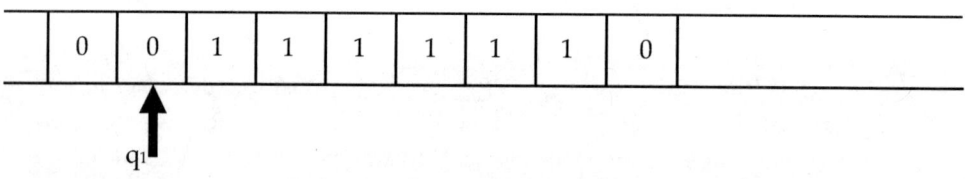

4. T_A is still at the q_1-cell, but the pointer now reads 0. The left-hand arrow instructs the machine to move the pointer one square to the right and to change its state to q_2. We have this figure. (Notice that at this stage we have six 1's.)

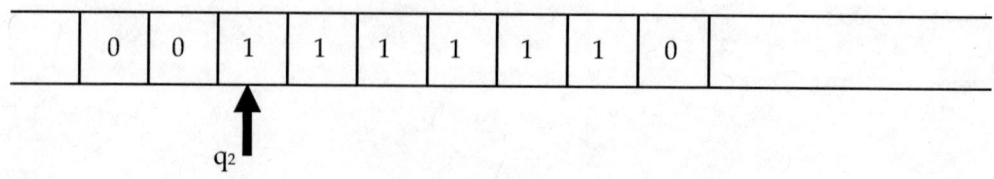

5. T_A is at the q_2-cell. The pointer reads 1. The right-hand arrow indicates the following instruction: replace 1 with 0 and remain at state q_2. The tape looks like this. (Now we have the right number of 1's.)

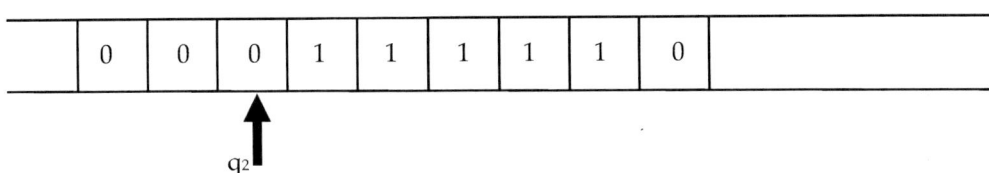

6. T_A is still at the q_2-cell. The pointer reads 0. Thus the left-hand arrow instructs the machine to move its pointer one square to the right and enter the terminal state q_e, which means that the machine halts. The pointer now is positioned at the leftmost 1 of a sequence of five 1's. The figure below describes the halting stage.

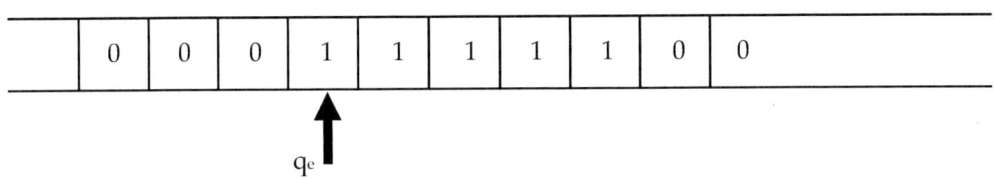

4.2.5 We revisit the notion of a partial numerical function. A **partial numerical function** F is a function from $\mathbb{N}^n$ into $\mathbb{N}$ such that for every n-tuple $\langle m_1, m_2, m_3, \ldots, m_n \rangle$, F assigns *at most* one natural number as a value. If F assigns a value to the n-tuple $\langle m_1, m_2, m_3, \ldots, m_n \rangle$, we say that $F(m_1, m_2, m_3, \ldots, m_n)$ is **defined**, and if F assigns no value to the n-tuple $\langle m_1, m_2, m_3, \ldots, m_n \rangle$, we say that $F(m_1, m_2, m_3, \ldots, m_n)$ is **undefined**. It should be clear that if $F(m_1, m_2, m_3, \ldots, m_n)$ is defined, then $F(m_1, m_2, m_3, \ldots, m_n)$ denotes the value of the function F at the argument $\langle m_1, m_2, m_3, \ldots, m_n \rangle$. A traditional notation for the preceding two cases is $F(m_1, m_2, m_3, \ldots, m_n) = \downarrow$ (F is defined) and $F(m_1, m_2, m_3, \ldots, m_n) = \uparrow$ (F is undefined). An equivalent way of defining a partial numerical function is to say that F is a function from $\mathbb{N}^n$ into $\mathbb{N}$ such that $dom(F) \subseteq \mathbb{N}^n$, where $dom(F)$ (the domain of F) is the set $\{\langle m_1, m_2, m_3, \ldots, m_n \rangle : F(m_1, m_2, m_3, \ldots, m_n) = \downarrow\}$, that is, $dom(F)$ is the set of all the arguments in $\mathbb{N}^n$ at which F has values. As stated previously, a **total numerical function** is a special type of a partial function: it is a partial function whose domain is the set $\mathbb{N}^n$. If $dom(F)$ is a *proper* subset of $\mathbb{N}^n$, F is said to be a **strictly partial function**. Writing n-tuples is cumbersome. In most cases the number of the coordinates of the ordered tuple is irrelevant. In this case, we use "vector" notation. We write $\vec{m}$ to denote the n-tuple $\langle m_1, m_2, m_3, \ldots, m_n \rangle$. An unfortunate aspect of this notation is that it cannot distinguish between an n-tuple that consists of distinct coordinates and an n-tuple that consists of all identical coordinates, that is, between $\langle m_1, m_2, m_3, \ldots, m_n \rangle$ and $\langle m, m, m, \ldots, m \rangle$ (where m is repeated n times). To maintain this distinction, we will denote an n-tuple that consists of the coordinate m repeated n times as $\bar{m}$. We use similar notation and terminology for Turing machines. T is a Turing machine on

$\mathbb{N}^n$ iff T is a Turing machine that takes n numbers $m_1, m_2, m_3, \ldots,$ and m_n as inputs, separated by single 0-squares, and produces at most one number as output. If T generates an output for the inputs $m_1, m_2, m_3, \ldots,$ and m_n, we say that $T(m_1, m_2, m_3, \ldots, m_n)$ is defined and we write $T(m_1, m_2, m_3, \ldots, m_n) = \downarrow$ (T halts), and if T does not generate an output, we say that $T(m_1, m_2, m_3, \ldots, m_n)$ is undefined and we write $T(m_1, m_2, m_3, \ldots, m_n) = \uparrow$ (T does not halt). Here too, if $T(m_1, m_2, m_3, \ldots, m_n)$ halts, $T(m_1, m_2, m_3, \ldots, m_n)$ denotes the output of T when the inputs are $m_1, m_2, m_3, \ldots,$ and m_n. Now we state an important definition.

Turing-Computable Function: A partial function F from $\mathbb{N}^n$ into $\mathbb{N}$ is Turing-computable iff there is a Turing machine T_F that computes F, that is:

$$\text{for every } \vec{m} \in \mathbb{N}^n, T_F(\vec{m}) = \begin{cases} F(\vec{m}) & \text{if } \vec{m} \in \text{dom}(F) \\ \uparrow & \text{if } \vec{m} \notin \text{dom}(F) \end{cases}.$$

Observe that Turing computability is a formal notion that has a precise definition, unlike the informal and intuitive notion of computable function. A computable function is a function that can be computed via a mechanical procedure such that when it is applied to an argument in the domain of the function, it yields, after a finite sequence of deterministic steps, the value of the function at that argument. A mechanical procedure is understood as a procedure that involves no creative steps. All of the notions involved — "mechanical," "deterministic," and "creative" — are informal, yet intuitive, notions. Thus it is a philosophically serious question to ask whether the notion of a Turing-computable function actually captures the intuitive notion of a computable function. Before trying to answer this question, the reader should be warned that many authors use the expressions 'mechanically computable function', 'algorithmically computable function', and 'effectively computable function' to describe what we have been simply calling "computable function." Many authors think that the notion of a computable function does not imply that the function is necessarily computable via an effective procedure. So for them, there might be computable functions that are not effectively computable. We will not make this distinction in this book. We use the term 'computable' to mean mechanically or effectively or algorithmically computable. For us, to say that a function is computable is to say that there is an effective procedure that computes the values of this function. Our usage is in harmony with the terminology usually employed in cognitive science and the philosophy of mind. For example, one reads in the literature of these fields that "computationalism" is the doctrine that mental processes are computational processes, where 'computational' is meant to describe a process that is effectively computational.

We return to our philosophical question regarding the relation between the informal notion of computable function and the formal notion of Turing-computable function. The American mathematician and logician Alonzo Church (1903–95) conjectured a thesis famously know as Church's Thesis. This thesis says that the class of computable numerical functions is identical with the class of Turing-computable functions. Here is the precise statement of this thesis.

Church's Thesis: For every partial function F from $\mathbb{N}^n$ into $\mathbb{N}$, F is computable iff F is Turing-computable.

Since the concept of computability is an informal concept that has no precise definition, this thesis cannot be proved. Any attempt to prove this thesis requires some precise definition of computability. But such a definition, like Turing computability, is just another formal representation of this intuitive concept. To prove that this new definition is equivalent to Turing computability does not prove the thesis itself; it only proves that those two formal representations of computable functions characterize the same class of numerical functions.

To be sure, we have several proofs of this sort. Turing computability is not the only formal notion of computability. In Section 4.4, we will define partial recursive functions. These functions are also proposed as a characterization of the class of computable functions. There are other formal characterizations of computability proposed in the twentieth century. For example, there is Church's λ-calculus. It is a striking result that all of these diverse formal representations of the informal notion of computability are demonstratively equivalent to each other; in other words, all of them characterize precisely the same class of numerical functions. Thus while Church's Thesis itself is unprovable demonstratively, all known formal representations of computable functions are demonstratively proven to be equivalent to each other. This remarkable fact lends support to Church's Thesis. One might argue that the reason behind this surprising mathematical result—namely, that all known formalizations of computability are equivalent—is that all of these formalizations are successful characterizations of the same informal concept of computability. This argument is an inference to the best explanation. It is not a demonstrative argument; it is a probabilistic argument that posits the success of these formalizations in capturing the informal concept of computability as the best explanation for this surprising mathematical fact. Another probabilistic argument for Church's Thesis is that all known computable functions are actually Turing-computable. So there hasn't been a counterexample to this thesis. A third argument for the correctness of Church's conjecture is that Turing machines are idealized computers: there are no limits to the complexity and length of their instruction sets, to their hardware, or to the time required for the computations. So these idealized computing devices have much more resources than actual computers, which have limited memory, hardware, and computing time. Hence it should be surprising if there is actually a computable function that is not Turing-computable. In the remainder of this book, we will accept Church's Thesis as true. Thus we will speak of computable functions without always specifying that they are Turing-computable. In other words, we will treat 'computable' and 'Turing-computable' as synonymous.

4.2.6   Having defined a Turing-computable function, we can now give precise definitions of decidable and semidecidable sets of n-tuples of natural numbers. Suppose that K is a subset of $\mathbb{N}^n$. Define the **characteristic function** of K as follows: χ_K is the characteristic function of K iff χ_K is the total function from $\mathbb{N}^n$ into $\mathbb{N}$ such that:

for every $\vec{m} \in \mathbb{N}^n$, $\chi_K(\vec{m}) = \begin{cases} 1 & \text{if } \vec{m} \in K \\ 0 & \text{if } \vec{m} \notin K \end{cases}$.

We also define the **listing function** of K: λ_K is the listing function of K iff λ_K is the partial function from $\mathbb{N}^n$ into $\mathbb{N}$ such that:

for every $\vec{m} \in \mathbb{N}^n$, $\lambda_K(\vec{m}) = \begin{cases} 1 & \text{if } \vec{m} \in K \\ \uparrow & \text{if } \vec{m} \notin K \end{cases}$.

A set K is **decidable** iff its characteristic function is computable and it is **semidecidable** iff its listing function is computable. If the characteristic function of K is computable, then we have an effective decision procedure for membership in K. We chose any member $\vec{m}$ of $\mathbb{N}^n$, and we compute $\chi_K(\vec{m})$. If $\chi_K(\vec{m})$ returns the value 1, $\vec{m}$ is a member of K; if $\chi_K(\vec{m})$ returns the value 0, $\vec{m}$ is not a member of K. A semidecidable set has a Yes-procedure. If we take any member $\vec{m}$ of $\mathbb{N}^n$ and compute $\lambda_K(\vec{m})$, we will obtain 1 iff $\vec{m}$ is a member of K; but if $\vec{m}$ is not a member of K, the procedure for computing $\lambda_K(\vec{m})$ will produce no answer.

Semidecidable sets are commonly referred to as **effectively enumerable**. The members of an effectively enumerable set A can be effectively listed in a numerical order e_0, e_1, e_2, ..., e_n, ... If A is finite, the list will terminate at some e_k. If A is infinite, the list will never terminate, but every member of A will eventually appear in the list. To be precise, a subset K of $\mathbb{N}^n$ is effectively enumerable iff there is a computable total function F from $\mathbb{N}$ into $\mathbb{N}^n$ such that ran(F) = K. (Recall that ran(F) is the range of F, which is the set of all the values of F.) We will give an informal proof establishing that both definitions are equivalent. However, before we can give this proof, we need to address a technical problem. We said that F is a computable function from $\mathbb{N}$ into $\mathbb{N}^n$. By Church's Thesis, F is Turing-computable, which means that there is a Turing machine T_F that computes F. But Turing machines do not generate n-tuples as outputs, only single natural numbers. So it is not clear how T_F computes F since F's values are n-tuples of natural numbers. Of course, one can complicate the definition of a Turing machine to allow it to generate n-tuples as outputs. But this is unnecessary. All n-tuples of natural numbers can be encoded as single numbers. There are several encoding procedures that can accomplish this task. One of them, for instance, encodes any n-tuple $\langle m_1, m_2, m_3, ..., m_n \rangle$ as the single number $2^{m_1} 3^{m_2} 5^{m_3} ... p_n^{m_n}$, where p_n is the nth prime number. The specific details of the encoding system are irrelevant. All that matters is that there is an effective encoding procedure that encodes every n-tuple of natural numbers into a unique single number such that no two distinct n-tuples have the same code, and that there is an effective decoding procedure that can be applied to any number, and if that number is the code of an n-tuple, it produces the original n-tuple. If $\vec{m}$ is an n-tuple of natural numbers, we let $[\vec{m}]$ denote the single natural number that encodes $\vec{m}$. Now we can say that a (partial) function F from $\mathbb{N}^n$ into $\mathbb{N}^k$ is computable iff there is a Turing machine T_F such that for every $\vec{m}$ in $\mathbb{N}^n$, $T_F(\vec{m}) = [F(\vec{m})]$ iff $\vec{m} \in \text{dom}(F)$ and $T_F(\vec{m})$ does not halt otherwise. In other words, when T_F is given $\vec{m}$ as an input and $\vec{m} \in \text{dom}(F)$, T_F generates as an output the numerical code of the value $F(\vec{m})$; and if $\vec{m} \notin \text{dom}(F)$, $T_F(\vec{m})$ does not halt. Now we are ready to prove the equivalence of the definitions of semidecidable and effectively enumerable sets.

Theorem 4.2.1: For every $K \subseteq \mathbb{N}^n$, K is semidecidable iff K is effectively enumerable.

We first prove the 'if' part. So assume that K is effectively enumerable, that is, there is a computable total function F from $\mathbb{N}$ into $\mathbb{N}^n$ such that ran(F) = K. Our goal is to use F to define a computable listing function of K. Since F is computable, there is a Turing machine T_F that computes F. More precisely, T_F computes the numerical code of F(m) when applied to the input m (m is a single natural number). We describe the following computational procedure. Let $\vec{m} \in \mathbb{N}^n$. We compute $T_F(0)$, $T_F(1)$, $T_F(2)$, $T_F(3)$, and so on until T_F generates $[\vec{m}]$. This establishes that $\vec{m} \in K$. We stop the process and let $\lambda_F(\vec{m}) = 1$. This process, however, might never terminate. If $\vec{m} \notin K$, then $\vec{m} \notin \text{ran}(F)$, which means that for any natural number q, $T_F(q)$ will never generate $[\vec{m}]$ as an output. So this procedure is only a Yes-procedure: if $\vec{m} \in K$, then $F(q) = \vec{m}$ for some natural number q; and hence $T_F(q)$ generates $[\vec{m}]$ as an output; when this happens, we let $\lambda_F(\vec{m}) = 1$; but if $\vec{m} \notin K$, the process of computing $T_F(0)$, $T_F(1)$, $T_F(2)$, $T_F(3)$, and so on will never terminate; in this case $\lambda_F(\vec{m})$ is undefined. Since every member of K will be computed by this process at some point, $\lambda_F(\vec{m}) = 1$ iff $\vec{m} \in K$, and it is undefined otherwise. Therefore, the listing function λ_F of K is computable. This proves that K is a semidecidable set.

Now we prove the 'only-if' part. Suppose that K is semidecidable, that is, its listing function λ_K is computable. Our goal is to define a computable total function F from $\mathbb{N}$ into $\mathbb{N}^n$ such that ran(F) = K. In other words, we want to find a computational procedure such that given any natural number q, this computational procedure assigns a member of K to q, and every member of K is assigned by this procedure to at least one natural number. We will describe a computational procedure that does exactly that, but we need to incorporate the notion of computation time. The first step is to enumerate all the members of $\mathbb{N}^n$. There are many effective procedures for enumerating the members of $\mathbb{N}^n$. For instance, we can use the previously discussed procedure of encoding n-tuples of natural numbers into single numbers, and then arrange these n-tuples by the magnitude of their numerical codes. Let the list $\vec{m}_0, \vec{m}_1, \vec{m}_2, \vec{m}_3, \ldots, \vec{m}_n, \ldots$ be exhaustive of the set $\mathbb{N}^n$. Since λ_K is computable, there is a Turing machine T_K that computes λ_K. Precisely speaking, for every $\vec{m}$ in $\mathbb{N}^n$, $T_K(\vec{m}) = 1$ iff $\vec{m} \in K$, and $T_K(\vec{m})$ does not halt otherwise. We assume that computation time is measured by some finite unit μ and that it is meaningful to speak of "running a Turing machine for some number of μ." We compute the total function F from $\mathbb{N}$ into $\mathbb{N}^n$ as follows. We apply T_K to the input $\vec{m}_0$ and we run T_K for 1 μ. If $T_K(\vec{m}_0)$ halts with an output 1 (if T_K halts, it always halts with an output 1), we let $F(0) = \vec{m}_0$. In this case, the procedure located the first member of K. Whether T_K halts after 1 μ or not, we apply T_K to the input $\vec{m}_1$ and run T_K for 1 μ. $T_K(\vec{m}_1)$ might halt or not. If it halts, we let $F(0) = \vec{m}_1$ if 0 has not been assigned yet, and we let $F(1) = \vec{m}_1$ if 0 has been already assigned to a member of K. Whether $T_K(\vec{m}_1)$ halts or not, we apply T_K to each of $\vec{m}_0$, $\vec{m}_1$, and $\vec{m}_2$ for 2 μ. If T_K halts for any of these inputs, we assign to that input the first natural number that has not been assigned yet. Now we apply T_K to each of $\vec{m}_0, \vec{m}_1, \vec{m}_2$, and $\vec{m}_3$, for 3 μ, and we assign the next available natural numbers to the inputs for which T_K halts. In general, we apply T_K to each of $\vec{m}_0, \vec{m}_1, \vec{m}_2, \vec{m}_3, \ldots$, and $\vec{m}_n$ for n μ. Any input for which T_K halts (with an output 1) is a member of K. Once this computational process locates a member of K, we assign to it the first natural number that has not been already assigned to a member of K.

This computational procedure enumerates effectively all the members of K. However, it enumerates every member of K infinitely many times. To see this, assume that T_K halts when applied to $\vec{m}_5$ after running for 11 μ. Say that 3 is the first available number. So we let $F(3) = \vec{m}_5$. The next time T_K is applied to $\vec{m}_5$, it will run for 12 μ. But if $T_K(\vec{m}_5)$ halts after 11 μ, it will remain at the halting position after running for 12 μ. The process assigns a new natural number to $\vec{m}_5$ and moves to the next n-tuple. After that, every time the process reencounters $\vec{m}_5$, it will assign to it a new natural number. So $\vec{m}_5$ is the value of the function F at infinitely many arguments. But this is all right. We did not require the enumerating function F to be one-to-one; we only required that it be total. The function F computed by the procedure just described is a total function from $\mathbb{N}$ into $\mathbb{N}^n$ such that ran(F) = K. Therefore, K is effectively enumerable.

Due to this theorem, we will use the terms 'semidecidable' and 'effectively enumerable' interchangeably. In section 4.4, we will introduce partial recursive functions. As we said previously, partially recursive functions are precisely those functions that are Turing-computable. We will call sets whose characteristic functions are recursive "recursive sets," and those whose listing functions are recursive "recursively enumerable sets." These are familiar concepts with different formulations: recursive and decidable sets are the same, and recursively enumerable and semidecidable sets are the same. It remains to define decidable relations. As explained in Chapter Two, an n-place relation on $\mathbb{N}$ is a set of n-tuples of natural numbers. In other words, R is an n-place relation on $\mathbb{N}$ iff $R \subseteq \mathbb{N}^n$. Hence there is no difference between a decidable n-place relation on $\mathbb{N}$ and a decidable subset of $\mathbb{N}^n$. R is a **decidable relation** iff it is a decidable set. Similarly, R is a **semidecidable relation** iff it is a semidecidable set.

We end this section by proving a theorem that is known as Kleene's Theorem.

Kleene's Theorem: For every $K \subseteq \mathbb{N}^n$, K is decidable iff K and $\mathbb{N}^n - K$ are effectively enumerable.

Recall from Chapter Two that $\mathbb{N}^n - K$ (the complement of K in $\mathbb{N}^n$) denotes the set of all the members of $\mathbb{N}^n$ that do not belong to K. We prove first the 'only-if' part. Suppose that K is decidable. We want to show that K and $\mathbb{N}^n - K$ are effectively enumerable. It is clear that if K is decidable, then it is semidecidable, since if the characteristic function of K, χ_K, is computable, then its listing function, λ_K, is computable as well. Furthermore, since the characteristic function χ_K of K assigns 0 to all and only the member of $\mathbb{N}^n - K$, the listing function of $\mathbb{N}^n - K$, $\lambda_{\mathbb{N}^n - K}$, can be defined as follows: for every $\vec{m} \in \mathbb{N}^n$, $\lambda_{\mathbb{N}^n - K}(\vec{m}) = 1$ iff $\chi_K(\vec{m}) = 0$. Given that χ_K is computable, $\lambda_{\mathbb{N}^n - K}$ is computable as well. It follows that $\mathbb{N}^n - K$ is semidecidable. We have proved that if K is decidable, then K and $\mathbb{N}^n - K$ are semidecidable. By Theorem 4.2.1, K and $\mathbb{N}^n - K$ are effectively enumerable.

We now prove the 'if' part. Let K and $\mathbb{N}^n - K$ be effectively enumerable. Our goal is to show that K is decidable. By Theorem 4.2.1, K and $\mathbb{N}^n - K$ are semidecidable. Let C be $\mathbb{N}^n - K$. By the definition of semidecidable set, the listing functions λ_K of K and λ_C of C are computable. Thus there are two Turing machines T_K and T_C such that T_K computes λ_K and T_C computes λ_C. In other words, $T_K(\overline{m}) = 1$ iff $\overline{m} \in K$ and $T_C(\overline{m}) = 1$ iff $\overline{m} \in C$. If $\overline{m} \in C$, T_K does not halt, and if $\overline{m} \in K$, T_C does not halt. We use the following computational procedure to compute the characteristic function χ_K of K. For any n-tuple $\overline{m}$ in $\mathbb{N}^n$, we apply T_K and T_C to $\overline{m}$ concurrently. If T_K halts, then $\overline{m}$ belongs to K; we let $\chi_K(\overline{m}) = 1$. On the other hand, if T_C halts, then $\overline{m}$ belongs to C; we let $\chi_K(\overline{m}) = 0$. One of these Turing machines must halt, since $\overline{m}$ either belongs to K or to $\mathbb{N}^n - K$. $\chi_K(\overline{m})$ will be either 1 or 0 depending on which Turing machine halts. This computational procedure shows that the function χ_K is computable. Therefore, K is a decidable set.

4.3 The Halting Problem

The Halting Problem is a mathematical theorem due to Alan Turing (1936). In this subsection we state and prove this theorem. Almost all functions that one encounters in a typical study of arithmetic or algebra are computable. And hence, by Church's Thesis, they are Turing-computable. Therefore one is inclined to think that all functions are computable—at least all functions for which there are precise definitions. It seems reasonable to affirm that if we can define a function precisely, then we should be able to compute it in principle. The Halting Problem establishes that this inclination is mistaken. There are functions that we can define precisely but which are not Turing-computable—by Church's Thesis, they are not computable. The Halting Problem, informally stated, says that there is no effective decision procedure that can decide for any given Turing machine T and any given natural number m whether $T(\overline{m})$ will halt or not, where $\overline{m}$ is the n-tuple whose coordinates are all m. The theorem needs to be stated in terms of computable functions rather than in terms of effective decision procedures, since we have a precise formalization of the notion of computable function. Computable functions are numerical functions. So to speak of a computable function that takes Turing machines as arguments requires that we encode Turing machines into numerical codes. Thus we begin by describing an effective encoding and decoding system for Turing machines. We will associate every Turing machine T with a unique numerical code [T].

Every Turing machine is uniquely characterized by its instruction set. Every line of such a set is of the form q_iXYq_j, where i is any natural number, j is any natural number or 'e' (q_e is the terminal state of every Turing Machine), X is either 0 or 1, and Y is either 0, 1, R, or L. It is easy to associate basic codes with these symbols. Here are the basic codes.

$[q_m] = 1\underbrace{000\ldots 0}_{m \text{ times}}$

$[0] = 2$

$[1] = 3$

$[R] = 4$

$[L] = 5$

$[q_e] = 7$

Let $\{l_1, l_2, l_3, \ldots, l_j\}$ be the instruction set of the Turing machine T. Suppose that l_i is $q_k XYq_s$. We define the numerical code $[l_i]$ to be $[q_k][X][Y][q_s]$. Finally we define the numerical code $[T]$ to be $[l_1][l_2][l_3]\ldots[l_j]$.

The Turing machine T_Z discussed in 4.2.4, which computes the zero function Z, has the four-line instruction set $\{l_1, l_2, l_3, l_4\}$. Its encoding procedure is as follows.

$[l_1] = [q_0 1 0 q_0] = 1321$

$[l_2] = [q_0 0 R q_1] = 12410$

$[l_3] = [q_1 1 1 q_0] = 10331$

$[l_4] = [q_1 0 0 q_e] = 10227$

$[T] = [l_1][l_2][l_3][l_4] = 1321124101033110227$

This is an effective procedure for encoding every Turing machine into a numerical code. The decoding is also an effective procedure. Given any natural number, we can see if its digits represent basic codes and whether the instruction set obtained is a valid Turing machine program. For instance the number above can be decoded as follows;

1 3   2 1 1 2 4 <u>1 0</u> 1 0   3 3 1 <u>1 0</u> 2 2 7

q_0 1 0 q_0 q_0 0 R q_1 q_1 1 1 q_0 q_1 0 0 q_e

Let F be the function from the set **TM** that consists of all the Turing machines into $\mathbb{N}$ such that for every T in **TM**, $F(T) = [T]$. This is clearly a function, since every Turing machine has a unique numerical code. It is also a one-to-one function; for no two distinct Turing machines have the same numerical code. It is, however, not an onto-function. Many natural numbers do not encode any Turing machines. For instance, no number smaller than 12211327 can encode any Turing machine, since 12211327 is the smallest number that encodes a Turing machine. (12211327 is the numerical code of the Turing machine whose instruction set is $q_0 0 0 q_0 q_0 1 0 q_e$.)

The **Halting Function** H is defined as follows: H is a total function from $\mathbb{N}^2$ into $\mathbb{N}$ such that

for all $\langle n, m \rangle$ in $\mathbb{N}^2$, $H(n, m) =$

$$\begin{cases} 1 & \text{if there is a Turing machine T such that } n = [T] \text{ and } T(\overline{m}) \text{ halts} \\ 2 & \text{otherwise} \end{cases}.$$

It is important to understand how this function works. H takes two arguments n and m. There are two cases: either n is the numerical code of a Turing machine or not. If n is not the numerical code of a Turing machine, H(n, m) = 2. On the other hand, if n is the numerical code of a Turing machine T, then we have two further cases: either T($\bar{m}$) halts or not. If it does not halt, H(n, m) = 2; and if does halt, H(n, m) = 1. We wrote T($\bar{m}$) instead of T(m) because T might be a k-place Turing machine, that is, a Turing machine that takes k inputs. In this case we apply T to k inputs each of which is the number m. In other words, T starts with its pointer positioned at the leftmost 1 of m consecutive 1's followed by a 0-square followed by m consecutive 1's followed by a 0-square, and so on, for k segments of m consecutive 1's separated by 0-squares. When T is applied to these k inputs, it might halt or not. If it halts, we assign 1 to H(n, m); and if does not, we assign 2 to H(n, m).

To illustrate the idea, we consider a simple example. H(12211321, 1) = 2, since 12211321 is not the numerical code of any Turing machine—if we decode 12211321, we obtain the string $q_0 00 q_0 q_0 10 q_0$, which is not an instruction set of any Turing machine since the instructions do not terminate with the terminal state q_e. On the other hand, H(12211327, 1) = 1 because 12211327 is the numerical code of the Turing machine whose instruction set is $q_0 00 q_0 q_0 10 q_e$ and because when T is applied to the input 1, it yields the output 0 and halts. However, H(12211327, 2) = 2 because when the same Turing machine is applied to an input consisting of two consecutive 1's, it replaces the first 1 with 0 and enters the terminal state q_e; but its pointer is positioned at the 0-square that is immediately to the left of a 1-square. This is not a valid halting position. So T(2) does not halt, and hence H(12211327, 2) = 2. In fact, given the way we defined Turing machines, T(0) also does not halt. T starts at state q_0 and reads 0; it enters into a loop $q_0 00 q_0$ in which its pointer remains stationary at a 0-square but the internal state of the machine is not q_e. We said that a valid halting position for a Turing machine requires that the internal state is q_e. In this case, H(12211327, 0) = 2.

Now we prove the central theorem of this section.

The Halting Problem: The Halting Function H is not computable.

By Church's Thesis, it is sufficient to prove that H is not Turing-computable. The proof is a reductio ad absurdum argument. So we assume for reductio that H is Turing-computable. Therefore there is a 2-place Turing machine T_H that computes H. Precisely speaking, for all n and m in $\mathbb{N}$,

$$T_H(n, m) = \begin{cases} 1 & \text{if } H(n, m) = 1 \\ 2 & \text{if } H(n, m) = 2 \end{cases}.$$

We construct a new Turing machine T* such that for each n in $\mathbb{N}$,

$$T^*(n, n) = \begin{cases} \uparrow & \text{if } H(n, n) = 1 \\ 2 & \text{if } H(n, n) = 2 \end{cases}.$$

T* is based on T_H. Let $\{l_1, l_2, l_3, ..., l_k\}$ be the instruction set of T_H. Suppose that q_g is the state with the greatest index g among the internal states of T_H. As any Turing machine, the terminal state of T_H is q_e. We do the following modifications to the instruction set $\{l_1, l_2, l_3, ..., l_k\}$. First, we replace q_e with q_{g+1}. Second, we add four more instruction lines:

l_{k+1}: $q_{g+1}00q_{g+1}$ This is a dummy line; it will never be executed, since T halts at an output 1 or 2.

l_{k+2}: $q_{g+1}1Rq_{g+2}$

l_{k+3}: $q_{g+2}0Lq_{g+1}$

l_{k+4}: $q_{g+2}1Lq_e$

At the halting position, the pointer of T_H is positioned at a 1-square surrounded by 0-squares or at a 1-square followed by another 1-square, and both squares are surrounded by 0 squares. In either position the internal state of the machine is the terminal state q_e. The first modification we made is that we replaced q_e with q_{g+1}, where g is the greatest index that appears in the instruction set of T_H. l_{k+1} is a dummy line: it is effective only if T* is at the state q_{g+1} and its pointer is reading 0. Since the pointer of T* initially reads 1 when the state is q_{g+1}, l_{k+1} will never be executed. l_{k+2} is the relevant line: l_{k+2} instructs T* to move its pointer to the right one square and enter state q_{g+2}. Now there are two possibilities. If the output of T_H is 1, the pointer of T* will read 0. In this case l_{k+3} is applicable. It instructs T* to move to the left one square and return to state q_{g+1}. Now the pointer reads 1 again and the state is q_{g+1}. Thus l_{k+2} will be re-executed: T* moves its pointer to the right and enters state q_{g+2}. Since the pointer reads 0, line l_{k+3} is re-executed. It is clear that T* has entered into a loop: it moves one square to the right; it reads 0; it moves one square to the left; it reads 1; it moves one square to the right; and so on. So if the output of T_H is 1, T* does not halt. It follows that if H(n, n) = 1, T*(n, n) does not halt. The second possibility is when the output of T_H is 2. In this case, after T* executes line l_{k+2}, its internal state is q_{g+2} and its pointer reads 1. Hence T* executes line l_{k+4}, which instructs T* to move its pointer one square to the left and halts. T* is in a valid halting position, where its pointer is positioned at the leftmost of 2 consecutive 1's. The output of T* is the same as the output of T_H, namely 2. Therefore, if H(n, n) = 2, T*(n, n) halts with an output 2.

In a certain sense T* reverses the effect of T_H. The latter halts with an output 1, when applied to inputs n and n, where n is the numerical code of a Turing machine T that halts for the input $\bar{n}$. Thus T_H(n, n) halts with an output 1 when T($\bar{n}$) halts. T*(n, n), on the other hand, does not halt when T($\bar{n}$) halts. Rather T*(n, n) halts when and only when n is not the numerical code of any Turing machine or n is the numerical code of a Turing machine T that does not halt for the input $\bar{n}$. If we suppose that n is the numerical code of a Turing machine T, then T*(n, n) halts when T($\bar{n}$) does not halt, and T*(n, n) does not halt when T($\bar{n}$) halts.

The preceding stage of the proof is known as "diagonalization." According to the definition of T*, T* acts on the diagonal of the function H(n, m). Suppose that n is the numerical code of a Turing machine T. Metaphorically speaking, H takes T and applies it to the input $\vec{m}$: if T($\vec{m}$) halts, H returns the value 1, and if T($\vec{m}$) does not halt, H returns the value 2. H(n, n) is the "diagonal" value of H. Again, suppose that n is the numerical code of a Turing machine T. Still metaphorically speaking, H takes T and applies it to its own numerical code: if T halts when applied to its own code, H returns the value 1; if T does not halt when applied to its own code, H returns the value 2. T* applies to these diagonal values: for any Turing machine whose code is n, if T halts when applied to n, T* does not halt; and if T does not halt when applied to n, T* halts. The question now concerns the diagonal value of T* itself. Since T* is a Turing machine, it must have a unique numerical code r, that is, r = [T*]. What is the diagonal value of r? That is, what is the value of H(r, r)? Is it 1 or 2? We will argue that either assumption leads to a contradiction.

Case 1: H(r, r) = 1. By the definition of the Halting Function H, since r is the numerical code of the Turing machine T*, T* must halt when applied to ⟨r, r⟩ (observe that T* is a 2-place Turing machine). Since when T* halts, it always halts with an output 2, T*(r, r) = 2. By the definition of T*, H(r, r) ≠ 1, which is a contradiction.

Case 2: H(r, r) = 2. Again, by the definition of the Halting Function H, since r is the numerical code of T*, T* must fail to halt when applied to the input ⟨r, r⟩. By the definition of T*, H(r, r) ≠ 2, which is a contradiction. Since H(r, r) can only be 1 or 2, the original Reductio Assumption is false; and hence the Halting Function H is not Turing-computable. By Church's Thesis, H is not a computable function. This completes our proof of the Halting Problem.

4.4 Partial Recursive Functions

Partial recursive functions are different formal representations of computable functions. However, as we stated above, the class of the partial recursive functions is identical with the class of the Turing-computable functions. This is a remarkable result, given that these formalizations are conceptually and historically independent of each other. They are both motivated by the same concern—namely, to give a precise formalization of the informal and intuitive concept of computable numerical function. Partial recursive functions are defined in a manner similar to how PL formulas are defined. We list the *basic* recursive functions and certain operations; the partial recursive functions are the basic functions and all functions that can be formed from the basic ones by finitely many applications of these operations. All the partial recursive functions are numerical functions with multiple arguments and a single value, that is, each partial recursive function is from $\mathbb{N}^n$, for some positive integer n, into $\mathbb{N}$, such that the domain of the function is a subset of $\mathbb{N}^n$. As before, if the domain of the function is $\mathbb{N}^n$, the function is total; and if it is a proper subset of $\mathbb{N}^n$, the function is *strictly* partial.

The **basic recursive functions** are infinitely many: the zero function, the successor function, and the projection functions. The operations are three: composition, primitive recursion, and minimization. The **zero function** Z is a total function from $\mathbb{N}$ into $\mathbb{N}$ such that for every $n \in \mathbb{N}$, Z(n) = 0. The **successor function** is a total function from $\mathbb{N}$ into $\mathbb{N}$ such that for every $n \in \mathbb{N}$, S(n) = the successor of n. It is incorrect to define S(n) = n+1. For the successor function is more basic than the addition function. In fact the addition function is a recursive function that can be generated from the basic ones. The successor function is a *primitive notion* that is understood intuitively. We know that $S(0) = 1$, $S(1) = 2$, $S(2) = 3$, and so on. No further definition is offered for the successor function. The **projection functions** are infinitely many. A projection function J_i^n takes as an argument an n-tuple $\langle m_1, m_2, ..., m_n \rangle$ of natural numbers and returns as a value the ith coordinate m_i. Precisely speaking, for all positive integers n and i where $n \geq i$, the projection function J_i^n is a total function from $\mathbb{N}^n$ into $\mathbb{N}$ such that for every n-tuple $\langle m_1, m_2, ..., m_i, ..., m_n \rangle$, $J_i^n(m_1, m_2, ..., m_i, ..., m_n) = m_i$. The simplest of the projection functions is the identity function J_1^1: for every $n \in \mathbb{N}$, $J_1^1(n) = n$. The second simplest projection functions are J_1^2 and J_2^2: for every $\langle m_1, m_2 \rangle \in \mathbb{N}^2$, $J_1^2(m_1, m_2) = m_1$ and $J_2^2(m_1, m_2) = m_2$.

The first recursive operation is **composition**. It is similar to function composition that we studied in Chapter Two but it is not identical with it. The difference has to do with the number of arguments that are to be replaced by other n-place functions. Let us first consider an example of this operation. Say we have two 3-place functions H_1 and H_2 and one 2-place function G, which are defined as follows: $H_1(m_1, m_2, m_3) = 2m_1+3m_2+m_3$, $H_2(m_1, m_2, m_3) = 5m_1 \times m_2 \times m_3$, and $G(k_1, k_2) = (k_1 \times k_2)+4k_1$. We can obtain a new 3-place function F by replacing k_1 in G with H_1 and k_2 with H_2. We express this function as $F(m_1, m_2, m_3) = G(H_1(m_1, m_2, m_3), H_2(m_1, m_2, m_3))$. F has the following formula: $F(m_1, m_2, m_3) = [(2m_1+3m_2+m_3) \times (5m_1 \times m_2 \times m_3)] + 4(2m_1+3m_2+m_3)$. This formula is obtained from $(k_1 \times k_2)+4k_1$ by replacing k_1 with $(2m_1+3m_2+m_3)$, which is H_1, and k_2 with $5m_1 \times m_2 \times m_3$, which is H_2. Observe that H_1, H_2, and F have the same number of places—namely 3. Of course, it is permissible to replace the arguments of G with the same function, say H_1. If G is an m-place function, then there should be at most m distinct n-place functions to replace the arguments of G. The resulting function F is n-place. Here is the general definition of the operation composition. If G is an m-place function, and H_1, H_2, ..., and H_m are at most m distinct n-place functions, then the composition of G and H_1, H_2, ..., and H_m is the n-place function F that is defined as follows:

$F(q_1, q_2, ..., q_n) = G(H_1(q_1, q_2, ..., q_n), H_2(q_1, q_2, ..., q_n), ..., H_m(q_1, q_2, ..., q_n))$.

The second recursive operation is **primitive recursion**. This operation is similar to a style of definition that we termed previously "inductive definition." The central idea is the same. The value of a function F is defined for the argument 0 and then its value is defined for the argument $S(q)$ in terms of its value for q; hence F is defined for every natural number. The logic that underlies this style of definition is the same logic that underlies the Principle of Mathematical Induction. Since $F(0)$ is defined and $F(S(q))$ is defined in terms of $F(q)$, $F(1)$ is defined in terms of $F(0)$; this means that $F(1)$ is defined. The same reasoning applies to $F(2)$, $F(3)$, $F(4)$ and so on. The outcome of this process is that $F(n)$ is defined for every natural number n. Primitive recursion generalizes this style of definition for functions with any number of arguments. For the sake of clarity, we will define the operation of primitive recursion for a non-zero number of arguments and for a zero number of arguments. So let n be any positive integer. Suppose that G is an n-place function and that

H is an n+2-place function. The n+1-place function F is defined by primitive recursion from the functions G and H iff

$F(m_1, m_2, \ldots, m_n, 0) = G(m_1, m_2, \ldots, m_n)$, and
$F(m_1, m_2, \ldots, m_n, S(k)) = H(m_1, m_2, \ldots, m_n, k, F(m_1, m_2, \ldots, m_n, k))$

If n = 0, the function F is defined by primitive recursion from the function H as follows:

$F(0) = p$ (where p is some fixed natural number), and
$F(S(k)) = H(k, F(k))$

Many arithmetical functions are defined by primitive recursion. We are, in fact, familiar with many of them. However, their standard definitions do not fit the exact format of primitive recursion. It is not hard to modify these definitions to make them exhibit the specific form of primitive recursion. Let us discuss two familiar examples: the sum and product of two numbers. The standard inductive definition of addition is this.

For all natural numbers k and d, $d+0 = d$, and $d+S(k) = S(d+k)$.

In familiar notation, the second (inductive) part says that $d+(k+1) = (d+k)+1$ (i.e., d + the successor of k = the successor of (d+k)). In fact, the definition above is a definition by primitive recursion for 1 argument (i.e., n = 1). We can reformulate it to fit exactly the form of primitive recursion. First let Sum(d, k) be the sum of d and k. We want to give a precise primitive recursive definition of it. So Sum is taking the place of the function F in the original formulation of primitive recursion. The identity function J_1^1 will assume the role of the 1-place function G in the definition. Thus the first clause is fulfilled:

$Sum(d, 0) = J_1^1(d) = d$.

Now we want a 3-place function H such that $H(d, k, Sum(d, k)) = S(Sum(d, k))$. The projection function J_3^3 returns the value Sum(d, k) when applied to the arguments ⟨d, k, Sum(d, k)⟩. But we want a function that returns the value S(Sum(d, k)) when applied to the arguments ⟨d, k, Sum(d, k)⟩. This can be achieved by using the operation composition. We take the successor function S(q), which is 1-place, and replace its argument q with the 3-place function $J_3^3(d, k, Sum(d, k))$. This fulfills the second clause:

$Sum(d, S(k)) = S(J_3^3(d, k, Sum(d, k))) = S(Sum(d, k))$.

Having defined Sum(d, k), we can now give a primitive recursive definition of Prod(d, k)—the product of d and k. The standard inductive definition of Prod(d, k) is this.

For all natural numbers d and k, $Prod(d, 0) = 0$, and $Prod(d, S(k)) = Sum(Prod(d, k), d)$.

In familiar notation, this definition says that $d \times 0 = 0$ and $d \times (k+1) = (d \times k)+d$. Prod(d, k) takes the place of the function F in the standard form of primitive recursion. We need a 1-place function to take the place of G. The zero function Z is the natural candidate for G. The first clause is now complete.

Prod(d, 0) = Z(d) = 0.

Our goal now is to find a 3-place function H such that H(d, k, Prod(d, k)) = Sum(Prod(d, k), d). The projection function J_3^3 returns the value Prod(d, k) when applied to the arguments ⟨d, k, Prod(d, k)⟩ and the projection function J_1^3 returns the value d when applied to the arguments ⟨d, k, Prod(d, k)⟩. Each of these functions is 3-place. Using the operation of composition, we can replace the arguments p and q of the function Sum(p, q) with J_3^3(d, k, Prod(d, k)) and J_1^3(d, k, Prod(d, k)), respectively. This gives us the 3-place function we are looking for. The inductive clause now has the standard form:

Prod(d, S(k)) = Sum(J_3^3(d, k, Prod(d, k)), J_1^3(d, k, Prod(d, k))) = Sum(Prod(d, k), d).

Since primitive recursion is a precise formulation of the notion of inductive definition, logicians call such definitions "recursive" rather than "inductive." We will adopt this usage in the remainder of this book.

The last recursive operation is **minimization**. Some functions have "zeros." A **zero** of a function is an argument whose value is zero. For instance, the function F(k, m) = (k+1)(m–3)(m–2) has zeros when m = 3, and m = 2. We have the following cases: (1) F(k, 3) = 0, but F(k, t) is not positive for every natural number t < 3, since F(k, 2) = (k+1)(2–3)(2–2) = 0; (2) F(k, 2) = 0, and F(k, t) is positive for every natural number t < 2, since F(k, 1) = (k+1)(1–3)(1-2) = 2(k+1), and F(k, 0) = (k+1)(0–3)(0–2) = 6(k+1). Hence, 2 is the smallest value of m at which F is 0. In other words, F(k, 2) = 0 and F(k, t) > 0 for every t < 2. In this case, we say that 2 is **the least zero** of F for all k. The operation minimization defines a new function μF on the basis of F: for all natural numbers k, μF(k) = 2. We can generalize this operation for any number of arguments. Let F be an n+1-place function, where n > 0. Suppose that for *some* k_1, k_2, ..., and k_n, there is a natural number m such that F(k_1, k_2, ..., k_n, m) = 0 and that for *every* natural number t < m, F(k_1, k_2, ..., k_n, t) is a positive integer. We define the n-place function μF such that μF(k_1, k_2, ..., k_n) = m. We say that the n-place function μF is defined by **minimization from** the n+1-place function F. Here is a compact formulation of this operation.

μF(k_1, k_2, ..., k_n) =
$$\begin{cases} m & \text{if } F(k_1, k_2, ..., k_n, m) = 0 \text{ and for every } t < m, F(k_1, k_2, ..., k_n, t) \text{ is positive} \\ \uparrow & \text{Otherwise} \end{cases}$$

In other words, m is the least zero of the function F(k_1, k_2, ..., k_n, m), where F(k_1, k_2, ..., k_n, t) is positive for every t < m. For this reason, minimization is sometimes called the "least-zero operation." If we apply minimization to the function Sum, we obtain μSum(0) = 0 and μSum(k) is undefined for every k > 0. The reason is that, in standard notation, Sum(k, m) = k+m = 0 iff k = m = 0. The minimization of the function Prod, on the other hand, yields the function μProd(k) = 0 for every k, since Prod(k, m) = k×m = 0 iff m = 0 or k = 0.

Below is the precise definition of a partial recursive function.

Partial Recursive Function: A function F from $\mathbb{N}^n$ into $\mathbb{N}$ is a partial recursive function iff F is a basic recursive function, that is, F is the zero function, the successor function, or a projection function; or F is obtained from the basic recursive functions by finitely many applications of one or more of the operations composition, primitive recursion, and minimization.

If we restrict the recursive operations to composition and primitive recursion the resulting functions are called **primitive recursive functions**. All primitive recursive functions are total, since all the basic recursive functions are total and since the operations of composition and primitive recursion yield total functions when applied to total functions. It is only minimization that might generate partial functions whether it is applied to total or partial functions. When minimization is included as one of the recursive operations, the resulting functions are called partial recursive functions. The term **recursive function** is often (but not always) used to refer to total recursive functions. In this case, minimization must be modified to ensure that it does not generate partial functions. Minimization may be modified as follows. Let $F(k_1, k_2, ..., k_n, m)$ be an n+1-place function such that for *all* natural numbers $k_1, k_2, ...,$ and k_n, there *exists* at least one m such that $F(k_1, k_2, ..., k_n, m) = 0$. Since $\mathbb{N}$ is well-ordered by $<$, there is always a least m such that $F(k_1, k_2, ..., k_n, m) = 0$. Now we define, for all natural numbers $k_1, k_2, ...,$ and k_n, $\mu F(k_1, k_2, ..., k_n) = m$, where m is the least number such that $F(k_1, k_2, ..., k_n, m) = 0$. This form of minimization yields only total functions. It is the same form of minimization above but it is restricted to total functions that have an m-zero for each n-tuple of arguments. As we indicated previously, the term 'partial recursive function' stands for total as well as strictly partial recursive functions. Partial recursive functions are equivalent to Turing-computable functions. We have the following theorem, which we state without a proof.

Theorem 4.4.1: For every numerical function F, F is a Turing-computable function iff F is a partial recursive function.

Decidable sets and relations and semidecidable sets and relations are called, when recursion theory is concerned, **recursive sets and relations**, and **recursively enumerable sets and relations**, respectively. These concepts have parallel definitions. A recursive set is a set whose characteristic function is recursive (characteristic functions are always total) and, similarly, a recursive relation is a relation whose characteristic function is recursive. A recursively enumerable set or a relation is a set or a relation whose listing function is a partial recursive function. Parallel to Theorem 4.2.1, a recursively enumerable set may also be defined as the range of a recursive function whose domain is $\mathbb{N}$. Both definitions are equivalent.

4.5 Exercises

Note: All answers must be adequately justified.

4.5.1* Let D be a function from ℕ into ℕ such that for every positive integer n, D(n) = 2n (D is the "doubling function").

    4.5.1a Describe a Turing-machine algorithm for computing this function.
    4.5.1b Generate the instruction set for this Turing machine.
    4.5.1c Draw the diagram of this Turing machine.

4.5.2 Define the total function G from ℕ² into ℕ as follows: for all ⟨n, m⟩ ∈ ℕ²,

$$G(n, m) = \begin{cases} n - m & \text{if } n \geq m \\ m - n & \text{if } n < m \end{cases}$$

    4.5.2a Describe a Turing-machine algorithm for computing G.
    4.5.2b Generate the instruction set of this Turing machine.
    4.5.2c Draw the diagram of this Turing machine.

4.5.3 Define the partial function Sub (for "subtraction") from ℕ² into ℕ as follows: for all ⟨n, m⟩ ∈ ℕ²,

$$\text{Sub}(n, m) = \begin{cases} n - m & \text{if } n \geq m \\ \uparrow & \text{if } n < m \end{cases}$$

    4.5.3a Describe a Turing-machine algorithm for computing Sub.
    4.5.3b Generate the instruction set of this Turing machine.
    4.5.3c Draw the diagram of this Turing machine.

4.5.4* Devise a Turing machine that computes the projection function J_2^3. Assume that all the arguments are positive integers.

4.5.5 Prove that if a function is Turing-computable, it is computable by infinitely many Turing machines.

4.5.6 Give a precise recursive definition of each of the following functions.

    4.5.6a* The doubling function 2n in terms of the addition function Sum, basic recursive functions, composition, and primitive recursion.
    4.5.6b The factorial function n!, where n! = n×(n–1)×(n–2)×(n–3)× ... ×2×1, in terms of the multiplication function Prod, basic recursive functions, composition, and primitive recursion.
    4.5.6c The exponentiation function m^n in terms of Prod, basic recursive functions, composition, and primitive recursion.

4.5.7 Let F(n, m) be a total recursive function. Prove that each of the following functions is also recursive.

    4.5.7a G(n, m) = F(m, n).
    4.5.7b G(n) = F(n, n).
    4.5.7c* G(n) = F(2, n).

4.5.8* Let $\chi_=$ be the characteristic function of the relation = on $\mathbb{N}$. Thus for all natural numbers n and m, $\chi_= = 1$ if n = m, and $\chi_= = 0$ if n ≠ m. Invoke Theorem 4.4.1 to demonstrate that $\chi_=$ is a recursive function

4.5.9 Let $\chi_\le$ be the characteristic function of the relation ≤ on $\mathbb{N}$. Thus for all natural numbers n and m, $\chi_\le = 1$ if n ≤ m, and $\chi_\le = 0$ if n > m.

    4.5.9a Invoke Theorem 4.4.1 to show that $\chi_\le$ is a recursive function.
    4.5.9b Let Max(n, m) = m if n ≤ m, and Max(n, m) = n if n > m. Prove that Max is a recursive function. You may invoke $\chi_\le$, Sum, and Prod.

4.5.10 Suppose that there are effective procedures for encoding and decoding partial recursive functions into and from natural numbers. Let [F] be the numerical code of the partial recursive function F. Define the 2-place function H as follows: H(n, m) = 1 if there is a partial recursive function F such that n = [F] and F(m) = ↓, and H(n, m) = 2 otherwise. Demonstrate that H is not a partial recursive function. You may invoke the fact that the 2-place function Sub is partial recursive, where Sub(n, m) = n–m if n ≥ m, and Sub = ↑ if n < m.

Solutions to the Starred Exercises

SOLUTIONS TO 4.5.1

    4.5.1a Let T_D be a Turing machine that computes the doubling function D. Here is an algorithm for T_D when the input is any natural number n.

    1. The pointer of T_D in the initial position is at the leftmost 1-square of a string of n 1's or at a 0-square.

    2. If the pointer reads 0, it does nothing, and halts.

    3. If the pointer reads 1, it moves to the left until it reaches the third 0-square after the leftmost 1-square. It replaces the 0's of the third and second 0-squares with 1's.

    4. Then it moves to the rightmost 1-square of the input, and replaces its 1 with 0. After that, it moves to the left until it reaches the leftmost 1-square of the input. It passes the 0-square that separates the input from the output and continues

moving to the left until it reaches the second 0-square to the left of the leftmost 1-square of the output. It replaces the 0's of these 0-squares with 1's, and repeats this step.

5. When the 1's of the input are all replaced with 0's, the pointer moves left until it reaches the leftmost 1-square of the output, and halts.

The main idea of the algorithm is simple: the pointer erases the rightmost 1 of the input tally and writes two 1's to the left of the input tally; it does this for every 1-square in the input.

4.5.1b Below is the instruction set of T_D.

$q_0 0 0 q_e$ $\quad q_0 1 1 q_1$ $\quad q_1 0 L q_2$ $\quad q_1 1 L q_1$ $\quad q_2 0 L q_3$ $\quad q_2 1 L q_2$ $\quad q_3 0 1 q_3$ $\quad q_3 1 R q_4$

$q_4 0 1 q_4$ $\quad q_4 1 1 q_5$ $\quad q_5 0 R q_6$ $\quad q_5 1 R q_5$ $\quad q_6 0 L q_7$ $\quad q_6 1 R q_6$ $\quad q_7 0 L q_8$ $\quad q_7 1 0 q_7$

$q_8 0 R q_9$ $\quad q_8 1 L q_8$ $\quad q_9 0 L q_{10}$ $\quad q_9 1 1 q_1$ $\quad q_{10} 0 L q_{10}$ $\quad q_{10} 1 1 q_{11}$ $\quad q_{11} 0 R q_e$ $\quad q_{11} 1 L q_{11}$

COMPUTABILITY 221

4.5.1c

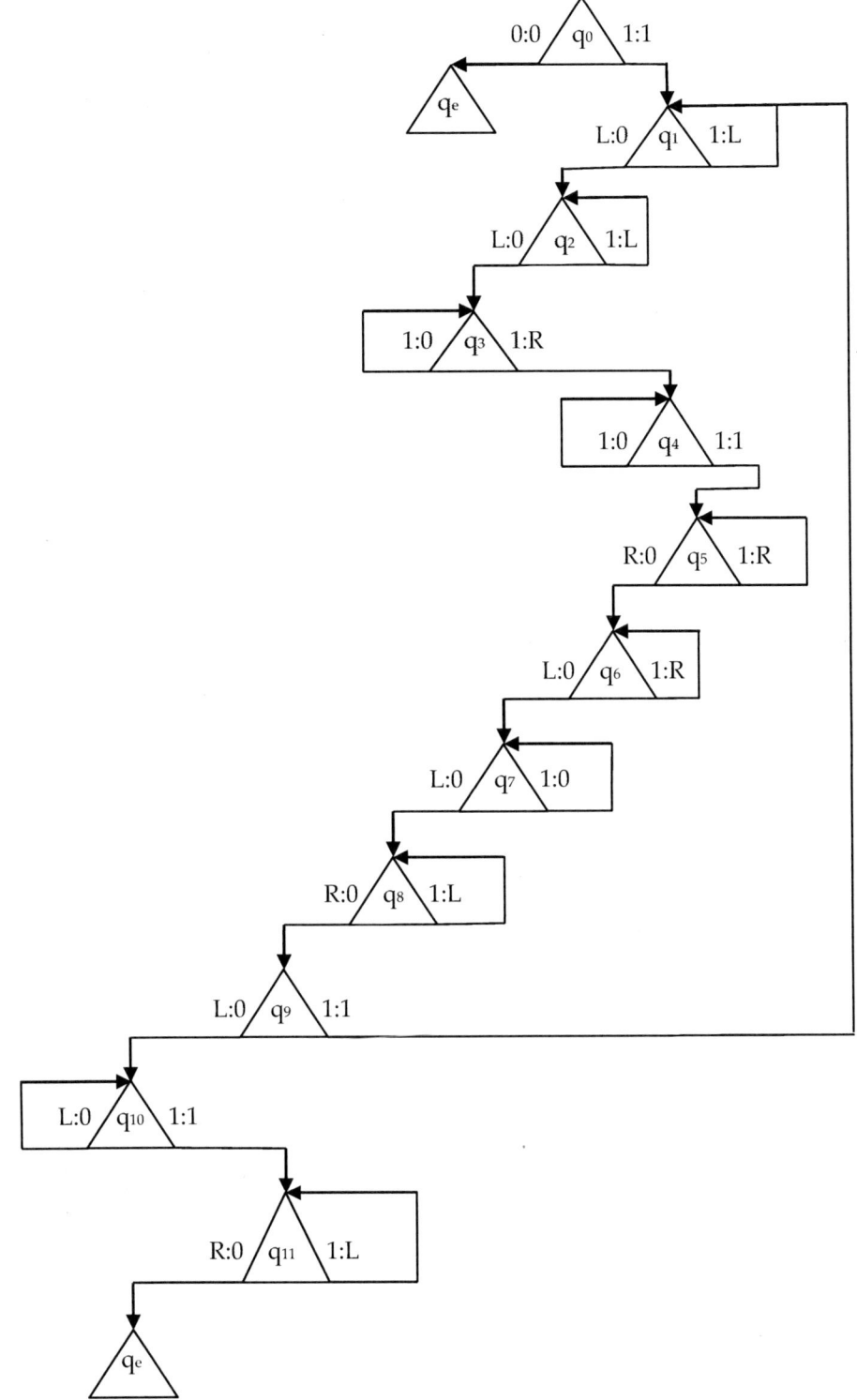

Solution to 4.5.4

Recall that $J_2^3(n, m, k) = m$, for all natural numbers n, m, and k. We are asked to give the instruction set of a Turing machine that computes this function when the arguments are positive integers. Hence the input of this Turing machine T consists of three strings of 1's: n 1's followed by a 0-square followed by m 1's followed by a 0-square followed by k 1's. T produces as an output a single string of m 1's. Here is the instruction set of T.

$q_0 0 R q_1$	$q_0 1 0 q_0$	$q_1 0 R q_2$	$q_1 1 1 q_0$	$q_2 0 R q_3$	$q_2 1 R q_2$	$q_3 0 R q_4$	$q_3 1 0 q_3$
$q_4 0 0 q_5$	$q_4 1 1 q_3$	$q_5 0 L q_5$	$q_5 1 1 q_6$	$q_6 0 R q_e$	$q_6 1 L q_6$		

Solution to 4.5.6

4.5.6a Let D be the doubling function. Below is a precise recursive definition of D.

$D(0) = 0$

$D(S(k)) = H(k, D(k)) = \text{Sum}(G_1(k, D(k)), G_2(k, D(k))) = \text{Sum}(J_2^2(k, D(k)), J_1^2(S(S(Z(k))), D(k)))$

In standard notation: $2 \times (k+1) = (2 \times k) + 2 = (2 \times k) + S(S(0)) = J_2^2(k, (2 \times k)) + J_1^2(S(S(0)), (2 \times k)) = J_2^2(k, (2 \times k)) + J_1^2(S(S(Z(k))), (2 \times k))$

Solution to 4.5.7

4.5.7c   We define G(n) in terms of F(2, n) to show that G(n) is recursive, given that F is a total recursive function.

$G(n) = F(H_1(m, n), H_2(m, n)) = F(J_2^2(m, S(S(Z(n)))), J_2^2(m, n))$

Solution to 4.5.8

We will prove that $\chi_=$ is Turing-computable; by Theorem 4.4.1, $\chi_=$ is recursive. So we need to devise a Turing machine that computes $\chi_=$. Our goal is to give the instruction set of a Turing machine T, such that if the input of T is two equal numbers, T produces 1, and if the input of T is two unequal numbers, T produces 0. Here is the instruction set of T. The lines $q_1 1 1 q_1$ and $q_7 1 1 q_7$ are dummy lines; they will not be executed.

$q_0 0 R q_1$	$q_0 1 1 q_5$	$q_1 0 R q_2$	$q_1 1 1 q_1$	$q_2 0 1 q_e$	$q_2 1 1 q_3$	$q_3 0 R q_4$
$q_3 1 0 q_3$	$q_4 0 0 q_e$	$q_4 1 0 q_3$	$q_5 0 R q_6$	$q_5 1 R q_5$	$q_6 0 L q_7$	$q_6 1 1 q_9$
$q_7 0 L q_8$	$q_7 1 1 q_7$	$q_8 0 0 q_e$	$q_8 1 0 q_7$	$q_9 0 L q_{10}$	$q_9 1 R q_9$	$q_{10} 0 L q_{11}$
$q_{10} 1 0 q_{10}$	$q_{11} 0 L q_{16}$	$q_{11} 1 1 q_{12}$	$q_{12} 0 L q_{13}$	$q_{12} 1 L q_{12}$	$q_{13} 0 R q_{14}$	$q_{13} 1 L q_{13}$
$q_{14} 0 R q_{15}$	$q_{14} 1 0 q_{14}$	$q_{15} 0 0 q_1$	$q_{15} 1 1 q_0$	$q_{16} 0 L q_{16}$	$q_{16} 1 1 q_{13}$	

Chapter Five

The Incompleteness Theorems

5.1 Peano Arithmetic

In this chapter we will prove a number of results, most of which concern the scope and limitation of PL theories. The central result of this chapter is Gödel's First Incompleteness Theorem. This theorem establishes that a certain important PL theory is incomplete. As defined in Chapter Three, a PL set Σ is a theory iff Σ contains all its logical consequences that are composed of Voc(Σ) (that is, the vocabulary of Σ). A PL set Σ is complete iff for every PL sentence that is composed of Voc(Σ), either it or its negation is a theorem of Σ. By the Soundness and Completeness Theorems, a PL theory may also be defined proof-theoretically as a PL set that contains all its theorems that are composed of its vocabulary; and a complete PL set may also be defined semantically as a PL set such that for every sentence **X** that is composed of the vocabulary of the set, either **X** or ¬**X** is a logical consequence of the set. For the most part, in this chapter, we will move freely between the semantical concepts and their proof-theoretic counterparts. We will make the distinction clear only where there is a risk of misunderstanding.

In this section we will develop the PL theory for which the First Incompleteness Theorem is proved. This theory is an axiomatizable number theory, whose set of axioms is infinite. We call this theory **Peano Arithmetic**,[1] and we denote its set of axioms as "PA." In other words, Peano Arithmetic is Th(PA), where, as defined in Chapter Three, Th(PA) = {**X**: **X** is a PL sentence composed of Voc(PA) that is a logical consequence of PA} = {**X**: **X** is a PL sentence composed of Voc(PA) that is a theorem of PA}. We begin by describing the vocabulary of PA. We will refer to Voc(PA) as the **standard arithmetical vocabulary**, and we will denote it as "AV." A PL sentence or formula that is composed of AV is an **AV sentence** or an **AV formula**, a set of PL sentences that are composed of AV is an **AV set**, and a PL theory whose vocabulary is AV is an **AV theory**. AV consists of the logical vocabulary of PL (i.e., the two connectives and quantifier, the identity symbol, the variables, and the parentheses), and the following extra-logical vocabulary: one name **0**, one 1-place

---

[1] This theory is named after the Italian mathematician Giuseppe Peano (1858–1932). He described its set of axioms. Peano was not the first to discover these axioms. The German mathematician Richard Dedekind (1831–1916) described a close version of the same axioms.

function symbol s, and two 2-place function symbols $+$ (a boldfaced '+') and $\bullet$.[2] The intended interpretation of PA is the structure of the natural numbers N. It is described below. We refer to this structure as the **standard interpretation of arithmetic**.

The PL Interpretation N

UD: $\mathbb{N}$: $\{0, 1, 2, 3, \ldots\}$
LN: **0**, $c_1, c_2, c_3, \ldots$

Semantical assignments

$N(\mathbf{0})$: 0; $N(c_1)$: 1; $N(c_2)$: 2; $N(c_3)$: 3; …; $N(c_n)$: n; …
$N(sx)$: "the successor of x": $S(x)$
$N(+xy)$: "the sum of x and y": $x + y$
$N(\bullet xy)$: "the product of x and y": $x \times y$

To keep with standard notation, we will write '$(x + y)$' instead of '$+xy$' and '$(x \bullet y)$' instead of '$\bullet xy$'. To simplify our expressions, we will write Sx instead of $S(x)$.

Now we list the PA axioms.

The AV Set PA

Ax1 $(\forall x)\mathbf{0} \neq sx$
Ax2 $(\forall x)(\forall y)(sx = sy \rightarrow x = y)$
Ax3 $(\forall x)(x + \mathbf{0}) = x$
Ax4 $(\forall x)(\forall y)(x + sy) = s(x + y)$
Ax5 $(\forall x)(x \bullet \mathbf{0}) = \mathbf{0}$
Ax6 $(\forall x)(\forall y)(x \bullet sy) = ((x \bullet y) + x)$

IS If **X**[**z**] is an AV formula that contains exactly one free variable **z**, and it does not contain any occurrences of the variables v and y, then the following is an axiom:

X[**0**]$\rightarrow$(($\forall v$)(**X**[v]$\rightarrow$**X**[sv])$\rightarrow$($\forall y$)**X**[y])

where **X**[**0**], **X**[v], and **X**[y] are obtained from **X**[**z**] by replacing all the occurrences of **z** with **0**, v, and y, respectively.

It is important to note that when we say that a PL formula **X** contains a free variable **z**, we mean that **z** has *only* free occurrences in **X**, that is, **X** contains occurrences of **z** but no **z**-quantifiers. We will follow this usage throughout this chapter. If a formula contains exactly one free variable, we describe it as a 1-variable formula; if it contains exactly two free variables, we say that it is a 2-variable formula; and similarly for any number of variables. 'IS' stands for "Induction Schema." It is not a single axiom. It is an "axiom schema," that is, it is a schema that generates infinitely many axioms. IS is the PL representation of the Principle of Mathematical Induction. Observe that PA is a decidable

---

[2] Precisely speaking, the extra-logical vocabulary of AV does not belong to the PL standard extra-logical vocabulary introduced in Chapter One. To keep with standard notation, we will not modify the symbols of AV; rather we will assume that the PL extra-logical vocabulary has been expanded to include the symbols of AV.

set: there is an effective decision procedure for determining membership in this set. The list Ax1–Ax6 presents no problem since it is a finite list of axioms, and the axiom schema IS is decidable, since the set of AV formulas **X** is decidable and the set of all sentences that have the form "**X**[0]→((∀v)(**X**[v]→**X**[sv])→(∀y)**X**[y])" is also decidable.

We will presuppose a substantive claim, which is not standardly presupposed when the Incompleteness Theorems are presented. We will presuppose that the standard interpretation of arithmetic, N, is a model of PA; in other words, every sentence of PA is true on N. This presupposition is sufficiently intuitive, but it is a substantive mathematical assumption. To see the intuitive appeal of this presupposition, we will give the natural readings of the PA axioms on N.

The natural reading of $(\forall x) 0 \neq sx$ (Ax1) on N:
Zero is not the successor of any natural number.
Ax1 is true on N.

The natural reading of $(\forall x)(\forall y)(sx = sy \rightarrow x = y)$ (Ax2) on N:
Any two natural numbers that have the same successor are identical.
Ax2 is true on N.

The natural reading of $(\forall x)(x + \mathbf{0}) = x$ (Ax3) on N:
For any natural number n, (n + 0) = n.
Ax3 is true on N.

The natural reading of $(\forall x)(\forall y)(x + sy) = s(x + y)$ (Ax4) on N:
For all natural numbers n and m, (n + Sm) = S(n + m).
If Sm is understood to mean (m + 1), this statement becomes (n + (m +1)) = ((n + m) + 1). Of course, we are not supposed to replace Sm with (m + 1), since addition is to be defined in terms of S. S is a primitive notion: it is intuitively understood but not explicitly defined. The interpretations of Ax3 and Ax4 represent a recursive definition of addition.
Ax4 is true on N.

The natural reading of $(\forall x)(x \bullet \mathbf{0}) = \mathbf{0}$ (Ax5) on N:
For every natural number n, (n × 0) = 0.
Ax5 is true on N.

The natural reading of $(\forall x)(\forall y)(x \bullet sy) = ((x \bullet y) + x)$ (Ax6) on N:
For all natural numbers n and m, (n × Sm) = ((n × m) + n). Again if we allow ourselves to understand Sm as (m + 1), this statement becomes (n × (m + 1)) = ((n × m) + (n × 1)) = ((n × m) + n), which is an instance of the distributive property of multiplication. As explained above, we are not supposed to understand Sm as (m + 1). The interpretations of Ax5 and Ax6 offer a recursive definition of multiplication.
Ax6 is true on N.

The Induction Schema is the reason why the presupposition that N is a model of PA is substantive. Since **X**[z] can be any AV formula with one free variable, we do not know the interpretation of every instance of IS in advance. We can decide whether a specific instance of IS is true on N once we know what **X**[z] is. However, the situation is not as bad as these remarks might make it appear. The logic of IS is that of the Principle of Mathematical Induction, and we accepted that PMI is a valid principle of inference. IS does not ask us to accept that $(\forall y)$**X**[y] is true on N based on the interpretation of $(\forall y)$**X**[y]. The instances of IS are conditionals. So the truth of $(\forall y)$**X**[y] is supposed to follow from the truth of two sentences: **X**[0] and $(\forall v)($**X**[v]$\to$**X**[sv]$)$. If these sentences are true on N, then $(\forall y)$**X**[y] is true on N. It is not necessary to know the specific interpretation of the formula **X**[z]. It is understood that once **X**[z] is interpreted by N, the resulting interpretation is a metalinguistic one-variable formula that attributes some property to natural numbers. The truth of **X**[0] and $(\forall v)($**X**[v]$\to$**X**[sv]$)$ on N entails that this metalinguistic formula is true of 0 and is true of Sk whenever it is true of k. By PMI, it follows that it is true of every natural number n. Thus the truth of the instances of IS on N is justified by the validity of PMI for the structure of the natural numbers N.

The above reasoning shows that the presupposition that N is a model of PA, although mathematically substantive, is intuitively justifiable. Accepting this presupposition means that we have assumed that N is a model of Peano Arithmetic, Th(PA). The members of Th(PA) are all logical consequences of PA. Hence if the PA axioms are true on N, then, by the definition of logical consequence, all their logical consequences are true on N. It follows that N is a model of Th(PA). This establishes that Th(PA) is satisfiable. By the Soundness Theorem, Th(PA) is consistent, that is, no contradiction is derivable from Th(PA).

The question now is whether Th(PA) is complete. Since every sentence in Th(PA) is a theorem of PA, if a PL sentence **X** is a theorem of Th(PA), then it is a theorem of PA. This follows from the lemma below.

Lemma 5.1.1: If a PL set $\Gamma \vdash$ **X** for every **X** in Σ, and $\Sigma \vdash$ **Z**, then $\Gamma \vdash$ **Z**.

One can prove this lemma from the Soundness and Completeness Theorems: if $\Gamma \vdash$ **X** for every **X** in Σ and $\Sigma \vdash$ **Z**, then by the Soundness Theorem, $\Gamma \vDash$ **X** for every **X** in Σ and $\Sigma \vDash$ **Z**; it is obvious that if M is any model of Γ that is relevant to **Z**, then it is a model of Σ as well, and hence **Z** is true on M; so **Z** is true on every model of Γ that is relevant to it; this shows that $\Gamma \vDash$ **Z**;[3] but now by the Completeness Theorem, $\Gamma \vdash$ **Z**. However, invoking the Soundness and Completeness Theorems to prove this simple lemma is like using a machine gun to hunt a hare. This lemma is a consequence of the definition of PL derivation. If $\Sigma \vdash$ **Z**,

---

3 The preceding reasoning presupposes that M is also relevant to the members of Σ. If M is not relevant to some members of Σ, it can always be expanded into another interpretation M* that is identical with M except that M* also interprets the additional PL vocabulary in Σ. Since M and M* agree on their interpretations of the vocabulary of Γ and **Z**, they assign identical truth values to **Z** and to the members of Γ. Thus since M is a model of Γ, M* is also a model of Γ; and since **Z** is true on M*, it is also true on M.

then there is a finite $\Sigma^* \subseteq \Sigma$ such that $\Sigma^* \vdash Z$. Since $\Gamma \vdash X$ for every X in Σ, every member of Σ^* is a theorem of Γ. This implies that for every X in Σ^*, there is a derivation D_X of X from Γ. Given that Σ^* is finite, there are only finitely many of these derivations. There is also a derivation D_Z of Z from Σ^*. All of these derivations can be combined into one derivation of Z from Γ. The idea is simple: for every X in Σ^*, we list its derivation D_X, which has X as its conclusion. This allows us to introduce all the members of Σ^* in one main derivation. Once we have obtained all the members of Σ^*, we list the derivation D_Z, whose conclusion is Z. This combined derivation is a derivation of Z from Γ. Therefore, $\Gamma \vdash Z$.

By Lemma 5.1.1, the question whether Th(PA) is complete is equivalent to the question whether PA is complete. Is it the case that for every sentence X that is composed of the vocabulary AV, PA $\vdash X$ or PA $\vdash \neg X$? This is a very important question. Peano Arithmetic is a paradigm of an axiomatic mathematical system. Mathematicians work with Peano axioms, trying to prove from those axioms true arithmetical propositions, that is, AV sentences that are true on the standard interpretation of arithmetic, N. One would expect that such an activity is in principle always fruitful. If a sentence is true on N, then it should be possible to prove it from the axioms, and if it is false on N, then it should be possible to disprove it from the axioms, that is, to prove its negation from the axioms. Part of this expectation is that if one discovers an arithmetical proposition that can neither be proved nor disproved from the axioms, then this only shows that Peano axioms are not sufficiently strong: there should be a set of arithmetical axioms that can in principle settle every question about the truth or falsity of arithmetical propositions.

Alas! The Austrian mathematician and logician Kurt Gödel (1906–78) proved in 1931 that this natural expectation could not be met: Peano Arithmetic is incomplete. There is an AV sentence G_{PA} such that neither PA $\vdash G_{PA}$ nor PA $\vdash \neg G_{PA}$. This is the First Incompleteness Theorem. Furthermore, we will argue that G_{PA} is true on N. This shows that Th(PA) is a *proper subset* of Th$_{AV}$(N). Recall that Th$_{AV}$(N) = {X: X is a PL sentence composed of AV and X is true on N}. A consequence of Corollary 3.5.1e is that Th$_{AV}$(N) is a consistent complete theory whose vocabulary is AV. G_{PA} is constructed on the basis of a certain "diagonal" procedure. This procedure also establishes that every consistent extension of Th(PA), including Th(PA) itself, is not a decidable set. In other words, for any consistent theory Σ, if Th(PA) is a subset of Σ, then there is no effective decision procedure for determining whether any given sentence belongs to Σ or not. Since Th$_{AV}$(N) is a consistent extension of Th(PA), Th$_{AV}$(N) is also not a decidable set. Theorem 3.5.1 of Chapter Three affirms that every complete axiomatizable PL theory is decidable. Since Th$_{AV}$(N) is complete and undecidable, it cannot be axiomatizable. So it is impossible to meet that expectation no matter how the axioms are strengthened. We will prove this fact again later in this chapter. It is standard to define **Arithmetic** as the complete theory Th$_{AV}$(N). Therefore, the preceding result can be restated by saying that Arithmetic is not axiomatizable. One frequently reads that the First Incompleteness Theorem shows that Arithmetic is not axiomatizable. In this chapter we will not give a detailed proof of the First Incompleteness Theorem, but we will give a detailed outline of the proof, describing its main components. In the remainder of this chapter, we will presuppose the fact that for every natural number n, the successor of n is n+1, that is, Sn = n+1. This will simplify our expressions.

5.2 Representability in Peano Arithmetic

In this section, we present the first major component of the proof of the First Incompleteness Theorem. We will state a theorem that asserts that all (total) recursive functions are representable in Th(PA). Of course, we will have to explain first what representability in Th(PA) amounts to. We begin our development of this component by defining a sequence of AV singular terms (i.e., singular terms that are composed of AV—the standard arithmetical vocabulary). These singular terms are called AV numerals. An **AV numeral** is obtained from the name **0** and the function symbol s by iterating s any finite number of times to the left of **0**. We introduce the following defined notation. Recall that defined notations are metalinguistic symbols that abbreviate strings of symbols of the object language. We define $s^0\mathbf{0}$ as an abbreviation for $\mathbf{0}$, $s^1\mathbf{0}$ as an abbreviation for $s\mathbf{0}$, $s^2\mathbf{0}$ as an abbreviation for $ss\mathbf{0}$; and in general we define $s^n\mathbf{0}$ as an abbreviation for $\underbrace{sss \ldots}_{n \text{ times}}\mathbf{0}$. We can give a precise recursive definition of this notation: $s^0\mathbf{0} = \mathbf{0}$ and $s^{n+1}\mathbf{0} = ss^n\mathbf{0}$. There is a close relation between the natural numbers and the AV numerals. First, we prove a simple theorem.

Theorem 5.2.1: For all natural numbers n and m, n = m iff PA $\vdash s^n\mathbf{0} = s^m\mathbf{0}$, and n ≠ m iff PA $\vdash s^n\mathbf{0} \neq s^m\mathbf{0}$.

Proof

We will only prove the first part, that is, n = m iff PA $\vdash s^n\mathbf{0} = s^m\mathbf{0}$. The proof is a PMI argument on n. Assume that n = 0. If n = m, m = 0. PA $\vdash \mathbf{0} = \mathbf{0}$. By definition of $s^0\mathbf{0}$, PA $\vdash s^0\mathbf{0} = s^0\mathbf{0}$. On the other hand, if PA $\vdash s^0\mathbf{0} = s^m\mathbf{0}$, m must be 0; otherwise we would have PA $\vdash s^0\mathbf{0} = ss^{m-1}\mathbf{0}$, which implies that PA $\vdash \mathbf{0} = ss^{m-1}\mathbf{0}$, and this contradicts Ax1: $(\forall x)\mathbf{0} \neq sx$. We proved so far that if n = 0, then n = m iff PA $\vdash s^n\mathbf{0} = s^m\mathbf{0}$. So the Base Step is established. The Induction Hypothesis is that for every m, k = m iff PA $\vdash s^k\mathbf{0} = s^m\mathbf{0}$. The goal is to prove that for every m, k+1 = m iff PA $\vdash s^{k+1}\mathbf{0} = s^m\mathbf{0}$. So let k+1 = m. Thus k = m–1. By the Induction Hypothesis, PA $\vdash s^k\mathbf{0} = s^{m-1}\mathbf{0}$. It follows that PA $\vdash ss^k\mathbf{0} = ss^{m-1}\mathbf{0}$, which implies, by the definition of AV numerals, that PA $\vdash s^{k+1}\mathbf{0} = s^m\mathbf{0}$. Now suppose that PA $\vdash s^{k+1}\mathbf{0} = s^m\mathbf{0}$. We have that PA $\vdash ss^k\mathbf{0} = ss^{m-1}\mathbf{0}$ (observe that m ≠ 0; otherwise we would have PA $\vdash ss^k\mathbf{0} = s^0\mathbf{0}$, which contradicts Ax1). Ax2, which is $(\forall x)(\forall y)(sx = sy \rightarrow x = y)$, entails that PA $\vdash s^k\mathbf{0} = s^{m-1}\mathbf{0}$. By the Induction Hypothesis, k = m–1, which gives us k+1 = m. This establishes the Inductive Step. By PMI, for all natural numbers n and m, n = m iff PA $\vdash s^n\mathbf{0} = s^m\mathbf{0}$.

We also prove another close parallel between the natural numbers and the AV numerals.

Theorem 5.2.2: For every AV numeral $s^n\mathbf{0}$, its referent on N is the natural number n, that is, $N(s^n\mathbf{0})$ = n, for every natural number n.

Proof

The proof is by PMI on n. Let n = 0. The AV numeral is $s^0 0$, which by definition is **0**. By the construction of N, N(**0**) = 0. This establishes the Base Step. The Induction Hypothesis is that $N(s^k 0) = k$. Our goal is to prove that $N(s^{k+1} 0) = k+1$. By definition, $N(s^{k+1} 0) = N(s s^k 0)$. By clause 1.2.1c of the definition of PL interpretation (Chapter One), $N(s s^k 0) = N(s)N(s^k 0)$. By the Induction Hypothesis and the construction of N, we have $N(s)N(s^k 0) = Sk = k+1$ (recall that S is the successor function of N and N(s) = S). This establishes the Inductive Step. Hence by PMI, $N(s^n 0) = n$, for every natural number n.

The preceding two theorems show that the set of AV numerals mirrors the set ℕ: the AV numerals are the formal copies of the natural numbers. Because of this close relation between the VA numerals and the natural numbers, we will introduce a new defined notation: for each natural number n, define the boldfaced **n** to be the AV numeral $s^n 0$. Hence **1** is $s^1 0$, **2** is $s^2 0$, **3** is $s^3 0$, and so on.

Most of the properties and relations of the natural numbers have counterparts in Th(PA). Below is a small list of these properties and relations. Later in this section, we will state a general proposition about the representability of all recursive functions, sets, and relations in Th(PA). To simplify matters, we will employ logical connectives and quantifiers that belong to the full version of PL. However, we will think of them as defined notations. So our PL will remain the economical version, but we now have metalinguistic abbreviations of some of its strings. We explained before how to express all the binary connectives and the existential quantifier in terms of ¬, →, and ∀. We repeat these definitions here for the convenience of the reader. **X**∧**Y** abbreviates ¬(**X**→¬**Y**), **X**∨**Y** abbreviates ¬**X**→**Y**, **X**↔**Y** abbreviates ¬((**X**→**Y**)→¬(**Y**→**X**)), and (∃z)**W** abbreviates ¬(∀z)¬**W**. Also we will avail ourselves of the rules of inference that pertain to these connectives and quantifier. We begin by defining order predicates as abbreviations of AV formulas.

5.2a   For all AV terms **t** and **s**, **t** < **s** abbreviates (∃z)(z ≠ 0 ∧ (**t** + z) = **s**). (Recall that a term is a variable, a singular term, or a functional term.)

5.2b   For all AV terms **t** and **s**, **t** ≤ **s** abbreviates **t** < **s** ∨ **t** = **s**.

5.2c   For all natural numbers n and m, if n < m, then PA ⊢ **n** < **m**.

5.2d   For all natural numbers n and m, if n ≤ m, then PA ⊢ **n** ≤ **m**.

5.2e   For every natural number n, if n is even, then PA ⊢ (∃x)(x ≤ **n** ∧ **n** = (x + x)).

5.2f   For every natural number n, if n is odd, then PA ⊢ (∃x)(x ≤ **n** ∧ **n** = ((x + x) + 1)).

5.2g   For all natural numbers n, m, and k, if (n + m) = k, then PA ⊢ (**n** + **m**) = **k**.

5.2h   For all natural numbers n, m, and k, if (n × m) = k, then PA ⊢ (**n** • **m**) = **k**.

Now we explain what is meant by the statement that all recursive sets, relations, and functions are representable in Th(PA). We define first the concept of **representability in Th(PA)**.

Definition 5.2.1

5.2.1a For every set of natural numbers B, B is representable in Th(PA) iff there is an AV formula **X**[z] with one free variable such that for each natural number k, if k ∈ B, then PA ⊢ **X**[k], and if k ∉ B, then PA ⊢ ¬**X**[k].

5.2.1b For every n-place relation R on ℕ (i.e., R ⊆ ℕⁿ), R is representable in Th(PA) iff there is an AV formula **X**[z_1, z_2, z_3, ..., z_n] with n free variables such that for each n-tuple ⟨k_1, k_2, k_3, ..., k_n⟩ of natural numbers, if ⟨k_1, k_2, k_3, ..., k_n⟩ ∈ R, then PA ⊢ **X**[k_1, k_2, k_3, ..., k_n], and if ⟨k_1, k_2, k_3, ..., k_n⟩ ∉ R, then PA ⊢ ¬**X**[k_1, k_2, k_3, ..., k_n].

5.2.1c For every total n-place function F from ℕⁿ into ℕ, F is representable in Th(PA) iff there is an AV formula **X**[z_1, z_2, z_3, ..., z_n, z_{n+1}] with n+1 free variables such that for each n-tuple ⟨k_1, k_2, k_3, ..., k_n⟩ of natural numbers and for each natural number k, if F(k_1, k_2, k_3, ..., k_n) = k, then PA ⊢ **X**[k_1, k_2, k_3, ..., k_n, k], and PA ⊢ ($\forall x$)(**X**[k_1, k_2, k_3, ..., k_n, x] → x = k).

The formulas described in the definitions above are said to **represent** in Th(PA) the set B, the relation R, or the function F. These definitions indicate a relationship between membership in certain sets and provability in Th(PA). The first definition states that if a number belongs to B, then it is a theorem of PA that the formula **X** holds for its AV numeral, and if it does not belong to B, then it is a theorem of PA that **X** does not hold for its AV numeral. Hence membership in B corresponds to the provability in Th(PA) of a certain set of sentences. The same explanation applies to relations and functions. After all, according to set theory, relations and functions are sets of n-tuples of objects. An n-tuple of numbers belongs to a relation R only if it is provable in Th(PA) that the formula **X** holds for the AV numerals of these numbers, and it does not belong to R only if it is provable in PA that the formula **X** does not hold for the AV numerals of these numbers. Functions require an additional clause. We note that a total n-place function is an n+1 relation that assigns a unique value for every n-tuple of numbers. So if F(k_1, k_2, k_3, ..., k_n) = k, it must be provable in Th(PA) that the formula **X** holds for the n+1-tuple ⟨k_1, k_2, k_3, ..., k_n, k⟩ and it must also be provable that **k** is unique. The second clause asserts the uniqueness condition. It is a theorem of PA that if **X** holds for the n+1-tuple ⟨k_1, k_2, k_3, ..., k_n, x⟩, then x is identical with **k**.

The representability of functions, like of sets and relations, also has a negative clause—namely that if F(k_1, k_2, k_3, ..., k_n) ≠ k, then PA ⊢ ¬**X**[k_1, k_2, k_3, ..., k_n, k]. This clause, however, is provable. It need not be included in the definition. We prove this clause below.

Theorem 5.2.3: Suppose that the formula **X**[z_1, z_2, z_3, ..., z_n, z_{n+1}] represents in Th(PA) the total n-place function F. For each n-tuple ⟨k_1, k_2, k_3, ..., k_n⟩ of natural numbers and for each natural number k, if F(k_1, k_2, k_3, ..., k_n) ≠ k, then PA ⊢ ¬**X**[k_1, k_2, k_3, ..., k_n, k].

Proof

1. Assume that the formula $X[z_1, z_2, z_3, \ldots, z_n, z_{n+1}]$ represents in Th(PA) the n-place function F. Assume further that $F(k_1, k_2, k_3, \ldots, k_n) \neq k$.
2. Since F is a total function, there is a natural number h such that $F(k_1, k_2, k_3, \ldots, k_n) = h$.
3. From 1 and 2: $k \neq h$.
4. From 3 by Theorem 5.2.1: PA $\vdash \mathbf{k} \neq \mathbf{h}$.
5. From 1 by Definition 5.2.1c: PA $\vdash (\forall x)(X[\mathbf{k}_1, \mathbf{k}_2, \mathbf{k}_3, \ldots, \mathbf{k}_n, x] \to x = \mathbf{h})$.
6. From 5 by Universal Instantiation: PA $\vdash X[\mathbf{k}_1, \mathbf{k}_2, \mathbf{k}_3, \ldots, \mathbf{k}_n, \mathbf{k}] \to \mathbf{k} = \mathbf{h}$.
7. From 4 and 6 by Modus Tollens: PA $\vdash \neg X[\mathbf{k}_1, \mathbf{k}_2, \mathbf{k}_3, \ldots, \mathbf{k}_n, \mathbf{k}]$.

We will state later that all (total) recursive functions are representable in Th(PA). This is sufficient to establish that all recursive sets and relations are representable in Th(PA). The following theorem makes this connection between the representability of functions and the representability of sets and relations clear. (See Subsection 4.2.6 for the definition of characteristic function.)

Theorem 5.2.4: Let D be any subset of $\mathbb{N}^n$, where $n \geq 1$ (we take $\mathbb{N}^1$ to be $\mathbb{N}$). D is representable in Th(PA) iff its characteristic function is representable in Th(PA).

The next theorem is the central theorem of this component of the Incompleteness Proof.

The Representability Theorem: Every (total) recursive function is representable in Peano Arithmetic.

An immediate corollary of the preceding two theorems is that all recursive sets and relations are representable in Th(PA). Let D be a recursive subset of $\mathbb{N}^n$. By definition, D is recursive iff its characteristic function χ_D is recursive. By the Representability Theorem, χ_D is representable in Th(PA). Hence by Theorem 5.2.4, D is representable in Th(PA). Since we will need to invoke this fact often, we state this corollary below.

Corollary 5.2.1: All recursive sets and relations are representable in Peano Arithmetic.

5.3 Arithmetization of the Metatheory

In this section we explain the second major component of the Incompleteness proof. A large portion of the metatheory of Peano Arithmetic, Th(PA), can be "mirrored" in the structure of the natural numbers N. The vocabulary AV is encoded into numerical codes, and on the basis of these codes numerical analogues are constructed that in some sense mirror certain sets and relations of the metatheory of Th(PA). Many of these numerical analogues are recursive. Hence by the Representability Theorem, they are representable in Th(PA).

Metaphorically speaking, through this 2-step procedure, Th(PA) can be made to "speak" about its own metatheory, including its own syntax and proof theory. It is standard to use single quotation marks to create metalinguistic names of the expressions of the object language. But it has been our convention in this book to drop these single quotation marks and rely on the context to distinguish between using an expression and mentioning it. The process of encoding the metatheory in N is based on encoding the syntax of Th(PA) into numerical codes. There are several numerical coding systems that can serve the purpose of arithmetizing the metatheory of Peano Arithmetic equally well. The numerical codes that these systems generate are traditionally referred to as **gödel numbers**, even if these systems differ from each other and even if they differ from the system that Gödel originally devised. We will present one of these systems, and following the tradition we will refer to the numerical codes generated by this system as "gödel numbers." In fact we will have two types of numbers: basic numbers and gödel numbers. If θ is an AV expression, we will denote its basic or gödel number as [θ].

We begin by assigning numerical codes to the basic symbols of AV. These numerical codes are called the **basic numbers**. The basic numbers are the odd natural numbers.

Extra-Logical Symbol	**0**	s	+	•
Its Basic Number	1	3	5	7

Logical Symbol	(	)	¬	→	∀	=
Its Basic Number	9	11	13	15	17	19

Logical Symbol (Variable)	u	v	w	x	y	z	u_1	v_1	w_1	x_1	y_1	z_1	...
Its Basic Number	21	23	25	27	29	31	33	35	37	39	41	43	...

The task now is to create sets of natural numbers that mirror sets of AV grammatical expressions. First, we define the arithmetical analogue of the set of all the AV terms. We call this arithmetical set TERM. The members of TERM are the basic numbers of **0** and of the PL variables, and all the numbers that are generated from these numbers by applying one or more of the following operations finitely many times: if n and m belong to TERM, then so do $2^3 \times 3^n$ (the code for st), $2^9 \times 3^n \times 5^5 \times 7^m \times 11^{11}$ (the code for (t + s)), and $2^9 \times 3^n \times 5^7 \times 7^m \times 11^{11}$ (the code for (t • s)). This set contains the numerical codes of the AV terms. Each AV term is associated with a unique number in TERM and every number in TERM is associated with a unique AV term. The unique number in TERM that is associated with the AV term **t** is the gödel number of **t**. As we stated above, we denote this number as [t]. The encoding and decoding procedures are effective. Here are a few examples of encoding and decoding of AV terms.

1. $[ssx] = 2^3 \times 3^{[sx]} = 2^3 \times 3^{2^3 \times 3^{27}}$

    $[ssx]$ can be decoded by reversing the encoding procedure. So we prime-factor $[ssx]$. We obtain $2^3 \times 3^k$. We prime factor k; we get $2^3 \times 3^{27}$. All the gödel numbers are even numbers and all the basic numbers are odd numbers. So long as we have even powers of prime numbers, we continue the process of prime-factorization. Once we reach odd numbers, we stop and decode these basic numbers. Thus once we obtain the powers 3, 3, and 27, we stop the process of prime-factorization and decode these basic numbers in the order of their prime-factorization. We obtain the string ssx, which is an AV term.

2. $[sss0] = 2^3 \times 3^{[ss0]} = 2^3 \times 3^{2^3 \times 3^{[s0]}} = 2^3 \times 3^{2^3 \times 3^{2^3 \times 3^1}}$.

    As above, the decoding of $[sss0]$ reverses the order of the encoding procedure. We prime-factor $[sss0]$; we get $2^3 \times 3^k$. k is even; so we prime-factor k; we obtain $2^3 \times 3^h$. Again, since h is even, we prime-factor it; we get $2^3 \times 3^1$. The powers are odd numbers. So we stop the process of prime-factorization, and decode the basic numbers in the order of their prime-factorization. The resulting expression is sss0.

3. $[(s0 + sv)] = 2^9 \times 3^{[s0]} \times 5^5 \times 7^{[sv]} \times 11^{11} = 2^9 \times 3^{2^3 \times 3^1} \times 5^5 \times 7^{2^3 \times 3^{23}} \times 11^{11}$.

4. $[(sw \bullet (0 + sv))] = 2^9 \times 3^{[sw]} \times 5^7 \times 7^{[(0+sv)]} \times 11^{11} =$
    $2^9 \times 3^{2^3 \times 3^{25}} \times 5^7 \times 7^{2^9 \times 3^1 \times 5^5 \times 7^{[sv]} \times 11^{11}} \times 11^{11} =$
    $2^9 \times 3^{2^3 \times 3^{25}} \times 5^7 \times 7^{2^9 \times 3^1 \times 5^5 \times 7^{2^3 \times 3^{23}} \times 11^{11}} \times 11^{11}$.

Of course, if we begin with an arbitrary natural number and follow the decoding procedure described above, we will most likely end up with a prime-factorization that is not the gödel number of any AV term. However, determining whether a number is a gödel number of an AV term or not is an effective decision procedure. So it is clear that the characteristic function of TERM is recursive. Hence TERM is a recursive set, which means that it is representable in Th(PA). By Definition 5.2.1a, there is an AV formula **term**$[x]$ with one free variable such that for every natural number k, if k ∈ TERM, then PA ⊢ **term**$[k]$, and if k ∉ TERM, then PA ⊢ ¬**term**$[k]$. We can interpret the fact that PA ⊢ **term**$[k]$ by saying that Th(PA) "declares" that k is the gödel number of an AV term, or that Th(PA) "declares" that the AV expression whose gödel number is k is a term of its language. So metaphorically speaking, we made Th(PA) "speak" about its own syntax.

The next stage in the arithmetization of the syntax of Th(PA) is to arithmetize the category of AV atomic formulas. Every AV atomic formula is of the form **s = t**, where **s** and **t** are any AV terms. The arithmetical analogue of this category is the set AFORM that consists of all the gödel numbers of the AV atomic formulas. AFORM is easily defined. Every member of AFORM is a number of the form $2^n \times 3^{19} \times 5^m$ (the code for **t = s**) where n and m ∈ TERM. It is clear that AFORM is a recursive set. Hence there is an AV formula **aform**$[x]$ with one free variable such that **aform** represents the set AFORM in Th(PA). The AV formulas are constructed from the atomic ones by finite applications of one or more of the negation, conditional, and universal quantifier formation rules. The arithmetical analogue of the category of AV formulas is the set FORM of all the gödel numbers of the AV formulas. In order to construct the set FORM, we need to introduce the notion of prime-powers of a number. For any positive natural number m, we define inductively the set of the prime-powers of n, PP(n), to be the *smallest* set that consists of all the positive powers of the prime numbers that constitute the prime-factorization of n, and of all the prime-powers of the members of PP(n). For example, PP(145152) = {8, 4, 1 3, 2}, since $145152 = 2^8 \times 3^4 \times 7^1$, $8 = 2^3$, and $4 = 2^2$. We construct the set FORM as follows. FORM consists of all the numbers in AFORM and all the numbers that can be generated from those numbers by applying one or more of the following operations finitely many times: if n and m are members of FORM, then so are $2^{13} \times 3^n$ (the code for $\neg X$), $2^9 \times 3^n \times 5^{15} \times 7^m \times 11^{11}$ (the code for $(X \to Y)$), and $2^9 \times 3^{17} \times 5^k \times 7^{11} \times 11^n$ (the code for $(\forall z)Y$) provided that (1) k is an odd number greater than 20, that (2) k belongs to PP(n), and that (3) no member of PP(n) has the prime-factorization $2^9 \times 3^{17} \times 5^k \times 7^{11}$ (the code for $(\forall z)$, where k is [z]). The last three conditions are needed to ensure that the variable z occurs in the formula Y and that no z-quantifier occurs in Y. FORM is also a recursive set. Hence it is representable in Th(PA) by an AV formula **form**[x] with one free variable.

To complete the arithmetization of the syntax of Th(PA), we need to construct the set SENT, which consists of all the gödel numbers of the AV sentences. An AV sentence is an AV formula that contains no free variables. So we only need to isolate a proper subset of FORM whose members are the gödel numbers of the AV formulas that contain no free variables. Consider an AV formula X and a PL variable z. An occurrence of z in X is bound iff it is an occurrence anywhere in a subformula $(\forall z)Y$ of X. This suggests the following construction of SENT. Let k be the basic number of the PL variable z and g be the gödel number of the AV formula X. If g or any of its prime-powers has a prime-factorization of the form $2^9 \times 3^{17} \times 5^k \times 7^{11} \times 11^m$ (the code for $(\forall z)Y$), we say that k represents a bound occurrence of z in X. It is possible for k to represent bound and free occurrences of z in X. Let k_1, k_2, k_3, ..., and k_n be *all* the odd numbers greater than 20 that are prime-powers of g. In other words, these numbers are the basic numbers of *all* the variables that occur in X. We take each k_j to be the basic number of z_j. If every k_j represents *only* bound occurrences of z_j in X, then g is the gödel number of an AV sentence. Thus for every natural number g, g $\in$ SENT iff g $\in$ FORM, and either (1) no prime-power of g is an odd number greater than 20 (i.e., the formula contains no variables) or (2) each odd number k greater than 20 that is a prime-power of g represents only bound occurrences of a variable in the formula whose gödel number is g. SENT is clearly a recursive set. Therefore there is an AV formula **sent**[x] with one free variable that represents SENT in Th(PA).

We have represented the syntax of Th(PA) within Th(PA) itself. The procedure is a two-stage process. We first arithmetize the vocabulary and syntactical categories of PA by encoding them into natural numbers and recursive sets of natural numbers. Second, the natural numbers are represented in Th(PA) as AV numerals, and the recursive sets of numbers are represented in Th(PA) by certain AV formulas. Via this two-stage procedure,

Th(PA) is made to "declare" certain facts about its own syntax. We illustrated the idea previously, but will discuss one more example here. Consider the AV sentence $(\forall x)0 \neq sx$, which is the first axiom of PA. The gödel number of this sentence is $[(\forall x)0 \neq sx]$. Say, m = $[(\forall x)0 \neq sx]$. m ∈ SENT. The AV numeral that represents m is **m** (which is $s^m 0$). Since the formula **sent**$[x]$ represents the set SENT in Th(PA) and since m ∈ SENT, PA ⊢ **sent**[**m**]. We can be more direct and assert that since $(\forall x)0 \neq sx$ is a sentence of the language of Th(PA), PA ⊢ **sent**[**m**], where m = $[(\forall x)0 \neq sx]$. This informally says that Th(PA) "declares" that $(\forall x)0 \neq sx$ is a sentence of its language.

The arithmetization of the metatheory of Th(PA) is not yet complete. We also need to arithmetize PA proofs, that is, PL derivations of AV sentences from PA. PA proofs are sequences of AV sentences such that the last sentence in the sequence is the conclusion of the proof and every sentence in the sequence is either a PA axiom or is licensed by one of the inference rules of MDS. We will not explain the details of the arithmetization of PA proofs; but we will outline the procedure. Sequences of AV sentences can also have gödel numbers. If $X_1 X_2 X_3 \ldots X_n$ is a sequence of n AV sentences, then we can define the gödel number of this sequence, $[X_1 X_2 X_3 \ldots X_n]$, to be the number $2^{[X_1]} \times 3^{[X_2]} \times 5^{[X_3]} \times \ldots \times p_n^{[X_n]}$, where p_n is the nth prime number. It might be thought that this encoding procedure confuses gödel numbers of AV terms and formulas with gödel numbers of sequences of AV sentences, since both of them utilize the sequence of prime numbers. It is not so. No two distinct items have the same gödel number. The powers of the prime numbers that constitute the prime-factorization of a gödel number of a sequence of AV sentences are the numerical codes of *complete* sentences without any *connectives* between them. This is not the case with the gödel numbers of AV sentences. The powers of the prime numbers that constitute the prime-factorization of the gödel number of an AV term or formula do *not* encode complete sentences that are not connected to other sentences. The encoding procedures described above ensure that every grammatical AV expression and every sequence of AV sentences has a unique numerical code and that no different items have the same numerical code, whether it is a basic or a gödel number. Also these encoding procedures and their accompanying decoding procedures are all effective procedures.

PA proofs are a special kind of sequence of AV sentences. Every member of a sequence that is a PA proof is either a PA axiom or is licensed by one of the MDS rules. Hence to encode PA proofs, we need to encode the PA axioms and we need to describe certain arithmetical operations that correspond to the applications of the MDS rules. The details of this encoding procedure are tedious and hardly illuminating. The examples of encoding procedures we gave above are sufficient to illustrate the arithmetization of the metatheory of Th(PA). So we simply assume that the task of encoding PA proofs is carried out successfully. Let PROOF be the binary relation on ℕ such that for every ordered pair ⟨m, k⟩ in ℕ², ⟨m, k⟩ ∈ PROOF iff k is the gödel number of an AV sentence **X** and m is the gödel number of a PA derivation of **X**. PROOF is a recursive relation. We present an argument for this claim that relies on Church's Thesis. Given any finite sequence D of AV sentences and any AV sentence **X**, there is an effective decision procedure for determining whether D is a

PL derivation of **X** from PA or not. We observe that the set of PA axioms is decidable. Decidability is a condition we previously placed on every set of axioms, and PA is no exception. So there is an effective decision procedure determining whether any AV sentence is a PA axiom or not. Since D is finite, we examine every sentence in D. Let D be the sequence $X_1 X_2 X_3 \ldots X_n$. We first examine the terminal sentence X_n to see if it is the desired conclusion **X** or not. If it is not, we determine that D is not a derivation of **X**. If it is, we examine the first sentence X_1 to see if it is a PA axiom or is introduced by one of the rules of MDS. As we said, the set PA is decidable, but so is the set MDS. There are finitely many rules in MDS. Of course, every inference rule has infinitely many applications. However, the applicability of each rule to any sentence is an effective decision procedure. For instance, we know that if X_1 is not a PA axiom, then it is either introduced by the rule Identity or by one of the hypothetical rules. If it is introduced by one of the hypothetical rules, then we know what to look for in D. For example, if it is introduced as the assumption of Conditional Proof, then we expect to see later a sentence **Y** followed by a sentence of the form $X_1 \rightarrow Y$; if it is introduced by Reductio Ad Absurdum, we expect to see later two sentences of the forms **Y** and $\neg Y$ followed by a sentence of the form $\neg X_1$ or of the form **Z** if X_1 is $\neg Z$. If we determine that X_1 is neither a PA axiom nor the result of applying one of the MDS rules, then we determine that D is not a PA proof. On the other hand, if X_1 passes the test, we examine X_2 and determine whether it is a PA axiom or is introduced by one of the MDS rules. Again, if it is, we proceed to X_3; if it is not, we determine that D is not a PA proof. Since D is finite, we only have to apply this procedure finitely many times. If every sentence in D passes the test, that is, it is either a PA axiom or is licensed by one of the MDS rules, and if X_n is **X**, then we have determined that D is a PL derivation of **X** from PA. In fact, there are in existence computer programs that can check the correctness of many PL derivations.

This effective decision procedure can be implemented as part of a decision procedure for determining whether any ordered pair $\langle m, k \rangle$ of natural numbers is a member of PROOF or not. We first apply the decoding procedure to see if m is the gödel number of a sequence of AV sentences and if k is the gödel number of an AV sentence. As explained previously, this decoding procedure is an effective procedure. If m and k are indeed the gödel numbers of a sequence of AV sentences and an AV sentence, respectively, then this decoding procedure will produce this sequence and this sentence. If they are not, our decoding procedure can determine that they are not appropriate numerical codes. Once the decoding procedure determines that either one of these numbers is not an appropriate numerical code, we place $\langle m, k \rangle$ outside PROOF. On the other hand, if the decoding procedure produces a sequence D of AV sentences and an AV sentence **X**, we apply our previous procedure to determine whether D is a PA proof of **X**. If it is, we place $\langle m, k \rangle$ inside PROOF; if it is not, we place $\langle m, k \rangle$ outside PROOF. So now we have an effective decision procedure for determining whether the relation PROOF holds between any given two numbers or not. This shows that PROOF is a decidable relation. By Church's Thesis, the characteristic function of PROOF is Turing-computable, which implies, by Theorem 4.4.1, that it is recursive. Hence PROOF is a recursive set. Of course, it is possible to construct PROOF directly as an arithmetical relation, just as we did for TERM, AFORM, and FORM. In this case, one can show directly that PROOF is recursive without invoking Church's Thesis.

Since PROOF is recursive, it is representable in Th(PA). Thus there is an AV formula **proof**[x, y] with two free variables such that for all natural numbers m and k, if ⟨m, k⟩ ∈ PROOF, then PA ⊢ **proof**[**m**, **k**]; and if ⟨m, k⟩ ∉ PROOF, then PA ⊢ ¬**proof**[**m**, **k**]. It follows that D is a PA proof of **X** iff PA ⊢ **proof**[**m**, **k**], where m = [D] and k = [**X**]. The 'only-if' part is obvious: if D is a PA proof of **X**, then ⟨[D], [**X**]⟩ ∈ PROOF; and hence PA ⊢ **proof**[**m**, **k**], where m = [D] and k = [**X**]. For the 'if' part, suppose that PA ⊢ **proof**[**m**, **k**]. Since PA is consistent (we assumed that the natural number structure N is a model of PA), PA ⊬ ¬**proof**[**m**, **k**]. Given that if ⟨m, k⟩ ∉ PROOF, then PA ⊢ ¬**proof**[**m**, **k**], it follows that ⟨m, k⟩ ∈ PROOF. Since m = [D] and k = [**X**], we conclude that D is a PA derivation of **X**.

On the basis of the binary relation PROOF, we can define the set PROV that consists of all the gödel numbers of the AV sentences that are *provable* from PA. Since Th(PA) = {**X**: **X** is an AV sentence such that PA ⊢ **X**}, PROV is really the set of all the gödel numbers of the members of Th(PA). PROV is the arithmetical analogue of Th(PA). PROV can easily be defined from PROOF: for any natural number k, k ∈ PROV iff there exists a natural number m such that ⟨m, k⟩ ∈ PROOF. This definition suggests that PROV is not a recursive set but only recursively enumerable. We will give a rigorous proof later that PROV is not decidable. However, for now we will explain why the definition we gave of PROV does not constitute an effective decision procedure for membership in PROV. The problem lies with the *unbounded* existential quantifier "there exists a natural number m." Consider the following effective procedure. Suppose we are given a number k, and we would like to determine whether k is a member of PROV or not. This task amounts to finding another natural number m, if there is one, such that the ordered pair ⟨m, k⟩ ∈ PROOF. If we can find such a number, then we know that k is the gödel number of a theorem of PA; and hence k ∈ PROV. Of course, there might be no such number. In this case, k is not the gödel number of any theorem of PA; and so k ∉ PROV. We have to search for such a number, if there is one. We start with number 0. PROOF is recursive, which means there is an effective decision procedure for determining membership in PROOF. Thus we run this procedure to see if the ordered pair ⟨0, k⟩ ∈ PROOF. If it does, then we have determined that k ∈ PROV. If it does not, then we examine the ordered pair ⟨1, k⟩: we run the procedure for PROOF, and see whether this ordered pair belongs to PROOF or not. Again, if it does, then we have determined that k ∈ PROV; and if it doesn't, we run the procedure for PROOF for the next ordered pair ⟨2, k⟩. We repeat this process for successive ordered pairs. If there is a number m such that ⟨m, k⟩ ∈ PROOF, this procedure will locate it eventually; and thus we know that k ∈ PROV. What would happen if there is no such number? The procedure would run continually without ever locating an ordered pair ⟨m, k⟩ that is a member of PROOF. In this case, we would never be able to know that k is not a member of PROV.

Observe that if the existential quantifier in the definition of PROV has an upper bound, this effective procedure would be a decision procedure. To see this, assume that the upper bound is some (large) positive number b. In this case PROV would be defined as follows: for every natural number k, k ∈ PROV iff there is a natural number m ≤ b such that ⟨m, k⟩ ∈ PROOF. Thus we would have to run the effective decision procedure for PROOF for only finitely many ordered pairs: at most we would have to run it for ⟨0, k⟩, ⟨1, k⟩, ⟨2, k⟩, ..., and ⟨b, k⟩. If any of these ordered pairs belongs to PROOF, then we would know that k ∈ PROV; and if none of them belongs to PROOF, we would know that k ∉ PROV. It is clear that this is an effective decision procedure. In this case, PROV would be decidable, that is, recursive, and hence representable in Th(PA). But there is no reasonable way of placing an upper bound for the existential quantifier in the definition of PROV. The number m we are looking for is supposed to be the gödel number of a PA proof of **X**, where k = [**X**]. But a PL derivation can be of any length—there is no upper bound on the length of a PL derivation. This means that for any given number b, there are PA proofs whose gödel numbers are greater than b (assuming that the gödel number of a PA proof is correlated with its length). It follows that the existential quantifier in the definition of PROV is *essentially unbounded*. However, the effective procedure we just described for PROV is only a Yes-procedure: it would tell us when k is a member of PROV, but it would produce no answer if k is not a member of PROV. This implies that the listing function of PROV is computable; and hence PROV is effectively (i.e., recursively) enumerable. We can enumerate the members of PROV as an infinite list. PROV is also not representable in Th(PA). We will prove this fact later in this chapter. However, there is an AV formula **prov**[y] with one free variable y such that if k ∈ PROV, then PA ⊢ **prov**[k], but it is *not* true that if k ∉ PROV, then PA ⊢ ¬**prov**[k]. The formula **prov**[y] is defined to be the formula (∃x)**proof**[x, y]. Informally speaking, (∃x)**proof**[x, y] says that there is a PA proof of y, which is the same as saying that y is provable from PA. Let k ∈ PROV. Thus there is a number m such that ⟨m, k⟩ ∈ PROOF. Since **proof**[x, y] represents PROOF in Th(PA), PA ⊢ **proof**[m, k]. By Existential Generalization PA ⊢ (∃x)**proof**[x, k], that is, PA ⊢ **prov**[k]. This short proof establishes our claim that if k ∈ PROV, then PA ⊢ **prov**[k].

5.4   Diagonalization and the First Incompleteness Theorem

5.4.1   The third and last major component of the incompleteness proof is diagonalization. The First Incompleteness Theorem follows from a lemma called the Diagonalization Lemma. In this section, we will prove this lemma and a few of its consequences, including Gödel's First Incompleteness Theorem. Our approach to diagonalization is less general than what is standardly presented in introductory textbooks on metalogic. Our approach is motivated by the fact that only a small portion of diagonalization is needed for proving the Diagonalization Lemma and its consequences. So we will limit our presentation to this portion. Our presentation, however, has the virtue of simplicity and economy. Let **X**[z] be an AV formula that contains one free variable z but no z-quantifiers. The AV sentence **X**[k], where k is the gödel number of **X**[z], is called the **diagonalization** of **X**[z]. The reason for the name 'diagonalization' is that, informally speaking, **X**[k] says that **X** is true of its own gödel number. If we think of the gödel number of **X**[z] as a "name" of **X**[z], then **X**[k] says that **X** is true of itself. When a formula is applied to its own gödel number (more precisely, to the AV numeral that represents its gödel number), it is natural to think of this substitutional instance as "the diagonal instance" of **X**[z]. The table below suggests a justification for this natural thought. Let $X_1, X_2, X_3, \ldots, X_n, \ldots$ be a complete list of all the AV formulas with only one free variable, and let $k_1, k_2, k_3, \ldots, k_n, \ldots$ be the gödel numbers of these formulas, respectively, that is, for every positive natural number n, $k_n = [X_n]$. We list the formulas at the top of the table and their gödel numbers to the left side of the table. Every entry of the table is the result of substituting the AV numeral that represents the ith gödel number k_i for the free variable in the jth formula X_j. Hence, the entry of the third row and sixth column is the substitutional instance $X_6[k_3]$.

	X_1	X_2	X_3	X_4	...	X_n	...
k_1	$X_1[k_1]$	$X_2[k_1]$	$X_3[k_1]$	$X_4[k_1]$	...	$X_n[k_1]$	...
k_2	$X_1[k_2]$	$X_2[k_2]$	$X_3[k_2]$	$X_4[k_2]$	...	$X_n[k_2]$	...
k_3	$X_1[k_3]$	$X_2[k_3]$	$X_3[k_3]$	$X_4[k_3]$	...	$X_n[k_3]$	...
k_4	$X_1[k_4]$	$X_2[k_4]$	$X_3[k_4]$	$X_4[k_4]$	...	$X_n[k_4]$	...
⋮	⋮	⋮	⋮	⋮	⋮	⋮	⋮
k_n	$X_1[k_n]$	$X_2[k_n]$	$X_3[k_n]$	$X_4[k_n]$	...	$X_n[k_n]$	...
⋮	⋮	⋮	⋮	⋮	⋮	⋮	⋮

It is clear from this table that the diagonal consists of entries of the form $X_m[k_m]$, where a formula is applied to the AV numeral that represents its own gödel number. These are the diagonal instances. This explains why we called $X_m[k_m]$ the "diagonalization of $X_m[z]$."

We state the following theorem without proof.

Theorem 5.4.1: There is a (total) recursive function DIAG from $\mathbb{N}$ into $\mathbb{N}$ such that for every natural number n, if n is the gödel number of an AV formula **X**[z] with exactly one free variable z, then DIAG(n) is the gödel number of the diagonalization of **X**[z], and if n is not the gödel number of an AV formula with only one free variable, then DIAG(n) = 0.

Let us consider an example of an ordered pair that belongs to this function. Let k be the gödel number of the formula $(x + ss0) = (ss0 \bullet x)$, that is, $k = [(x + ss0) = (ss0 \bullet x)]$. The diagonalization of this formula is the sentence $(k + ss0) = (ss0 \bullet k)$. Let $m = [(k + ss0) = (ss0 \bullet k)]$. Hence the ordered pair $\langle k, m \rangle \in$ DIAG. In standard function notation, DIAG(k) = m. Since DIAG is a recursive function, it is representable in Th(PA). By Definition 5.2.1c, there is an AV formula **diag**[x, y] with two free variables such that for all natural numbers k and m, if DIAG(k) = m, then PA $\vdash$ **diag**[k, m], and PA $\vdash (\forall x)(\mathbf{diag}[k, x] \rightarrow x = m)$. Now we are ready to state and prove the central theorem of this component of the proof of the First Incompleteness Theorem.

The Diagonalization Lemma: If **W**[z] is an AV formula with only one free variable z, then there is an AV sentence G such that PA $\vdash$ G↔**W**[**g**], where g = [G].

Before proving this lemma, we would like to explain its import. It is a theorem of PA that G is equivalent to a sentence that states, informally, that **W** is true of the of the gödel number of G. If we treat the gödel number of G as a name of G, then it is a theorem of PA that G is equivalent to a sentence that states that **W** is true of G. If we further think of equivalent sentences as making the same assertion, then it is a theorem of PA that G asserts that **W** is true of G itself. Metaphorically speaking, it is a theorem of PA that G asserts "I am **W**." Although the last statement is a metaphor, it is rather a useful metaphor: it will aid our understanding of the First Incompleteness Theorem.

Proof of the Diagonalization Lemma

1. Let **W**[z] be any AV formula with one free variable z. Take **diag**[x, y] to be the AV formula that represents DIAG in Th(PA). Assume that the variable y does not occur in **W**[z] (if it does, we choose a different variable). Let **W**[y] be the AV formula that results from replacing all the occurrences of z in **W**[z] with y.

2. From 1: The AV formula (∃y)(**diag**[x, y]∧**W**[y]) contains only one free variable x. Define G to be the diagonalization of (∃y)(**diag**[x, y]∧**W**[y]). Thus G is the AV *sentence* (∃y)(**diag**[**n**, y]∧**W**[y]), where n is the gödel number of the AV formula (∃y)(**diag**[x, y]∧**W**[y]). We will prove that PA $\vdash$ G↔**W**[**g**], where g is the gödel number of G.

3. From 2: DIAG(n) = g, where, as stated above, n = [(∃y)(**diag**[x, y]∧**W**[y])] and g = [G].

4. From 1 and 3: PA $\vdash$ **diag**[**n**, **g**] and PA $\vdash$ (∀v)(**diag**[**n**, v]→v = **g**).

5. From 4: There are PA derivations D_1 and D_2 of **diag**[**n**, **g**] and of (∀v)(**diag**[**n**, v]→v = **g**), respectively. Let Σ_1 and Σ_2 be the finite subsets of PA that consist of the PA axioms invoked in D_1 and D_2, respectively. Hence Σ_1 $\vdash$ **diag**[**n**, **g**] and Σ_2 $\vdash$ (∀v)(**diag**[**n**, v]→v = **g**).

6. From 5: Let $\Sigma = \Sigma_1 \cup \Sigma_2$. It is clear that Σ is a finite subset of PA, and Σ $\vdash$ **diag**[**n**, **g**] and Σ $\vdash$ (∀v)(**diag**[**n**, v]→v = **g**).

7. From 5 and 6: D_1 is a PL derivation of **diag**[**n**, **g**] from Σ and D_2 is a PL derivation of (∀v)(**diag**[**n**, v]→v = **g**) from Σ.

8. From 7: it is possible to construct a PL derivation of G↔**W**[**g**] from Σ. Here is such a derivation.

THE INCOMPLETENESS THEOREMS 241

[0	0	Σ	Premises (recall that Σ is a finite subset of PA)
	D₁		This derivation is given
	i	**diag[n, g]**	Conclusion of D₁
	D₂		This derivation is given
	j	$(\forall v)(\mathbf{diag[n}, v] \to v = \mathbf{g})$	Conclusion of D₂
[1	j+1	$(\exists y)(\mathbf{diag[n}, y] \wedge \mathbf{W}[y])$	CPA (this is G)
[2	j+2	**diag[n, t]**∧**W[t]**	EIA (**t** is a PL name that does not occur in any member of Σ nor does it occur in $(\exists y)(\mathbf{diag[n}, y] \wedge \mathbf{W}[y])$. Since Σ is finite, there is always such a name.)
	j+3	**diag[n, t]**	j+2, Simp
	j+4	**diag[n, t]** → **t** = **g**	j, UI
	j+5	**t** = **g**	j+3, j+4, MP
	j+6	**W[t]**	j+2, Simp
2]	j+7	**W[g]**	j+5, j+6, Sub (since **t** does not occur in **W**[y], **W[g]** may be considered as obtained from **W**[y] by replacing y with **g**.)
1]	j+8	**W[g]**	(j+2) – (j+7), EI
	j+9	$(\exists y)(\mathbf{diag[n}, y] \wedge \mathbf{W}[y]) \to \mathbf{W[g]}$	(j+1) – (j+8), CP
[3	j+10	**W[g]**	CPA
	j+11	**diag[n, g]**	i, Reit
	j+12	**diag[n, g]** ∧ **W[g]**	j+10, j+11, Conj
3]	j+13	$(\exists y)(\mathbf{diag[n}, y] \wedge \mathbf{W}[y])$	j+12, EG
	j+14	$\mathbf{W[g]} \to (\exists y)(\mathbf{diag[n}, y] \wedge \mathbf{W}[y])$	(j+10) – (j+13), CP
	j+15	$((\exists y)(\mathbf{diag[n}, y] \wedge \mathbf{W}[y]) \to \mathbf{W[g]}) \wedge (\mathbf{W[g]} \to (\exists y)(\mathbf{diag[n}, y] \wedge \mathbf{W}[y]))$ j+9, j+14, Conj	
0]	j+16	$(\exists y)(\mathbf{diag[n}, y] \wedge \mathbf{W}[y]) \leftrightarrow \mathbf{W[g]}$	j+15, Bc

9. From 2, 6 and 8: Since Σ is a finite subset of PA, PA ⊢ $(\exists y)(\mathbf{diag[n}, y] \wedge \mathbf{W}[y]) \leftrightarrow \mathbf{W[g]}$, that is, PA ⊢ G ↔ **W[g]**, where g is the gödel number of G.

5.4.2 There are many important consequences of this lemma, the most famous and important of which is Gödel's First Incompleteness Theorem. We first define the notion of ω-consistency. A PL set Σ is said to be **ω-consistent** if and only if there is no formula **X**[z] composed of Voc(Σ) with one free variable **z** such that Σ ⊢ ¬**X[n]**, for each natural number n, and Σ ⊢ $(\exists \mathbf{z})\mathbf{X}[\mathbf{z}]$. Informally speaking, Σ is ω-consistent if and only if there is no formula **X** such that it is provable from Σ that **X** is false of every natural number and yet it is provable from Σ that **X** is true of something. We assumed that PA is consistent, since we assumed that the standard interpretation of arithmetic, N, is a model of PA. But ω-consistency is a stronger assumption: ω-consistency entails consistency but the converse is not true. To see that ω-consistency entails consistency, consider the contrapositive statement—namely that if Σ is inconsistent, then it is ω-inconsistent. If Σ is inconsistent, then a contradiction is derivable from Σ. By the rule Explosion, every sentence is a theorem of Σ. Take the formula **X** to be $0 = sx$. Hence the sentences $(\exists x)0 = sx$, $0 \neq s0$, $0 \neq s1$, $0 \neq s2$, ..., $0 \neq sn$, ... are all theorems of Σ, which implies that Σ is ω-inconsistent. Therefore if Σ is inconsistent, then it is ω-inconsistent. Exercise 5.8.10 describes a case of a consistent set that is ω-inconsistent. Now we prove the First Incompleteness Theorem.

Gödel's First Incompleteness Theorem: If Peano Arithmetic is ω-consistent, then it is incomplete.

Proof

1. Suppose that Th(PA) is ω-consistent. We will prove that there is an AV sentence G$_{PA}$ such that Th(PA) $\nvdash$ G$_{PA}$ and Th(PA) $\nvdash$ ¬G$_{PA}$. Since by definition Th(PA) consists of all the theorems of PA that are composed of AV, by Lemma 5.1.1 it suffices to show that PA $\nvdash$ G$_{PA}$ and PA $\nvdash$ ¬G$_{PA}$. (We can also show the preceding statement without relying on Lemma 5.1.1: if PA $\nvdash$ **X**, then Th(PA) $\nvdash$ **X**, since if **X** were a theorem of Th(PA), then, given that Th(PA) is a theory, **X** would belong to Th(PA); and hence it would be a theorem of PA, which contradicts the fact that PA $\nvdash$ **X**.)

2. Since the relation PROOF is recursive, it is representable in Th(PA). Hence there is an AV formula **proof**[x, y] with two free variables that represents PROOF in Th(PA).

3. Let **prov**[y] be the AV formula $(\exists x)$**proof**[x, y], which contains only one free variable y. Informally speaking, **prov**[y] says that there is a PA proof of y, that is, y is provable from PA.

4. By the Diagonalization Lemma, there is an AV sentence G$_{PA}$ such that PA $\vdash$ G$_{PA} \leftrightarrow$ ¬**prov**[g], where g is the gödel number of G$_{PA}$. Using our previous metaphorical language, we say that it is a theorem of PA that G asserts "I am not provable from PA."

5. First Reductio Assumption: PA $\vdash$ G$_{PA}$.

6. From 5: there is a PL derivation D of G$_{PA}$ from PA. Let d be the gödel number of D.

7. From 4 and 6: since g = [G$_{PA}$] and d = [D], the ordered pair $\langle d, g \rangle \in$ PROOF.

8. From 2 and 7 by Definition 5.2.1b: PA $\vdash$ **proof**[d, g].

9. From 4 and 5 by Biconditional Modus Ponens: PA $\vdash$ ¬**prov**[g].

10. From 3 and 9: PA $\vdash$ ¬$(\exists x)$**proof**[x, g].

11. From 10 by Negated Conditional: PA $\vdash$ $(\forall x)$¬**proof**[x, g].

12. From 11 by Universal Insanitation: PA $\vdash$ ¬**proof**[d, g].

13. From 8 and 12: PA is inconsistent. This contradicts our supposition that PA is consistent. So we reject the first Reductio Assumption, and conclude that PA ⊬ G$_{PA}$.

14. Second Reductio Assumption: PA ⊢ ¬G$_{PA}$.

15. From 4 and 14 by Biconditional Modus Tollens: PA ⊢ ¬¬**prov**[g]. By Double Negation, PA ⊢ **prov**[g].

16. From 3 and 15: PA ⊢ (∃x)**proof**[x, g].

17. From 13: there is no PL derivation of G$_{PA}$ from PA.

18. From 17 by the construction of PROOF: since g = [G$_{PA}$], for any natural number m, the ordered pair ⟨m, g⟩ ∉ PROOF.

19. From 2 and 18 by Definition 5.2.1b: PA ⊢ ¬**proof**[m, g], for any natural number m.

20. From 16 and 19: since PA ⊆ Th(PA), Th(PA) ⊢ ¬**proof**[m, g], for every natural number m, and Th(PA) ⊢ (∃x)**proof**[x, g].

21. From 20 by the definition of ω-consistency: Th(PA) is ω-inconsistent, which contradicts 1. Thus we reject the second Reductio Assumption and conclude that PA ⊬ ¬G$_{PA}$.

22. From 13 and 18: PA ⊬ G$_{PA}$ and PA ⊬ ¬G$_{PA}$.

23. From 1 and 19: Th(PA) ⊬ G$_{PA}$ and Th(PA) ⊬ ¬G$_{PA}$.

24. From 23 by the definition of completeness: since G$_{PA}$ is composed of AV, Th(PA) is incomplete.

Observe that we did not need the assumption of ω-consistency for the first part. The consistency of PA is sufficient for proving that PA ⊬ G$_{PA}$. We invoked the supposition that Th(PA) is ω-consistent in proving the second part—namely that PA ⊬ ¬G$_{PA}$. It is possible to prove the First Incompleteness Theorem without the supposition that Th(PA) is ω-consistent; but the choice of G$_{PA}$ will have to be more complicated. Exercise 5.8.17 is devoted to proving the Incompleteness Theorem without the assumption of ω-consistency. By the Completeness Theorem of PL, Th(PA) ⊬ G$_{PA}$ and PA ⊬ ¬G$_{PA}$. By the definition of logical consequence, there are two models of Th(PA), one of which makes G$_{PA}$ true and the other makes it false. Hence these two models are not elementarily equivalent. By Theorem 3.4.1, those models are not isomorphic. Hence Th(PA) is not a categorical theory. In fact, every incomplete consistent PL theory is not categorical. Is Th(PA) at least ℵ$_0$-categorical? That is, are all the models of Th(PA) whose size is ℵ$_0$ isomorphic? We will answer this question negatively in the last exercise of this chapter; we will show that Arithmetic, which is Th$_{AV}$(N) = {X: X is an AV sentence that is true on N}, is not ℵ$_0$-categorical. It immediately follows that Th(PA) is not ℵ$_0$-categorical (recall that Th(PA) ⊆ Th$_{AV}$(N), and Voc(PA) = AV). It is traditional to refer to the sentence G$_{PA}$ as the **Gödel sentence of** PA.

Under the supposition that the standard interpretation of arithmetic, N, is a model of PA, it is possible to show that G$_{PA}$ is true on N. If we employ our earlier metaphor, it is clear why G$_{PA}$ is true on N. G$_{PA}$ says of itself that it is not provable from PA. The First

Incompleteness Theorem shows that indeed G_PA is not provable from PA. Thus G_PA says something that is the case, and hence it is true. We can turn this metaphorical argument into a rigorous argument. We have the following theorem.

Theorem 5.4.2: G_PA is true on N.

Proof
1. Suppose that N is a model of PA.
2. From the proof of the First Incompleteness Theorem: PA ⊢ G_PA ↔ ¬**prov**[g], where g = [G_PA].
3. From 2 by the definition of **prov**[y]: PA ⊢ G_PA ↔ ¬(∃x)**proof**[x, g].
4. From 3 by the Soundness Theorem: PA ⊨ G_PA ↔ ¬(∃x)**proof**[x, g].
5. From 1 and 4: since N is a model of PA, G_PA ↔ ¬(∃x)**proof**[x, g] is true on N.
6. By the First Incompleteness Theorem: PA ⊬ G_PA.
7. From 6: there is no PL derivation of G_PA from PA.
8. From 7 by the construction of PROOF: since g = [G_PA], for every natural number m, the ordered pair ⟨m, g⟩ ∉ PROOF.
9. From 8: since **proof**[x, y] represents PROOF in Th(PA), PA ⊢ ¬**proof**[m, g], for every natural number m.
10. From 9 by the Soundness Theorem: PA ⊨ ¬**proof**[m, g], for every natural number m.
11. From 1 and 10: since N is a model of PA, the AV sentence ¬**proof**[m, g] is true on N for every VA numeral **m**.
12. By the definition of PL interpretation: the list of names (LN) of N consists of infinitely many names such that every name **t** refers to exactly one natural number and every natural number has at least one name in LN. We argued previously that every AV numeral refers to exactly one natural number and that every natural number is named by one AV numeral. Hence for every name **t** in LN of N, there is an AV numeral **m** such that **t** = **m** is true on N.
13. From 11 and 12: ¬**proof**[t, g] is true on N for every name **t** in LN. In other words, every basic substitutional instance of (∃x)**proof**[x, g] is false on N.
14. From 13 by the truth conditions of the existential quantifier: (∃x)**proof**[x, g] is false on N; and hence ¬(∃x)**proof**[x, g] is true on N.
15. From 5 and 14 by the truth conditions of the biconditional: G_PA is true on N.

5.5 Consequences of Diagonalization and Incompleteness

5.5.1   There are several important consequences of the Diagonalization Lemma and the First Incompleteness Theorem. In this section, we will cover some of them. We present these consequences in a series of theorems.

Theorem 5.5.1: $Th(PA) \subset Th_{AV}(N)$.

Proof

1. By definition: $Th_{AV}(N) = \{X: X \text{ is an AV sentence that is true on } N\}$.
2. From 1 by Theorem 5.4.2: $G_{PA} \in Th_{AV}(N)$.
3. By definition: $Th(PA) = \{X: PA \vdash X\}$.
4. From 3 by the proof of the First Incompleteness Theorem: $G_{PA} \notin Th(PA)$.
5. Since N is a model of $Th(PA)$, $Th(PA) \subseteq Th_{AV}(N)$.
6. From 2, 4, and 5: $Th(PA) \subset Th_{AV}(N)$.

The same result follows from the facts that $Th_{AV}(N)$ is a complete theory (whose vocabulary is AV) and that $Th(PA)$ is an incomplete theory (whose vocabulary is also AV). Hence $Th_{AV}(N) \neq Th(PA)$. Since $Th(PA) \subseteq Th_{AV}(N)$, $Th(PA) \subset Th_{AV}(N)$.

We now prove a very important theorem that establishes a sufficient condition for a theory's being undecidable. We will see that a consequence of this theorem is that a theory that is consistent and in which all recursive functions are representable is undecidable. Almost all important mathematical theories meet this condition, and hence they are undecidable. We will give later an example of a fairly weak theory that meets this condition.

Theorem 5.5.2: If Σ is a consistent PL theory in which all recursive functions are representable, then the set of the gödel numbers of the sentences in Σ is not representable in Σ.

Proof

1. Let Σ be a consistent theory in which all recursive functions are representable.
2. Let $THRM_\Sigma = \{[X]: X \in \Sigma\}$. Since Σ is a theory, it consists of all its theorems that are composed of $Voc(\Sigma)$, that is, for every sentence **X** that is composed of $Voc(\Sigma)$, **X** is a theorem of Σ iff it is a member of Σ. Thus $THRM_\Sigma$ is the set of all the gödel numbers of the theorems of Σ that are composed of $Voc(\Sigma)$. Our goal is to show that $THRM_\Sigma$ is not representable in Σ.
3. Reductio Assumption: $THRM_\Sigma$ is representable in Σ.

4. From 3: there is a formula $X[z]$ composed of $Voc(\Sigma)$ with one free variable such that $X[z]$ represents $THRM_\Sigma$ in Σ, that is, for every natural number k, if k $\in$ $THRM_\Sigma$, then $\Sigma \vdash X[k]$, and if k $\notin$ $THRM_\Sigma$, then $\Sigma \vdash \neg X[k]$.

5. From 1: since the function DIAG is recursive, it is representable in Σ. (See Theorem 5.4.1 for the definition of the function DIAG.) Hence the Diagonalization Lemma holds for Σ.

6. From 4 and 5: there is a sentence G_Σ composed of $Voc(\Sigma)$ such that $\Sigma \vdash G_\Sigma \leftrightarrow \neg X[g]$, where g is the gödel number of G_Σ.

7. Reductio Assumption: G_Σ is not a theorem of Σ.

8. From 2 and 7: since g = $[G_\Sigma]$, g $\notin$ $THRM_\Sigma$.

9. From 4 and 8: since $X[z]$ represents $THRM_\Sigma$ in Σ, $\Sigma \vdash \neg X[g]$.

10. From 6 and 9 by Biconditional Modus Ponens: $\Sigma \vdash G_\Sigma$. Thus G_Σ is a theorem of Σ, which contradicts 7. Therefore the Reductio assumption is false. We conclude that $\Sigma \vdash G_\Sigma$.

11. From 2 and 10: since g = $[G_\Sigma]$, g $\in$ $THRM_\Sigma$.

12. From 4 and 11: since $X[z]$ represents $THRM_\Sigma$ in Σ, $\Sigma \vdash X[g]$. By Double Negation, $\Sigma \vdash \neg\neg X[g]$.

13. From 6 and 12 by Biconditional Modus Tollens: $\Sigma \vdash \neg G_\Sigma$.

14. From 10 and 13: Σ is inconsistent, which contradicts 1.

15. From 3 through 14: the original Reductio Assumption is false; and hence $THRM_\Sigma$ is not representable in Σ.

This theorem fulfills the promise we made earlier to show that the arithmetical set PROV is not representable in Th(PA) and not a recursive set. By definition, PROV = {[X]: X is an AV sentence that is provable from PA}. Hence PROV is just the set of the gödel numbers of the theorems of PA. Equivalently, PROV = {[X]: X $\in$ Th(PA)}. Since Th(PA) is a consistent theory in which all recursive functions are representable, by Theorem 5.5.2, PROV is not representable in Th(PA). It immediately follows that PROV is not a recursive set (if it were, it would be representable in Th(PA)).

Theorem 5.5.3: If Σ is a consistent PL theory in which all recursive functions are representable, then Σ is undecidable.

Proof

1. Suppose that Σ is a consistent theory in which all recursive functions are representable.

2. Let $THRM_\Sigma$ = {[X]: X $\in \Sigma$}. Since Σ is a theory, it consists of all its theorems that are composed of $Voc(\Sigma)$, that is, for every sentence X that is composed of $Voc(\Sigma)$, X is a

theorem of Σ iff it is a member of Σ. Thus THRM_Σ is the set of all the gödel numbers of the theorems of Σ that are composed of Voc(Σ).

3. Reductio Assumption: Σ is decidable.
4. From 3: there is an effective decision procedure P for determining membership in Σ.
5. From 4: we use the decision procedure P to construct an effective decision procedure for THRM_Σ. Let k be any natural number. Determining whether or not k = [X] for some sentence **X** composed of Voc(Σ) is clearly an effective decision procedure, since the decoding procedure is effective, and since, as is always the case, the set of the PL sentences composed of Voc(Σ) is assumed to be decidable. If k is not the gödel number of a sentence composed of Voc(Σ), we place k outside THRM_Σ. If k is the gödel number of a sentence composed of Voc(Σ), we apply P to **X** to determine whether **X** ∈ Σ or not. If **X** ∈ Σ, we place k inside THRM_Σ, and if **X** ∉ Σ, we place k outside THRM_Σ. This procedure is an effective decision procedure for determining membership in THRM_Σ.
6. From 5 by Church's Thesis: the characteristic function of THRM_Σ is Turing-computable. By Theorem 4.4.1, the characteristic function is recursive. Hence THRM_Σ is a recursive set.
7. From 1: since all recursive functions are representable in Σ, all recursive sets and relations are also representable in Σ.
8. From 6 and 7: THRM_Σ is representable in Σ.
9. From 1 and 2 by Theorem 5.5.2: THRM_Σ is not representable in Σ, which contradicts 8.
10. From 3 through 9: The Reductio Assumption is false. Hence Σ is undecidable.

In this proof, we described a procedure that is clearly an effective decision procedure. We invoked Church's Thesis to conclude that this procedure can be made into a Turing-computable function. Observe that this claim also presupposes that all effective procedures can be converted into numerical functions. We explained why this assumption is true in Chapter Four. Finally we employed Theorem 4.4.1 to argue that the Turing-computable function is recursive. In the remaining part of this chapter, we will only describe effective procedures. We won't bother with citing Church's Thesis and Theorem 4.4.1. We will assume that the reader would fill in the necessary details. There is a very interesting consequence of Theorem 5.5.3 above. The phenomenon of incompleteness is not restricted to Peano Arithmetic and similar theories. In fact every consistent axiomatizable PL theory that meets a certain reasonable condition is incomplete. The condition is that all recursive functions are representable in that theory. We will see later that this condition is not too demanding. There are quite weak theories that meet this condition. We state this fact in the following theorem.

Theorem 5.5.4: If Σ is a consistent axiomatizable PL theory in which all recursive functions are representable, then Σ is incomplete.

Proof

This theorem is an immediate consequence of Theorems 5.5.3 and 3.5.1. Since Σ is a consistent PL theory in which all recursive functions are representable, by Theorem 5.5.3 it is undecidable. By theorem 3.5.1 if Σ is complete and axiomatizable, it is decidable. Since Σ is axiomatizable and undecidable, it must be incomplete.

The next theorem is also an immediate consequence of Theorem 5.5.3. We define a **consistent extension** of Th(PA) to be any set Σ of AV sentences such that Σ is a consistent theory and Th(PA) $\subseteq$ Σ.

Theorem 5.5.5:

5.5.5a No consistent extension of Th(PA) is decidable.

5.5.5b Neither Th(PA) nor Th$_{AV}$(N) is a decidable theory.

Proof

5.5.5a Let Σ be any consistent extension of Th(PA). Since all recursive functions are representable in Th(PA), they are also representable in any extension of Th(PA); and hence they are representable in Σ. By Theorem 5.5.3, Σ is undecidable.

5.5.5b Both Th(PA) and Th$_{AV}$(N) are consistent extensions of Th(PA). (Every consistent theory is a consistent extension of itself.) By 5.5.5a, they are undecidable.

Theorem 5.5.6: Th(PA) is semidecidable.

Proof

1. By definition, the members of Th(PA) are precisely the theorems of PA.
2. Let Proof$_{PA}$ be the set of all the PA proofs, that is, D ∈ Proof$_{PA}$ iff D is a PL derivation from PA.
3. There is an effective decision procedure for determining membership in Proof$_{PA}$. For any finite sequence D of AV sentences, we can effectively determine whether D is a PA proof or not: we simply examine every sentence of D and determine whether it is an axiom of PA (PA is a decidable set) or it is licensed by one of the MDS rules. We arrange the members of Proof$_{PA}$ according to the magnitude of their gödel numbers: D precedes D* iff the gödel number of D is less than the gödel number of D*.
4. For any given AV sentence Z, Z is a theorem of PA iff there is a PA proof whose conclusion is Z. In other words, Z ∈ Th(PA) iff there is D in Proof$_{PA}$ such that Z is the last sentence of D.
5. From 3 and 4: by systematically examining the members of Proof$_{PA}$, we will be able to locate a PA proof of Z, if Z is a theorem of PA. However, if Z is not a theorem of

PA, the process of examining the members of Proof$_{PA}$ will never terminate. Hence, this procedure is an effective Yes-procedure for determining membership in Th(PA).

6. From 5: Th(PA) is semidecidable.

The following theorem demonstrates an important limitation of the axiomatic method. It implies that no matter how Peano axioms might be strengthened by adding new axioms to them, it is impossible to produce a decidable set of arithmetical axioms that are adequate for proving all and only the true AV sentences.

Theorem 5.5.7: Arithmetic, i.e., Th$_{AV}$(N), is not axiomatizable.

Proof

1. Reductio Assumption: Th$_{AV}$(N) is axiomatizable.
2. From 1 by Definition 3.5.1: there is a decidable set Ω such that Th$_{AV}$(N) = Th(Ω).
3. By definition: Th$_{AV}$(N) = {**X**: **X** is an AV sentence that is true on N}.
4. From 3 by Corollary 3.5.1e: Th$_{AV}$(N) is a consistent complete theory whose vocabulary is AV.
5. We describe an effective decision procedure for determining membership in Th$_{AV}$(N). There is an effective decision procedure for determining whether or not any given finite sequence D of AV sentences is a PL derivation from Ω. This procedure works in the usual way: every sentence of D can be effectively determined whether it is an axiom in Ω, as Ω is a decidable set, or it is licensed by one of the MDS rules. Hence the set {D: D is a PL derivation from Ω} is a decidable set. Call this set Proof$_\Omega$. The members of Proof$_\Omega$ can be arranged sequentially such that D$_1$ precedes D$_2$ iff the gödel number of D$_1$ is less than the gödel number of D$_2$. Given any AV sentence **Z**, we can effectively determine its membership in Th$_{AV}$(N) by examining the members of Proof$_\Omega$ in order. Since Th$_{AV}$(N) is complete, Th$_{AV}$(N) $\vdash$ **Z** or Th$_{AV}$(N) $\vdash \neg$**Z**. Given that Th$_{AV}$(N) = Th(Ω), by Lemma 5.1.1, $\Omega \vdash$ **Z** or $\Omega \vdash \neg$**Z**. Thus by systematically examining the derivations in Proof$_\Omega$ we will locate a derivation D whose conclusion is **Z** or a derivation D* whose conclusion is $\neg$**Z**. If we locate the first derivation, then **Z** $\in$ Th$_{AV}$(N); and if we locate the second derivation, then **Z** $\notin$ Th$_{AV}$(N). This is an effective decision procedure for determining membership in Th$_{AV}$(N).
6. From 5: Th$_{AV}$(N) is decidable, which contradicts Theorem 5.5.5b.

   [numbered 7 in source]
7. From 1 through 6: the Reductio Assumption is false; and hence Arithmetic is not axiomatizable.

   [numbered 8 in source]

We chose to give a direct proof of Theorem 5.5.7. But in fact it is an immediate consequence of Theorem 3.5.1 and Theorem 5.5.5b. The former asserts that any PL theory that is complete and axiomatizable must be decidable. As we explained in Chapter Three, this means that the triad of (1) completeness, (2) axiomatizability, and (3) undecidability is an inconsistent combination. It follows that since Arithmetic is complete and, by Theorem 5.5.5b, undecidable, it cannot be axiomatizable. The First Incompleteness Theorem also follows from Theorems 3.4.2 and 5.5.5b: since Peano Arithmetic is axiomatizable and undecidable, it cannot be complete. Let us call an axiomatizable AV theory for which the standard interpretation of arithmetic is a model a **number theory**. The following theorem says, in effect, that no number theory is complete.

Theorem 5.5.8: For any decidable set Ω of AV sentences, if $\Omega \subseteq \text{Th}_{AV}(N)$, then $\text{Th}(\Omega)$ is incomplete.

Proof

Since $\Omega \subseteq \text{Th}_{AV}(N)$ and $\text{Th}_{AV}(N)$ is a theory, $\text{Th}(\Omega) \subseteq \text{Th}_{AV}(N)$. Given that $\text{Th}_{AV}(N)$ is a complete theory, it follows that if $\text{Th}(\Omega)$ is complete, then $\text{Th}(\Omega) = \text{Th}_{AV}(N)$. To see this, suppose that $\text{Th}(\Omega)$ is complete and take X to be any AV sentence such that $X \in \text{Th}_{AV}(N)$. $\neg X$ cannot be a theorem of $\text{Th}(\Omega)$; for otherwise $\neg X \in \text{Th}(\Omega)$ since $\text{Th}(\Omega)$ is a theory; and hence $\neg X \in \text{Th}_{AV}(N)$, making $\text{Th}_{AV}(N)$ inconsistent. It follows, by the completeness of $\text{Th}(\Omega)$, that X must be a theorem of $\text{Th}(\Omega)$, which entails that $X \in \text{Th}(\Omega)$. In this case, we end up with $\text{Th}(\Omega) \subseteq \text{Th}_{AV}(N)$ and $\text{Th}_{AV}(N) \subseteq \text{Th}(\Omega)$, and by the Principle of Extensionality $\text{Th}(\Omega) = \text{Th}_{AV}(N)$. Hence, as we stated previously, if $\text{Th}(\Omega)$ is complete, $\text{Th}_{AV}(N) = \text{Th}(\Omega)$; and since Ω is decidable, $\text{Th}_{AV}(N)$ would be axiomatizable, contradicting Theorem 5.5.7. Therefore $\text{Th}(\Omega)$ is incomplete.

5.5.2   Is the concept of "arithmetical truth" definable in Peano Arithmetic? We need to clarify what we mean by "arithmetical truth" and "definable in Peano Arithmetic." It is reasonable to assume that 'arithmetical truth' refers to any AV sentence that is true on the standard interpretation of arithmetic—namely, N. If "definable in Peano Arithmetic" means that the set of the gödel numbers of the arithmetical truths is representable in Th(PA), then the answer is clearly No. By definition, $\text{Th}_{AV}(N)$ is the set of all the AV sentences that are true on N. Since all recursive functions are representable in Th(PA), and $\text{Th}_{AV}(N)$ is an extension of Th(PA), all recursive functions are representable in $\text{Th}_{AV}(N)$. Thus by Theorem 5.5.2, the set of the gödel numbers of all the sentences in $\text{Th}_{AV}(N)$ is not representable in $\text{Th}_{AV}(N)$ (recall that $\text{Th}_{AV}(N)$ is a consistent theory). But if this set is not representable in $\text{Th}_{AV}(N)$, it cannot be representable in Th(PA); for since Th(PA) is a subset of $\text{Th}_{AV}(N)$, any set that is representable in Th(PA) is representable in $\text{Th}_{AV}(N)$.

Let us give a weaker understanding of what is meant by "arithmetical truth being definable in Peano Arithmetic." Let us say that "definable in Peano Arithmetic" does not necessarily mean "representable in Peano Arithmetic"; rather we say that whatever this notion might mean, it has to satisfy the following necessary condition: a 1-variable AV formula **true**[x] "defines" arithmetical truth in Peano Arithmetic only if it satisfies the following condition:

TS   For every AV sentence **X**, PA ⊢ **true**[k]↔**X**, where k is the gödel number of **X**.

TS (for "Tarskian Schema") is the formal counterpart of an important condition that is called **Convention-T**. This convention was proposed by the Polish logician and mathematician Alfred Tarski (1901–83) in 1933 as a condition on any *adequate* definition of truth. The **Tarskian Schema** is the schema whose instances are all the biconditionals of the form "x is true iff S" where 'S' is to be replaced by a declarative sentence of the object language and 'x' by a name or a description of the sentence S. For instance, all the biconditionals below are instances of the Tarskian Schema.

> 'Snow is white' is true iff snow is white.
> The sentence 'Johanna loved Bruce' is true iff Johanna loved Bruce.
> The sentence that 1 = 0 is true iff 1 = 0.
> The sentence Sammy just uttered is true iff the economy is in recession. (Assuming that Sammy just said that the economy was in recession.)

Convention-T asserts that any adequate definition of truth for some language L must entail all the instances of the Tarskian Schema for L. Tarski argued that if one does not restrict the scope of this schema in some fashion, then an adequate definition of truth could entail contradictory biconditionals; and of course any definition that entails a contradictory sentence is inconsistent. A contradictory biconditional can be generated from the Tarskian Schema by instantiating the schema for the following sentence:

> The boxed sentence in this book is not true.

This sentence is an instance of a **Liar Sentence**, which is any sentence that asserts of *itself* that it is not true. Let us refer to this sentence as λ. Applying the Tarskian Schema to this sentence, we obtain the following biconditional.

λ is true iff the boxed sentence in this book is not true.

But λ refers to the boxed sentence in this book. By substitution, we get:

> The boxed sentence in this book is true iff the boxed sentence in this book is not true.

The last biconditional is contradictory in classical logic. Hence if we require an adequate definition of truth for at least a portion of English to fulfill Convention-T, and if it is possible to formulate a Liar Sentence in this portion of English, then this definition of truth must be inconsistent, assuming that the laws of classical logic hold for this portion of English. This problem is referred to as the **Liar Paradox**.

Tarski's solution to the Liar Paradox is to restrict the scope of the Tarskian Schema for a language L by restricting what one can assert in L. In simple terms, Tarski's solution is to prevent the sentences of L from talking about their own truth. Tarski proved that there can be no formula **true**[x] that satisfies the condition TS if PA is consistent. This statement is a version of a theorem that is known as Tarski's Indefinability Theorem. In Exercise 5.8.15, we will give a different version of this theorem. Tarski proved this theorem without the aid of Gödel's Incompleteness Proof. Tarski's Theorem and his definition of truth in formalized languages (1933) were an important accomplishment in the history of the semantics of PL. However, if one invokes the Diagonalization Lemma, Tarski's Theorem is easy to prove.

Tarski's Indefinability Theorem: The notion of arithmetical truth is not definable in Peano Arithmetic.

Proof

1. As usual, we presuppose that PA is satisfiable.
2. Reductio Assumption: arithmetical truth is definable in Peano Arithmetic. Thus there is an AV formula **true**[x], with only one free variable, such that for every AV sentence **X**, PA $\vdash$ **true**[k]$\leftrightarrow$**X**, where k is the gödel number of **X**.
3. From 2 by the Diagonalization Lemma: there is an AV sentence λ such that PA $\vdash$ $\lambda\leftrightarrow\neg$**true**[l], where l is the gödel number of λ, that is, l = [λ]. (λ is a Liar Sentence in Th(PA).)
4. From 2 and 3: PA $\vdash$ **true**[l]$\leftrightarrow\lambda$.
5. From 3 and 4 by Hypothetical Syllogism (applied to biconditionals): PA $\vdash$ **true**[l]$\leftrightarrow\neg$**true**[l].
6. From 5 by the Soundness Theorem for PL: PA $\vDash$ **true**[l]$\leftrightarrow\neg$**true**[l].
7. From 6: since **true**[l]$\leftrightarrow\neg$**true**[l] is contradictory, PA is not satisfiable, which contradicts 1.
8. From 2 through 7: the Reductio Assumption is false; and hence the notion of arithmetical truth is not definable in Peano Arithmetic

5.5.3   We proved in Theorem 5.5.3 that any consistent PL theory in which all the recursive functions are representable is undecidable. In the Representability Theorem we stated that all recursive functions are representable in Th(PA). PA is an infinite set of axioms. It is clear that any proof of the Representability Theorem would invoke only finitely many PA axioms. After all, like PL derivations, metaproofs are finite sequences of sentences. Thus any proof in the metatheory of PA would never need to invoke infinitely many PA axioms. Let **Q** be the *finite* set of PA axioms that are invoked in the proof of the Representability Theorem. Hence all recursive functions are representable in Th(**Q**). If one would like to see an explicit example of a finite set of AV sentences that is adequate for representing all recursive functions, then the set of Robinson's axioms is such a set. Robinson's axioms, which are due to the American mathematician Raphael Robinson (1911–95), are only 7 axioms. The set of these axioms is denoted as "RA." The theory whose set of axioms is RA, i.e., Th(RA), is known as **Robinson Arithmetic** and is standardly denoted as

THE INCOMPLETENESS THEOREMS

Q. However, we will use 'Q' to refer to the finite fragment of PA that is needed to prove the Representability Theorem. The RA axioms are the following.

RA1 $(\forall x)0 \neq sx$
RA2 $(\forall x)(\forall y)(sx = sy \rightarrow x = y)$
RA3 $(\forall x)(x \neq 0 \rightarrow (\exists y)x = sy)$
RA4 $(\forall x)(x + 0) = x$
RA5 $(\forall x)(\forall y)(x + sy) = s(x + y)$
RA6 $(\forall x)(x \bullet 0) = 0$
RA7 $(\forall x)(\forall y)(x \bullet sy) = ((x \bullet y) + x)$

Every member of RA is derivable from PA. All of the RA axioms are PA axioms except for RA3. However RA3 is a theorem of PA. Here is a PL derivation of $(\forall x)(x \neq 0 \rightarrow (\exists y)x = sy)$ from PA. The first sentence of the derivation is a PA axiom obtained by applying the Induction Schema to the AV formula $x \neq 0 \rightarrow (\exists y)x = sy$.

PA Proof of $(\forall x)(x \neq 0 \rightarrow (\exists y)x = sy)$

```
[0    1   (0 ≠ 0 → (∃y)0 = sy) → ((∀z)((z ≠ 0 → (∃w)z = sw) → (sz ≠ 0 → (∃w)sz = sw)) →
          (∀x)(x ≠ 0 → (∃y)x = sy))                                IS
[1    2   0 ≠ 0                                                    CPA
[2    3   ¬(∃y)0 = sy                                              RA
     4   0 = 0                                                     Id
2]   5   0 ≠ 0                                                     2, Reit
1]   6   (∃y)0 = sy                                                3–4, RAA
     7   0 ≠ 0 → (∃y)0 = sy                                        2–6, CPA
[3    8   a ≠ 0 → (∃w)a = sw                                       CPA
[4    9   sa ≠ 0                                                   CPA
     10  sa = sa                                                   Id
4]   11  (∃w)sa = sw                                               10, EG
3]   12  sa ≠ 0 → (∃w)sa = sw                                      9–11, CP
     13  (a ≠ 0 → (∃w)a = sw) → (sa ≠ 0 → (∃w)sa = sw)              8–11, CP
     14  (∀z)((z ≠ 0 → (∃w)z = sw) → (sz ≠ 0 → (∃w)sz = sw))        13, UG
     15  (∀z)((z ≠ 0 → (∃w)z = sw) → (sz ≠ 0 → (∃w)sz = sw)) → (∀x)(x ≠ 0 → (∃y)x = sy)
                                                                   1, 7, MP
0]   16  (∀x)(x ≠ 0 → (∃y)x = sy)                                  14, 15, MP
```

Since RA ⊆ Th(PA), Th(RA) ⊆ Th(PA). In fact RA is much weaker than PA. There are many theorems that are derivable from PA but not from RA. For example, one cannot prove from RA that + is commutative, that is, one cannot prove $(\forall x)(\forall y)(x + y) = (y + x)$. So Th(RA) ⊂ Th(PA). It is known that all recursive functions are representable in Th(RA). Th(RA) is consistent. All the RA axioms are true on the standard interpretation N. Since every member of Th(RA) is a logical consequence of RA, N is a model of Th(RA). Therefore, by the Soundness Theorem, Th(RA) is consistent. By Theorem 5.5.3, Th(RA) is undecidable. Also by Theorem 5.5.4, Th(RA) is incomplete.

It does not matter whether we use the finite fragment **Q** of PA, which is needed for proving that all the recursive functions are representable in Th(PA), or we use Robinson's axioms. Since RA is not powerful enough to establish Gödel's Second Incompleteness Theorem, we will continue to work with PA. However, in order to prove Church's Undecidability Theorem, we need the fact that there exists a finitely axiomatizable theory in which all recursive functions are representable. There are other proofs of Church's Undecidability Theorem that are not based on any finitely axiomatizable number theory. We will give a proof of Church's Undecidability Theorem that is based on the fact that all recursive functions are representable in Th(**Q**). Church's Undecidability Theorem asserts that the set of all the valid PL sentences is undecidable. In Subsection 3.5.2 we explained that the set of all the valid PL sentences is the theory Th(∅), whose set of axioms is the empty set. Given the Soundness and Completeness Theorems for PL, Th(∅) could be either defined as the set of all the valid PL sentences or the set of all the logical theorems of PL. Symbolically, Th(∅) = {**X**: ∅ ⊨ **X**} = {**X**: ∅ ⊢ **X**}. **X** in these definitions stands for any PL sentence, where the relevant vocabulary here is the complete standard vocabulary of PL. It is common to write ⊨ **X** and ⊢ **X** instead of ∅ ⊨ **X** and ∅ ⊢ **X**, respectively. We will follow this practice here to simplify our notation. Before we prove Church's Undecidability Theorem, we prove an easy lemma.

Lemma 5.5.1: For all PL sentences **Z** and **W**, **Z** ⊢ **W** iff ⊢ **Z**→**W**.

Proof

1. We first show the 'only if' part. So suppose that **Z** ⊢ **W**.
2. From 1: there is a PL derivation D of **W** from **Z**.
3. From 2: we can utilize D to produce a derivation of **Z**→**W** from the empty set (i.e., without premises). Using Conditional Proof, we list **Z** as the conditional proof assumption; then we invoke D. Since **W** is the conclusion of D, we obtain a derivation of **W** from the conditional proof assumption **Z**. Finally we close Conditional Proof, and conclude **Z**→**W**. This is a derivation of **Z**→**W** without premises. Hence **Z**→**W** is a logical theorem, that is, ⊢ **Z**→**W**.
4. From 1 through 3: if **Z** ⊢ **W**, then ⊢ **Z**→**W**.
5. Now we prove the 'if' part. Assume that ⊢ **Z**→**W**.
6. From 5: there is a PL derivation E of **Z**→**W** without premises.
7. From 6: employing E, we can construct a PL derivation of **W** from **Z**. We list **Z** as our only premise. Then we invoke E, which yields the conclusion **Z**→**W**. Given our premise **Z** and the conclusion **Z**→**W**, **W** follows by Modus Ponens. Thus **Z** ⊢ **W**.

8. From 5 through 7: if $\vdash Z \rightarrow W$, then $Z \vdash W$.
9. From 4 and 8: $Z \vdash W$ iff $\vdash Z \rightarrow W$, for all PL sentences Z and W.

Church's Undecidability Theorem: $Th(\varnothing)$ is undecidable.

Proof

1. Since **Q** is a finite set, we can combine all the members of **Q** in one conjunction. Let this conjunction be C. It is clear that for every AV sentence X, $\mathbf{Q} \vdash X$ iff $C \vdash X$.
2. From 1 by the definition of $Th(\mathbf{Q})$: for every AV sentence X, $X \in Th(\mathbf{Q})$ iff $\mathbf{Q} \vdash X$. Hence for every AV sentence X, $X \in Th(\mathbf{Q})$ iff $C \vdash X$.
3. From Lemma 5.5.1: for all PL sentences Z and W, $Z \vdash W$ iff $\vdash Z \rightarrow W$. In other words, W is a theorem of Z iff the conditional $Z \rightarrow W \in Th(\varnothing)$ (i.e., $Z \rightarrow W$ is a logical theorem).
4. From 2 and 3: for every AV sentence X, $X \in Th(\mathbf{Q})$ iff $C \rightarrow X \in Th(\varnothing)$.
5. Reductio Assumption: $Th(\varnothing)$ is decidable.
6. From 5: there is an effective decision procedure P for determining membership in $Th(\varnothing)$.
7. From 4 and 6: we can now construct an effective decision procedure for membership in $Th(\mathbf{Q})$. We take any AV sentence X (this step presupposes the standard fact that the set of all AV sentences is decidable). We form the conditional $C \rightarrow X$. Now we apply the procedure P to $C \rightarrow X$ to determine whether this conditional is a member of $Th(\varnothing)$ or not. Since P is an effective decision procedure, it must return either the answer Yes or No. If it is Yes, then we have determined that $C \rightarrow X \in Th(\varnothing)$, and hence $X \in Th(\mathbf{Q})$; if it is No, then we have determined that $C \rightarrow X \notin Th(\varnothing)$; and hence $X \notin Th(\mathbf{Q})$.
8. From 7: $Th(\mathbf{Q})$ is decidable.
9. Since $Th(\mathbf{Q})$ is a consistent theory in which all recursive functions are representable, by Theorem 5.5.3, $Th(\mathbf{Q})$ is undecidable, which contradicts 8.
10. From 5 through 9: the Reductio Assumption is false. We conclude that $Th(\varnothing)$ is undecidable.

Theorem 5.5.9: $Th(\varnothing)$ is semidecidable.

Proof

For every $X \in Th(\varnothing)$, there is a PL derivation D of X from the empty set, that is, D is a PL derivation without premises whose last sentence is X. Let LD (for "Logical Derivations") be the set {D: D is a PL derivation without premises}. LD is a decidable set. For any given finite sequence E of PL sentences, we can examine every sentence in E to determine whether it is licensed by one of the MDS rules or not. We arrange the members of LD sequentially as follows: D_1 precedes D_2 iff the gödel number of D_1 is less than the gödel number of D_2. In order to determine whether any sentence **X** is valid (i.e., a logical theorem), we examine the members of LD in order. If **X** is indeed valid, we will eventually locate a derivation D in LD whose terminal sentence is **X**. In this case, we determine that **X** belongs to Th(∅). It might also happen that a PL derivation of ¬**X** is located in the set LD. This means that **X** is contradictory; and thus **X** is not a member of Th(∅). However, if **X** is neither valid nor contradictory (i.e., **X** is contingent), this procedure will continue indefinitely without generating an answer. Since we would not know at any stage whether there is no PL derivation of **X** or of ¬**X** from the empty set or whether there is such a derivation but the procedure has not located it yet, we will not be able to determine that **X** is contingent. Thus this procedure is an effective Yes-procedure for the set Th(∅): it returns the answer Yes iff the correct answer is Yes, and it might not return an answer if the correct answer is No. Therefore, the set Th(∅) is semidecidable.

We have shown that the set of the valid PL sentences is only semidecidable. The same argument applies to the set of the contradictory PL sentences. In this case, we examine the members of LD to see whether the sentence ¬**X** appears as the terminal sentence of any derivation in LD. If it does, then **X** is contradictory. This is also an effective Yes-procedure for determining whether a PL sentence is contradictory. As is the case with Th(∅), the set of the contradictory PL sentences is not decidable. It immediately follows that the set of the contingent PL sentences is neither decidable nor semidecidable: there is no effective decision procedure for determining membership in this set nor is there an effective Yes-procedure for it. If there were a Yes-procedure P for the set of the contingent PL sentences, we would be able to combine this procedure with the Yes-procedure R for Th(∅) into a single effective decision procedure for Th(∅). We take any PL sentence **X** and we apply P and R to it *concurrently*. If P returns the answer Yes, we know that **X** is contingent, and hence is not a member of Th(∅). On the other hand, if R locates a PL derivation in LD whose terminal sentence is **X**, we know that **X** is valid, and so is a member of Th(∅). Alternatively, R might locate a derivation in LD whose terminal sentence is ¬**X**; in this case, we would know that **X** is contradictory and consequently does not belong to Th(∅). By applying P and R to **X** concurrently, we will be able to determine whether **X** is valid, contradictory, or contingent. This procedure, therefore, would be an effective decision procedure for determining membership in Th(∅). Since there is no such decision procedure, there is no effective Yes-procedure for determining whether a PL sentence is contingent.

In Chapter One we described *concepts* as decidable and semidecidable. We said that if there is an effective decision procedure for determining whether any object of the relevant kind is subsumed under a certain concept or not, then we call this concept "decidable." If there is only a Yes-procedure for the applicability of a concept, we call this concept "semidecidable." The discussion above implies that none of the concepts of valid sentence, contradictory sentence, and contingent sentence is decidable in PL. Furthermore, the first two concepts are semidecidable, but contingency is not even semidecidable. We also proved in this section that there are many decidable PL sets such that the sets of all their logical consequences are undecidable. For example, Th(PA) and Th(RA) are undecidable, yet PA and RA are decidable. This entails that the question of whether an argument whose premises are members of PA or RA and whose conclusion is some AV sentence is valid or not is undecidable. More precisely, there is no effective decision procedure for determining whether or not an argument Γ/X is valid, where Γ is any subset of PA or RA and X is any AV sentence. Generally speaking, the concepts of validity and invalidity are undecidable in PL: there is no effective decision procedure for determining whether any given PL argument is valid or invalid. To be sure, there are effective decision procedures for determining the validity and invalidity of the PL arguments that belong to some specific classes of arguments; but there is no effective decision procedure for determining the validity and invalidity of *every* PL argument. However, the concept of validity is semidecidable in PL, but the concept of invalidity is not even semidecidable (it does not have an effective Yes-procedure). Similarly the concept of logical equivalence is undecidable in PL; it is only semidecidable. The concepts of satisfiability and unsatisfiability have the same fate in PL: neither one is decidable. As expected, the concept of unsatisfiability is semidecidable, but satisfiability is not even semidecidable. Hence, none of the eight logical concepts we defined in Chapter One is decidable in PL.

5.6 The Incompleteness of Second-Order Predicate Logic

5.6.1 Consider the following valid argument.

P1 The Evening Star has every property that the Morning Star has.
P2 The Morning Star is a planet in the solar system.
P3 All space telescopes generated detailed images of every planet in the solar system.
P4 The Hubble Telescope is a space telescope.

C1 The Hubble Telescope generated a detailed image of the Evening Star.

The first premise of this argument contains second-order quantification, that is, quantification over properties and relations. We will present a faithful symbolic translation of this argument. However, before we do that, we need to discuss very briefly **second-order quantification and predication**. Logicians call sentences like P1 "second-order." The predicate logic we study here is commonly called "First-Order Logic." It is first-order because its quantifiers range only over individuals, and its predicates apply only to names that designate individuals. Individuals are considered *first-order objects*. First-order quantifiers do not range over properties and relations and first-order predicates do not apply to other predicates. Properties and relations *of individuals* are considered *second-order objects*. Second-order quantifiers range over properties and relations of individuals and second-order predicates apply to first-order predicates. For example, the sentences 'This book is red' and 'There is a red book' are first-order sentences. The first may be translated into the PL sentence 'Rb' and the second into '$(\exists x)(Bx \wedge Rx)$', where '$b$', '$Bx$', and '$Rx$' translate, respectively, 'This book', 'x is a book', and 'x is Red'. On the other hand, the sentences 'Red is a color' and 'This book has a color' are appropriately considered second-order sentences, since color is a property of red, green, blue, and so on, and red, green, and blue are properties of individuals. These sentences may be translated into the second-order sentences 'CR' and '$(\exists X)(CX \wedge Xb)$', where '$CX$' is a second-order predicate that translates 'X is a color' and 'X' is a second-order variable that ranges over properties (recall that R stands for the property of being red). Thus '$(\exists X)(CX \wedge Xb)$' says that there is X, such that X is a color and this book has X. P1 is a second-order sentence because it involves second-order quantification. To say that the Evening Star has every property that the Morning Star has is to say that for every property X, if the Evening Star has X, then the Morning Star has X as well.

Syntactically speaking, a first-order quantifier applies to a variable that occupies a "name-place," i.e., a place that could be occupied by a name if the quantifier were not present. For example, in the PL sentence $(\forall y)(Ay \rightarrow Lsy)$, the first-order universal quantifier applies to the second and third occurrences of the variable y. These occurrences occupy name-places. If we remove the quantifier, we obtain the open formula $Ay \rightarrow Lsy$, in which the variable y occurs freely in two places. Each of these places could be occupied by a name. We could write $Aa \rightarrow Lsa$, $Ab \rightarrow Lsb$, $At \rightarrow Lst$, and so on. A second-order quantifier applies to a variable that occupies a "predicate-place," i.e., a place that could be occupied by a (first-order) predicate if the quantifier were not present. It is a standard practice to use a different type of letters to denote second-order variables, i.e., variables that occupy predicate-places. Let us use the uppercase forms of PL (first-order) variables to denote second-order variables. In the sentence $(\exists Z)(Za \wedge Zb)$, the second-order existential quantifier applies to the second and third occurrences of the variable Z. If we drop the quantifier, we obtain the formula $Za \wedge Zb$, which contains two free occurrences of the second-order variable Z. Both of these occurrences could be occupied by 1-place predicates. We could write $Pa \wedge Pb$, $Ga \wedge Gb$, $Sa \wedge Sb$, and so on. The semantical content of the first-order sentence $(\forall y)(Ay \rightarrow Lsy)$ is that for every individual y, if y has the property A, then the individual s bears the relation L to y; while the semantical content of the second-order sentence $(\exists Z)(Za \wedge Zb)$ is that there is a property Z that both individuals a and b have.

As its name suggests, a second-order predicate applies to second-order variables as well as to first-order predicates. For instance, the three-place second-order predicate D may be applied to the second-order variables X and Y and to the first-order predicate R to form the second-order formula DXYR. On the other hand, a first-order predicate is used to form first-order formulas by appending to it a certain number of singular terms or first-order variables. For example, the three-place first-order predicate P may be applied to the names a and b and to the first-order variable z to form the first-order formula $Pabz$. If we augment First-Order Predicate Logic (PL) with second-order variables and predicates, the resulting logic is called **Second-Order Predicate Logic** (or simply, **Second-Order Logic**) and is usually denoted as PL². Of course, we must modify the syntax and semantics of PL in order to accommodate the second-order quantification and predication. We will not undertake this task here; we will only explain the semantics of second-order quantification over properties. But first we will give a faithful PL² translation of the argument above and argue that it is valid.

PL² Translation Key

e: The Evening Star
m: The Morning Star
h: The Hubble Telescope
Pz: z is a planet in the solar system
Tx: x is a space telescope
Gvw: v generated a detailed image of w

PL² Argument

S1	$(\forall Z)(Zm \rightarrow Ze)$	
S2	Pm	
S3	$(\forall x)(Px \rightarrow (\forall y)(Ty \rightarrow Gyx))$	
S4	Th	
S5	Ghe	

The reason this argument is valid is because the predicate P can be substituted for the second-order variable Z in the formula $Zm \rightarrow Ze$; we obtain the PL sentence $Pm \rightarrow Pe$. This sentence and S2 give us Pe. The truth of S3 entails that $Pe \rightarrow (\forall y)(Ty \rightarrow Gye)$ is true. Since Pe is true, we have that $(\forall y)(Ty \rightarrow Gye)$ is true. Again the truth of this universally quantified sentence entails that the substitutional instance $Th \rightarrow Ghe$ is true, which, together with S4, logically implies Ghe.

In a PL² interpretation, second-order quantification over properties is taken to be quantification over all the subsets of the universe of discourse (UD). UD is still considered a set of individuals. PL², like PL, is an extensional system, that is, properties are reduced to their extensions. A PL² interpretation allows quantification over individuals and over properties of and relations between individuals. Since properties in PL² are just subsets of UD, to quantify over properties is to quantify over the subsets of UD. Hence a second-order sentence such as $(\forall Z)(Zm \rightarrow Ze)$ is interpreted in a PL² interpretation J² as making an assertion about every subset of UD:[4] it says that for any subset S of UD, if the referent of m belongs to S, then the referent of e belongs to S. Thus this sentence is true on J² if and only if every subset of UD that contains J²(m) also contains J²(e). If there is at least one subset of UD

---

[4] The superscript "2" that is attached to symbols representing interpretations indicates that these interpretations are second-order; it does not designate a Cartesian product of these interpretations. Hence J² does not stand for the Cartesian product J×J; rather it simply indicates that J² is a second-order interpretation.

that contains $J^2(m)$ but not $J^2(e)$, then the sentence is false on J^2. On the other hand, the sentence $(\exists Z)(Zm \wedge Ze)$ is true on J^2 if and only if there is a subset of UD that contains both $J^2(m)$ and $J^2(e)$; if there is no such subset of UD, then $(\exists Z)(Zm \wedge Ze)$ is false on J^2.

5.6.2   We can give second-order axiomatization for Peano Arithmetic. We call the resulting theory **Second-Order Peano Arithmetic**, and we denote its set of axioms as "PA²." In other words, Second-Order Peano Arithmetic is Th(PA²). The vocabulary of PA² is the same as the standard arithmetical vocabulary, AV, except that the logical vocabulary of PA² also includes second-order variables. Thus the extra-logical symbols of PA² are the familiar **0**, **s**, **+**, and **•**. We denote the vocabulary of PA² as "AW." PA² axioms are the same as the PA axioms with one important difference. PA² does not contain the infinitely many PA axioms that are generated by the Induction Schema (IS). PA Induction Schema generates an axiom for each AV formula **X[z]** with one free variable **z**. If v and y do not occur in **X[z]**, then the relevant axiom is **X[0]**→((∀v)(**X[v]**→**X[sv]**)→(∀y)**X[y]**). PA² replaces all these axioms with a single axiom. We call this axiom the **Induction Axiom** (IA).

IA   (∀Z)(Z**0**→((∀v)(Zv→Zsv)→(∀y)Zy)

As IS, IA is intended as a formal counterpart of the Principle of Mathematical Induction. Informally speaking, IS asserts that any AV *formula* that is true of **0** and that is true of sv whenever it is true of v is true of everything. IA, on the other hand, asserts that any *property* that is instantiated by **0** and that is instantiated by sv whenever it is instantiated by v is instantiated by everything. On a PL² interpretation J^2, this assertion reduces to the following statement: any subset A of UD that contains $J^2(\mathbf{0})$ and that contains the successor of β whenever it contains β contains all the individuals in UD (i.e., A = UD).

In Section 5.1, we presupposed that the structure of the natural numbers N is a model of PA, and we argued that this presupposition was plausible. We do not need to make the same presupposition here: N², in fact, is a model of PA². N² is the second-order standard interpretation of arithmetic, which is identical with the (first-order) standard interpretation of arithmetic, N, with a single modification: N² allows for quantifications over the subsets of ℕ (the universe of discourse of N²). We showed in Section 5.1 that Ax1–Ax6 are all true on N; hence they are true on N². We need to show that IA is also true on N². To show this, we must show that for every subset A of ℕ, if A contains 0 (which is the referent of **0** on N²) and A contains Sk (the successor of k) whenever it contains k, then A = ℕ. So let A be any subset of ℕ that satisfies the following two conditions: (1) 0 ∈ A; and (2) for every natural number k, if k ∈ A, then Sk ∈ A. We take X[n] to be the metalinguistic formula 'n ∈ A', where n is a variable. According to the first condition, X[0]. According to the second condition, for every natural number k, if X[k], then X[Sk]. By PMI, for every natural number n, X[n], that is, every natural number belongs to A. Hence A = ℕ. This proof establishes that IA is true on the second-order standard interpretation of arithmetic, N².

PA² is a categorical set of axioms: all its models are isomorphic to one other. Since N² is a model of PA², the categoricity of PA² is equivalent to the fact that all the models of PA² are isomorphic to N². We prove this claim.

Theorem 5.6.1: PA² is categorical.

Proof

1. We will prove that every model M² of PA² is isomorphic to N² with respect to AW, which consists of the extra-logical symbols **0**, *s*, **+**, and **•** and the logical vocabulary of PL² (i.e., ¬, →, ∀, =, the first-order and second-order variables, and the parentheses). So let M² be any model of PA², and let the universe of discourse of M² be **M**.

2. M² assigns a referent to the name **0**. Let M²(**0**) = o. M² also assigns a 1-pace function F to the function symbol *s*, a 2-place function ⊕ to the function symbol **+**, and a 2-place function * to the function symbol **•**.

3. Since M² is a model of PA², Ax1 and Ax2 are true on M². Ax1 is the sentence $(\forall x)\mathbf{0} \neq sx$. The truth of Ax1 on M² entails that o is not a possible value of the function F. In other words, o ∉ ran(F). Ax2 is the sentence $(\forall x)(\forall y)(sx = sy \rightarrow x = y)$. The truth of Ax2 on M² entails that for all individuals α and β in **M**, if Fα = Fβ, then α = β. Hence, F is a one-to-one function from **M** into **M**.

4. Define the following sequence of individuals in **M**: F^0o = o, F^1o = Fo, F^2o = FFo, F^3o = FFFo, ..., F^no = $\underbrace{F ... F}_{n \text{ times}}$ o. Let A be the set whose members are the individuals defined above, that is, A = {F^no: n is a natural number}.

5. As explained previously, since IA is true on M², every subset of **M** that contains o and contains Fα whenever it contains α is identical with **M**.

6. From 4 by the definition of A: o ∈ A; and for every α, if α ∈ A, then Fα ∈ A.

7. From 5 and 6: A = **M**. Thus for every individual β, β ∈ **M** iff there is a natural number n such that β = F^no. In other words, **M** consists solely of o and every individual that is the result of applying F n times to o.

8. Define the function h from ℕ into **M** such that for every n ∈ ℕ, h(n) = F^no. Our goal now is to prove that h is an isomorphism between N² and M².

9. From 7 and 8: h is a function from ℕ *onto* **M**.

10. From 3 and 8: for all natural numbers n and m, if h(n) = h(m), then n = m. This is so because if h(n) = h(m), then F^no = F^mo; since F is a one-to-one function from **M** into **M**, if F^no = F^mo, then n = m. The last statement can be proved by PMI on n. Let n = 0, if F^0o = F^mo, then o = F^mo, which means that m = 0, since by Ax1, o cannot be the successor of any individual in **M**. This establishes the Base Step. The Induction Hypothesis is that for all m, if F^ko = F^mo, then k = m. We want to prove that if F^{k+1}o = F^mo, then k+1 = m. So let F^{k+1}o = F^mo. By the definition of F^no, FF^ko = FF^{m-1}o. (Observe that m ≠ 0; for otherwise, we would have F^{k+1}o = F^0o = o, which implies that o = FF^ko, contradicting Ax1.) Since, by Ax2, F is one-to-one, F^ko = F^{m-1}o. By the Induction

Hypothesis, k = m−1, which entails that k+1 = m. This establishes the Inductive Step. Hence by PMI, for all natural numbers n and m, if $F^n o = F^m o$, then n = m.

11. From 9 and 10: h is a one-to-one correspondence between ℕ and **M**.
12. From 4 and 8: h(0) = o. Since 0 is the referent of **0** on N^2 (i.e., $N^2(\mathbf{0}) = 0$) and o is the referent of **0** on M^2 (i.e., $M^2(\mathbf{0}) = o$), $h(N^2(\mathbf{0})) = M^2(\mathbf{0})$.
13. It remains to show that for every n and m, h(Sn) = Fh(n), h(n + m) = (h(n) ⊕ h(m)), and h(n × m) = (h(n) * h(m)). Since $N^2(s) = S$, $N^2(+) = +$, $N^2(\bullet) = \times$, $M^2(s) = F$, $M^2(+) = \oplus$, and $M^2(\bullet) = *$, the previous equalities can be fully expressed as $h(N^2(s)n) = M^2(s)h(n)$, $h(n\ N^2(+)\ m) = (h(n)\ M^2(+)\ h(m))$, and $h(n\ N^2(\bullet)\ m) = (h(n)\ M^2(\bullet)\ h(m))$.
14. From 4 and 8: for every n in ℕ, $h(Sn) = F^{Sn}o = FF^n o = Fh(n)$.
15. We prove h(n + m) = (h(n) ⊕ h(m)), for all n and m in ℕ, by mathematical induction on m. (1) Suppose that m = 0; h(n + 0) = h(n) = (h(n) ⊕ o) (because Ax3, which is $(\forall x)(x + \mathbf{0}) = x$, is true on M^2), and (h(n) ⊕ o) = (h(n) ⊕ h(0)); so the Base Step is established. (2) Suppose that h(n + m) = (h(n) ⊕ h(m)), for some m in ℕ. This is the Induction Hypothesis. h(n + Sm) = h(S(n + m)) = Fh(n + m) = F(h(n) ⊕ h(m)) by the Induction Hypothesis. Since Ax4, which is $(\forall x)(\forall y)(x + sy) = s(x + y)$, is true on M^2, F(h(n) ⊕ h(m)) = (h(n) ⊕ Fh(m)) = (h(n) ⊕ h(Sm)). This establishes the Inductive Step. By PMI, for all n and m in ℕ, h(n + m) = (h(n) ⊕ h(m)).
16. We prove h(n × m) = (h(n) * h(m)), for all n and m in ℕ, by mathematical induction on m. (1) Let m = 0; h(n × 0) = h(0) = o = (h(n) * o) (because Ax5, which is $(\forall x)(x \bullet \mathbf{0}) = \mathbf{0}$, is true on M^2), and (h(n) * o) = (h(n) * h(0)), which completes the Base Step. (2) Assume that h(n × m) = (h(n) * h(m)), for some m in ℕ. This is the Induction Hypothesis. h(n × Sm) = h((n × m) + n) = (h(n × m) ⊕ h(n)) = ((h(n) * h(m)) ⊕ h(n)) by the Induction Hypothesis. Given that Ax6, which is $(\forall x)(\forall y)(x \bullet sy) = ((x \bullet y) + x)$, is true on M^2, ((h(n) * h(m)) ⊕ h(n)) = (h(n) * Fh(m)) = (h(n) * h(Sm)). This establishes the Inductive Step. By PMI, for all n and m in ℕ, h(n × m) = (h(n) * h(m)).
17. From 11, 12, 14–16: h is an isomorphism between N^2 and M^2.
18. From 17: M^2 is isomorphic with N^2 with respect to AW.
19. From 18 by the definition of categoricity: PA^2 is categorical.

We said in Subsection 3.5.3 that a satisfiable categorical theory can be considered as defining the structure of its models relative to its vocabulary, since all the constituents of its models that interpret its vocabulary have the same structural relations to each other. For instance, every model M^2 of PA^2 consists of individuals that can be listed in an infinite sequence $\beta_0, \beta_1, \beta_2, \beta_3, \ldots, \beta_n, \ldots$, such that every individual appears once in this sequence and every β_n corresponds to the natural number n. Furthermore, for all natural numbers n and m, $M^2(\mathbf{n}) = \beta_n$, $M^2(s)\beta_n = \beta_{Sn}$, $(\beta_n\ M^2(+)\ \beta_m) = \beta_{n+m}$, and $(\beta_n\ M^2(\bullet)\ \beta_m) = \beta_{n \times m}$, where, as usual, $M^2(\mathbf{n})$ is the referent M^2 assigns to the AW numeral **n**, and $M^2(s)$, $M^2(+)$, and $M^2(\bullet)$ are the functions M^2 assigns to the function symbols s, +, and •, respectively. M^2, therefore, has exactly the same structure as the standard interpretation of arithmetic. This allows us to define precisely what we mean by "the structure of the natural numbers." We can say that the structure of the natural numbers is the structure that is exhibited by the constituents of any model of PA² that interpret the vocabulary AW. This is not available to us in First-Order Peano Arithmetic, Th(PA), or even First-Order Arithmetic, Th$_{AV}$(N), since, as we will see in the exercises, all of these theories have nonstandard models, that is, models that have very different structures from the standard model N. You may ask, Why is it important to be able to give a precise definition of the structure of the natural numbers N? It is important philosophically. There is a school of philosophy of mathematics that is called "structuralism." This school believes that the subject matter of mathematics consists of abstract structures, such as the structure of the natural numbers N. This position would not be very attractive if one is never able to define what these structures are. An adherent of this philosophical school would want to say that the subject matter of Th$_{AV}$(N) is the structure of the natural numbers N. But given that Th$_{AV}$(N) has many models that do not remotely resemble this structure, it is important to be able to single out N as the "intended" model of Th$_{AV}$(N). The second-order axiomatization of Peano Arithmetic gives us the means to distinguish this structure among all the others and say that this is the one that is the "real" subject matter of First-Order Peano Arithmetic.

As we will see later, there is no completeness theorem for PL². More precisely, any sound deduction system for PL² is incomplete. Thus in PL² the relations of logical consequence and derivability are not equivalent. Therefore it is important to make it clear that a PL² theory is a PL² set such that it contains all its *logical consequences* that are composed of its vocabulary. In PL², we cannot substitute "its theorems" for "its logical consequences." We have the following interesting consequences of Theorem 5.6.1. (It should be clear that M² is a model of Th(PA²) iff M² is a model of PA².) Th$_{AW}$(N²) is the set that consists of all the AW sentences that are true on the second-order structure of the natural numbers (N²).

Theorem 5.6.2:

5.6.2a Second-Order Peano Arithmetic, i.e., Th(PA²), is *semantically* complete, in the sense that for every AW sentence, either it or its negation is a logical consequence of Th(PA²).

5.6.2b Th(PA²) = Th$_{AW}$(N²).

5.6.2c Second-Order Arithmetic, i.e., Th$_{AW}$(N²), is finitely axiomatizable.

Proof of 5.6.2a

1. Let **X** be any AW sentence. Either **X** is true or false on N².
2. Assume that **X** is true on N².
3. By theorem 5.6.1: for every model M² of Th(PA²), M² is isomorphic to N² with respect to AW.
4. From 3 by Theorem 3.4.1: for every model M² of Th(PA²), M² is elementarily equivalent to N² with respect to AW.

5. From 2 and 4 by the definition of elementary equivalence: **X** is true on every model of Th(PA²).
6. From 5 by the definition of logical consequence: Th(PA²) ⊨ **X**.
7. From 2 through 5: if **X** is true on N², then Th(PA²) ⊨ **X**.
8. Now assume that **X** is false on N², that is, ¬**X** is true on N².
9. By similar reasoning to 4–5: ¬**X** is true on every model of Th(PA²).
10. From 8: Th(PA²) ⊨ ¬**X**.
11. From 8 through 10: if **X** is false on N², then Th(PA²) ⊨ ¬**X**.
12. From 1, 7, and 11: for every AW sentence **X**, Th(PA²) ⊨ **X** or Th(PA²) ⊨ ¬**X**.
13. From 12: Th(PA²) is a semantically complete theory.

Proof of 5.6.2b
1. Since N² is a model of Th(PA²), every sentence in Th(PA²) is true on N²; and hence every sentence in Th(PA²) belongs to Th$_{AW}$(N²).
2. Now let **X** be in Th$_{AW}$(N²). Hence it is true on N².
3. By Theorem 5.6.1: every model of Th(PA²) is isomorphic to N² with respect to AW.
4. From 3 by Theorem 3.4.1: every model of Th(PA²) is elementarily equivalent to N² with respect to AW.
5. From 2 and 4: **X** is true on every model of Th(PA²).
6. From 4: Th(PA²) ⊨ **X**.
7. From 6: since Th(PA²) is a theory, **X** belongs to Th(PA²).
8. From 2–7: if **X** ∈ Th$_{AW}$(N²), then **X** ∈ Th(PA²).
9. From 1 and 8 by the Principle of Extensionality: Th(PA²) = Th$_{AW}$(N²).

Proof 5.6.2c
1. By Theorem 5.6.2b: Th$_{AW}$(N²) = Th(PA²).
2. PA² is a finite set (and hence, decidable).
3. From 1 and 2 by Definition 3.5.1a: PA² is a set of axioms for Th$_{AW}$(N²).
4. From 3 by Definition 3.5.1c: Th$_{AW}$(N²) is finitely axiomatizable.

It is tempting to consider Theorem 5.6.2c as a fulfillment of the axiomatic enterprise. Since AV is a subset of AW, Th$_{AV}$(N) is a subset of Th$_{AW}$(N²). Hence PA² might be considered as a complete axiomatization of Th$_{AV}$(N): to determine whether an AV sentence is true or false, we only need to "derive" it or its negation from the second-order axioms PA². Unfortunately the situation isn't so simple. Whether PA² is a successful complete axiomatization of Arithmetic or not depends on what is meant by "derive it or its negation from the second-order axioms." If 'derive' here means its usual sense—namely, that there is a PL² derivation of **X** or of ¬**X** from PA²—then the preceding claim is false. Second-Order Arithmetic is not complete in this sense. To construct a PL² derivation of an AV sentence **X** from PA², one needs to specify a sound set of inference rules that is subject to certain normal constraints. We will prove later that any such set is incomplete in PL². This implies that there are logical consequences of PA² that cannot be derived from PA² no matter what rules of inference we stipulate for PL², so long as these rules meet certain reasonable conditions. It is true that Theorem 5.6.2a entails that for any AV sentence, either it or its negation is a logical consequence of PA². But unfortunately this type of completeness in PL² does not translate into proof-theoretic completeness. Just because **X** is a logical consequence of PA², it does not follow that **X** is proof-theoretically derivable from PA². This is very different from First-Order Logic: if **X** is a logical consequence of PA, then it is proof-theoretically derivable from PA. The difference between Second-Order and First-Order Peano Arithmetic is that the semantical powers of PA² are much more extensive than those of PA: many true arithmetical sentences that are not logical consequences of PA are logical consequences of PA². But these semantical powers do not correspond to equivalent proof-theoretic powers: there is no sound set of inference rules that meets certain reasonable conditions and that allows for the derivation from PA² of all the logical consequences of PA². Hence there is no resolution to the limitations of the axiomatic method.

5.6.3   The results of the previous subsection supply sufficient grounds for establishing that PL² is an incomplete Logic. We proved in Chapter Three that the Compactness Theorem follows from the Soundness and Completeness Theorems for PL. To remind the reader of this theorem, we restate it here. The Compactness Theorem asserts that for every PL set Σ and PL sentence **X**, if $\Sigma \vDash$ **X**, then there is a finite subset Γ of Σ such $\Gamma \vDash$ **X**. Our proof of the Compactness Theorem was general enough to apply to any logical system that has a well-defined notion of logical consequence, whose formal derivations are finite sequences of sentences, and that has a sound and complete proof theory. We will prove two incompleteness results about PL². The first one shows that the Compactness Theorem fails for PL². This establishes that PL² cannot have a sound and complete proof theory in which derivations are finite sequences. Let us refer to a proof theory in which derivations are finite sequences of sentences as a **finite proof theory**. We first rehearse our proof of the Compactness Theorem to show that it follows from the supposition that PL² has a sound and complete finite proof theory. Note that the relation of logical consequence is defined for PL² in the usual way: for every PL² set Σ and every PL² sentence **X**, $\Sigma \vDash^2$ **X** iff **X** is true on every (second-order) model of Σ that is relevant to **X**. We write '$\vDash^2$' to indicate that this is the relation of logical consequence in PL².

Theorem 5.6.3: If a relation of derivability $\vdash^2$ is definable for PL² such that for every PL² set Σ and every PL² sentence **X**, $\Sigma \vdash^2$ **X** iff $\Sigma \vDash^2$ **X**, where a PL² derivation is a finite sequence of PL² sentences, then the Compactness Theorem holds for PL².

Proof

1. Suppose that PL² has a well-defined derivability relation $\vdash^2$ that is sound, complete, and finite.
2. Let Σ be a PL² set and **X** be a PL² sentence such that $\Sigma \vDash^2 \mathbf{X}$. Our goal is to show that there is a finite subset of Σ that logically implies **X**.
3. From 1 and 2: since $\vdash^2$ is complete, $\Sigma \vdash^2 \mathbf{X}$.
4. From 3: there is a PL² derivation D of **X** from Σ.
5. From 1 and 4: D is a finite sequence of PL² sentences. Let Σ_D be the set of the sentences in Σ that are invoked in D.
6. From 5: Σ_D is a finite subset of Σ and $\Sigma_D \vdash^2 \mathbf{X}$.
7. From 1 and 6 by the soundness of $\vdash^2$: $\Sigma_D \vDash^2 \mathbf{X}$.
8. From 2–7: for all PL² set Σ and PL² sentence **X**, if $\Sigma \vDash^2 \mathbf{X}$, then there is a finite subset Σ^* of Σ such that $\Sigma^* \vDash^2 \mathbf{X}$.
9. From 8: the Compactness Theorem holds for PL².
10. From 1–9: if PL² has a sound and complete finite proof theory, the Compactness Theorem holds for PL².

We now prove that the Compactness Theorem does not hold for PL².

Theorem 5.6.4: There are a PL² set Σ and a PL² sentence **X** such that Σ logically implies **X** but no finite subset of Σ logically implies **X**.

Proof

1. Let W be the vocabulary of PA² plus a single additional name e. Recall that $s^n\mathbf{0}$ is an abbreviation for the singular term $\underbrace{sss \ldots s}_{n \text{ times}} \mathbf{0}$, and that $s^n\mathbf{0}$ refers to the number n in the standard model of Th(PA²), which is the second-order structure of the natural numbers N². Let Θ be the following set of W sentences: $\Theta = \{e \neq s^1\mathbf{0},\ e \neq s^2\mathbf{0},\ e \neq s^3\mathbf{0}, \ldots, e \neq s^n\mathbf{0}, \ldots\} = \{e \neq s^n\mathbf{0}: n \text{ is a positive integer}\}$. Since PA² is a finite set, we let C be the AW sentence that is the conjunction of these axioms. (Observe that AW is a subset of W.) Define Σ to be the PL² set = $\{C\} \cup \Theta$. We will establish that $\Sigma \vDash^2 e = \mathbf{0}$, and for every finite subset Γ of Σ, $\Gamma \nvDash^2 e = \mathbf{0}$.
2. First we show that Σ is satisfiable. Let N^e be the PL² interpretation that is identical with N² except that N^e contains the name e in its list of names and assigns to it the referent 0. We argued previously that N² is a model of PA², and hence it is a model of C. Since N^e does not change any of the interpretations made by N², N^e is a model of PA² as well. We also said previously that on N² every singular term $s^n\mathbf{0}$ refers to the number n. Hence the referent of e on N^e, which is the number 0, is distinct from the referent of any singular term $s^n\mathbf{0}$, where n is a positive integer. It follows that every sentence of the form $e \neq s^n\mathbf{0}$, where n is any positive integer, is true on N^e. This

shows that N^e is a model of Θ. Since $\Sigma = \{C\} \cup \Theta$, N^e is a model of Σ. We conclude that Σ is satisfiable.

3. Let M^2 be any model of Σ; thus M^2 is also a model of C and of Θ.

4. From 1 and 3: every sentence of the form $e \ne s^n\mathbf{0}$, where n is a positive integer, is true on M^2. Hence the referent of e on M is different from the referents of all the numerals $s^n\mathbf{0}$, where n is a positive integer.

5. From 1 and 3: since M^2 is a model of C (i.e., of PA^2), by the proof of Theorem 5.6.1, the universe of discourse of M^2, **M**, consists of every individual $F^k o$, where F is the function that M^2 assigns to the function symbol s, o is the referent that M^2 assigns to the name **0**, k is a natural number, and $F^k o$ is the result of applying the function F to o k times. (By stipulation, $F^0 o = o$.) We proved that $\mathbf{M} = \{F^k o:$ k is a natural number$\}$ by observing that since IA is true on M^2, every subset of **M** that contains o and that contains $F\alpha$ whenever it contains α is identical with **M**; and of course $\{F^k o:$ k is a natural number$\}$ meets these two conditions.

6. From 1 and 5: We now prove that $M^2(s^k\mathbf{0}) = F^k o$, for every natural number k. The proof is by PMI on k. Let k = 0. $M^2(s^0\mathbf{0}) = M^2(\mathbf{0}) = o = F^0 o$. So the Base Step is established. The Induction Hypothesis asserts that $M^2(s^h\mathbf{0}) = F^h o$. We want to show that $M^2(s^{h+1}\mathbf{0}) = F^{h+1} o$. By the definition of $s^k\mathbf{0}$, $M^2(s^{h+1}\mathbf{0}) = M^2(ss^h\mathbf{0})$. By the definition of PL^2 interpretation, M^2 satisfies all the conditions of a PL interpretation. Hence M^2 satisfies condition 1.2.1c (see Chapter One). It follows that $M^2(ss^h\mathbf{0}) = M^2(s)M^2(s^h\mathbf{0})$. By the Induction Hypothesis and since $M^2(s) = F$, $M^2(s)M^2(s^h\mathbf{0}) = FF^h o = F^{h+1} o$. This establishes the Inductive Step. By PMI, $M^2(s^k) = F^k o$, for all natural numbers k.

7. From 4 through 6: $M^2(e)$ can only be o, which is the referent of **0** on M^2. Thus $e = \mathbf{0}$ is true on M^2.

8. From 3 through 7: since M^2 is an arbitrary model of Σ, the sentence $e = \mathbf{0}$ is true on every model of Σ, that is, $\Sigma \vDash^2 e = \mathbf{0}$.

9. Now we prove that no finite subset of Σ logically implies $e = \mathbf{0}$. So let Γ be any finite subset of Σ. There are three cases: (1) Γ contains no members of Θ, that is, $\Gamma = \{C\}$; (2) $\Gamma \subseteq \Theta$, that is, $C \notin \Gamma$; (3) $C \in \Gamma$ and $\Gamma \cap \Theta \ne \emptyset$.

10. Case 1: $\Gamma = \{C\}$. Let N_e be the (second-order) structure of the natural numbers with e added to its list of names and e's referent on N_e is 1. It is clear that N_e is a model of Γ (since N_e is a model of PA^2 and C is the conjunction of the PA^2 axioms), and that $e = \mathbf{0}$ is false on N_e. Hence Γ does not logically imply $e = \mathbf{0}$.

11. Case 2: $\Gamma \subseteq \Theta$. Since Γ is finite, $\Gamma \subseteq \{e \ne s^1\mathbf{0}, e \ne s^2\mathbf{0}, e \ne s^3\mathbf{0}, ..., e \ne s^n\mathbf{0}\}$, where n is the greatest natural number such that the sentence $e \ne s^n\mathbf{0} \in \Gamma$ (if Γ is empty, take n = 0). We take J^2 to be the PL^2 interpretation whose UD is $\{0, 1, 2, 3, ..., n, n+1\}$, whose LN is $\{e, \mathbf{0}, s^1\mathbf{0}, s^2\mathbf{0}, s^3\mathbf{0}, ..., s^n\mathbf{0}\}$, and that assigns 0 to **0**, 1 to $s^1\mathbf{0}$, 2 to $s^2\mathbf{0}$, 3 to $s^3\mathbf{0}$, ..., n to $s^n\mathbf{0}$, and n+1 to e. It is obvious that J^2 is a model of Γ, and that the sentence $e = \mathbf{0}$ is false on J^2. Hence Γ does not logically imply $e = \mathbf{0}$.

12. Case 3: C ∈ Γ and Γ∩Θ ≠ ∅. Take I^2 to be the PL^2 interpretation that is identical with N^2 except e belongs to the list of names of I^2, and the referent of e on I^2 is the number n+1, where n is the greatest natural number such that the sentence $e \neq s^n0 \in \Gamma$. (There is always such n since Γ is finite.) Since N^2 is a model of PA^2, and since I^2 does not alter any of the interpretations made by N^2, I^2 is also a model of PA^2, that is, I^2 is a model of C. We previously proved that the referent of every s^m0 on N^2 is the number m. Given that I^2 makes the same interpretations as N^2, the referent of every s^m0 on I^2 is the number m. Since n is the greatest number such that the sentence $e \neq s^n0 \in \Gamma$, and since $I^2(e)$ = n+1, every sentence in Γ of the form $e \neq s^m0$ is true on I^2. It follows that I^2 is a model of Γ, and that $e = 0$ is false on I^2. Therefore Γ does not logically imply $e = \mathbf{0}$.

13. From 9 through 12: Γ does not logically imply $e = \mathbf{0}$.

14. From 9 through 13: since Γ is an arbitrary finite subset of Σ, no finite subset of Σ logically implies $e = \mathbf{0}$.

15. From 8 and 14: $e = \mathbf{0}$ is a logical consequence of Σ but not of any finite subset of Σ.

16. From 15: there are a PL^2 set Σ and a PL^2 sentence **X** such that Σ logically implies **X**, but no finite subset of Σ logically implies **X**.

The preceding theorem entails that the Compactness Theorem does not hold for PL^2. Our desired conclusion is an immediate consequence of the previous two theorems: if a derivability relation that is sound, complete, and finite is definable for PL^2, then by Theorem 5.6.3 the Compactness Theorem would hold for PL^2; since by Theorem 5.6.4 the Compactness Theorem fails in PL^2, no such derivability relation is definable for PL^2. We state this fact as the following theorem.

Theorem 5.6.5: There can be no relation of derivability for PL^2 that is sound, complete, and finite.

5.6.4 A natural reaction to the previous theorem is to wonder whether the restriction of finiteness can be relaxed. In other words, if we allow infinitely long derivations, could Second-Order Logic have a sound and complete proof theory? Before we can answer this question, we need to impose some restrictions on what count as permissible infinite derivations. Recall that derivations are meant to capture the notion of demonstrative proof. One essential feature of demonstrative proofs is their "verifiability." In other words, when we are presented with a demonstrative proof of a certain statement X from some set of premises Σ, we should be able to examine this proof and determine whether this is really a demonstrative proof of X from Σ.

But if we allow for infinitely long proofs, how are we supposed to examine such proofs and determine their conclusions? In fact, we just encountered an infinitely long proof and we had no difficulty determining its correctness and its conclusion. The set of the premises that consists of the seven PA^2 axioms and the infinitely many sentences of the form $e \neq s^n0$, where n is any positive integer, logically implied that $e = 0$. That was a semantical argument, but we can transform it into a formal derivation. Of course, we will need some inference rule that works with infinitely many antecedents. Let us devise such a rule. We call it the ω-rule. This rule takes infinitely many antecedents and produces a single conclusion of the form $(\forall z)X$. Here is the formal representation of this rule.

The ω-rule: Let **X** be any PL² formula that contains occurrences of the first-order variable **z** and that contains no z-quantifiers, let **X[t]** be the PL² sentence that is obtained from **X** by replacing every occurrence of **z** in **X** by the singular term **t**. This rule must be applied fully within an open block.

$$
\begin{array}{lll}
[n & & \\
& \vdots & \\
& i-(i+6) & \text{the seven PA}^2 \text{ axioms} \\
& \vdots & \\
& j & X[s^00] \\
& j+1 & X[s^10] \\
& j+2 & X[s^30] \\
& \vdots & \\
& j+k & X[s^k0] \\
& \vdots & \\
& \omega & (\forall z)X \qquad i-(i+6),\ j-\ldots,\ \omega\text{-rule} \\
& \vdots & \\
n] & &
\end{array}
$$

This rule is truth-preserving. Any PL² interpretation that satisfies the PA² axioms is isomorphic to the standard model of Th(PA²), N²; and hence its universe of discourse consists of infinitely many individuals each of which is the unique referent of some singular term s^k0. If on such an interpretation all of the referents of these singular terms satisfy the formula **X**, then **X** must be true of all the individuals in the universe of discourse. It follows that $(\forall z)X$ is true on that interpretation. Using this rule we can derive $e = 0$ from PA²∪Θ, where $\Theta = \{e \neq s^n0 : n \text{ is a positive integer}\}$.

$$
\begin{array}{llll}
[0 & 0.1\text{–}0.7 & \text{the seven PA}^2 \text{ axioms} & \\
[1 & 1 & e \neq s^00 & \text{RA} \\
& 2 & e \neq s^10 & \text{Premise} \\
& 3 & e \neq s^20 & \text{Premise} \\
& 4 & e \neq s^30 & \text{Premise} \\
& \vdots & & \\
& k & e \neq s^k0 & \text{Premise} \\
& \vdots & & \\
& \omega & (\forall x)e \neq x & (0.1\text{–}0.7),\ 1\text{–}\ldots,\ \omega\text{-rule} \\
& \omega+1 & e \neq e & \omega,\ \text{UI}
\end{array}
$$

1]	ω+2	$e = e$	Id
	ω+3	$e = s^0 0$	1–(ω+2), RAA
0]	ω+4	$e = \mathbf{0}$	ω+3, defined notation ($s^0\mathbf{0}$ is an abbreviation for $\mathbf{0}$)

The key idea is that although this derivation is infinite, an infinite sequence of its lines exhibits a decidable pattern: we know how every line k, where $1 \leq k < \omega$, looks. Since we also know what the PA^2 axioms are, we have an effective decision procedure for determining the applicability of the ω-rule: given any infinite sequence of PL^2 sentences, we can determine whether the inference rule ω is applicable to it or not. There is, however, an important point. As stated, the ω-rule presupposes that we can "locate" an infinite segment of sentences that consists of the seven PA^2 axioms and of an infinite sequence of sentences of the form $\mathbf{X}[s^k\mathbf{0}]$, where k is a natural number. It is possible, to have an infinitely long derivation, in which these sentences are scattered in an arbitrary, unpredictable fashion that exhibits no decidable pattern. In this case, we would not be able to decide whether we actually have all the required antecedents in order to apply the ω-rule.

We will stipulate certain restrictions on the proof theory of PL^2 that will allow for infinite derivations but without losing the ability to apply the rules of inference in a decidable way. We will try to keep these restrictions as minimal as possible. We stipulate the following three conditions on any proof theory for PL^2, that is, on any set of inference rules for PL^2. (1) If Σ is a decidable set of PL^2 sentences, then the set of all the PL^2 derivations that invoke only members of Σ as premises is effectively enumerable. In other words, if Σ is a decidable set of PL^2 sentences, then there is an effective procedure that generates an infinite sequence $D_1, D_2, D_3, \ldots, D_n, \ldots$ such that every D_j is a PL^2 derivation *from* Σ (i.e., all the premises that are invoked in D_j are members of Σ), and every PL^2 derivation from Σ appears as a member of this infinite sequence. (2) For any given PL^2 sentence $\mathbf{X}$ and any given PL^2 derivation D_j from Σ (where Σ is a decidable set), there is an effective *decision* procedure for determining whether $\mathbf{X}$ is the conclusion of D_j or not. (3) The set of PL^2 rules of inference are sound, that is, if $\mathbf{X}$ is derivable in PL^2 from Σ, then $\mathbf{X}$ is a logical consequence of Σ in PL^2. A relation of derivability that fulfills all these three conditions is denoted as "$\vdash^2$". More precisely, for any PL^2 set Σ and any PL^2 sentence $\mathbf{X}$, $\Sigma \vdash^2 \mathbf{X}$ iff there is a PL^2 derivation of $\mathbf{X}$ from Σ that satisfies the preceding three conditions. If there is such a derivability relation for PL^2 that is also *complete*, we say that PL^2 is a complete logic. Said differently, PL^2 is complete iff for all PL^2 sets Σ and all PL^2 sentences $\mathbf{X}$, if $\Sigma \vDash^2 \mathbf{X}$, then $\Sigma \vdash^2 \mathbf{X}$. We conclude this section by proving that PL^2 is an incomplete logic, even if a certain type of infinitely long derivations is allowed.

Theorem 5.6.4: PL^2 is incomplete.

Proof

1. Reductio Assumption: PL^2 is complete. Hence for every PL^2 set Σ and every PL^2 sentence $\mathbf{X}$, $\Sigma \vDash^2 \mathbf{X}$ only if $\Sigma \vdash^2 \mathbf{X}$.

THE INCOMPLETENESS THEOREMS   271

2. Let Σ be the set PA^2, which consists of the seven axioms of Second-Order Peano Arithmetic.

3. By Theorems 5.6.2a and 5.6.2b: $Th(PA^2)$ is semantically complete and $Th(PA^2) = Th_{AW}(N^2)$, where AW is $Voc(PA^2)$.

4. Let AV be the vocabulary of First-Order Peano Arithmetic, that is, $AV = Voc(PA)$. It is clear that $AV \subseteq AW$.

5. From 3: since $Th(PA^2)$ is a semantically complete theory and $Th(PA^2) = \{X: X$ is an AW sentence such that $PA^2 \vDash^2 X\}$, PA^2 is also semantically complete. (To see this let **X** be any AW sentence; either $Th(PA^2) \vDash^2 X$ or $Th(PA^2) \vDash^2 \neg X$; since $Th(PA^2)$ is a theory, either $X \in Th(PA^2)$ or $\neg X \in Th(PA^2)$; hence either $PA^2 \vDash^2 X$ or $PA^2 \vDash^2 \neg X$.)

6. From 5: for every AW sentence **X** either $PA^2 \vDash^2 X$ or $PA^2 \vDash^2 \neg X$.

7. From 6: given that $AV \subseteq AW$, for every AV sentence **X**, either $PA^2 \vDash^2 X$ or $PA^2 \vDash^2 \neg X$.

8. From 1 and 7: for every AV sentence **X**, $PA^2 \vDash^2 X$ or $PA^2 \vDash^2 \neg X$.

9. From 8: we describe an effective decision procedure for determining whether any AV sentence is a logical consequence of PA^2 or not. Since PA^2 is a decidable set, by the definition of the derivability relation $\vdash^2$, the set of all the PL^2 derivations from PA^2 is effectively enumerable. Let this set be the following infinite sequence: $D_1, D_2, D_3, \ldots, D_n, \ldots$. Also by the definition of $\vdash^2$, there is an effective decision procedure for determining whether any given PL^2 sentence **X** is the conclusion of any given PL^2 derivation D_j or not. So let **X** be any AV sentence. We apply this effective decision procedure for **X** and D_1. If **X** is the conclusion of D_1, we determine, by the soundness of $\vdash^2$, that $PA^2 \vDash^2 X$; if $\neg X$ is the conclusion of D_1, we determine, also by the soundness of $\vdash^2$, that $PA^2 \vDash^2 \neg X$; if neither is the conclusion of D_1, we examine D_2 and we repeat the same procedure. Since either $PA^2 \vDash^2 X$ or $PA^2 \vDash^2 \neg X$, this procedure eventually (after finitely many deterministic steps) will locate a PL^2 derivation D_j from PA^2 such that either **X** or $\neg X$ is its conclusion. If it is **X**, then we know that $PA^2 \vDash^2 X$; and if it is $\neg X$, then we know that $PA^2 \vDash^2 \neg X$, which entails that $PA^2 \nvDash^2 X$ because PA^2 is satisfiable, and thus it cannot logically imply **X** and $\neg X$.

10. From 3 and 9: the previous effective decision procedure yields an effective decision procedure for determining membership in $Th_{AV}(N)$, i.e., First-Order Arithmetic. Let **X** be any AV sentence. Apply the decision procedure described in 9 to determine whether $PA^2 \vDash^2 X$ or $PA^2 \nvDash^2 X$. If $PA^2 \vDash^2 X$, then we determine that $X \in Th(PA^2)$, and hence $X \in Th_{AW}(N^2)$ (because $Th(PA^2) = Th_{AW}(N)$). Given that **X** is an AV sentence and that AV is part of AW, it follows that $X \in Th_{AV}(N)$. On the other hand, if $PA^2 \nvDash^2 X$, then we determine that $X \notin Th(PA^2)$, which means that $X \notin Th_{AW}(N^2)$, and this in

turn entails that $\mathbf{X} \notin \text{Th}_{AV}(N)$. Thus we have an effective decision procedure for determining membership in First-Order Arithmetic, $\text{Th}_{AV}(N)$.

11. From 10: First-Order Arithmetic is decidable. This contradicts Theorem 5.5.5b.
12. From 1 through 11: the Reductio Assumption is false. We conclude that PL^2 is incomplete.

5.7 Gödel's Second Incompleteness Theorem

5.7.1 In Section 5.3, we defined PROOF to be the set of all the ordered pairs $\langle n, m \rangle$ such that n is the gödel number of a PL derivation from PA of an AV sentence whose gödel number is m, where AV is the standard arithmetical vocabulary. We said that PROOF is a recursive relation, and hence it is representable in Th(PA). We took **proof**$[x, y]$ to be an AV formula, with two free variables, that represents PROOF in Th(PA). This means that **proof**$[x, y]$ satisfies the following two conditions:

For all natural numbers n and m, if $\langle n, m \rangle \in$ PROOF, then PA $\vdash$ **proof**$[n, m]$, and
For all natural numbers n and m, if $\langle n, m \rangle \notin$ PROOF, then PA $\vdash \neg$**proof**$[n, m]$.

We also defined the "provability" formula **prov**$[y]$ as $(\exists x)$**proof**$[x, y]$. We showed there that if m is the gödel number of a theorem of PA, then PA $\vdash$ **prov**$[m]$. It is clear that since Th(PA) is not a decidable set, it is *not* true that if n is the gödel number of a sentence that is not a theorem of PA, then PA $\vdash \neg$**prov**$[n]$.

We did not specify the formula **proof**$[x, y]$. We only said that it is an AV formula with two free variables that represents the numerical relation PROOF. There are in fact more than one AV formula that represents PROOF in Th(PA). It is customary to refer to any AV formula that represents PROOF in Th(PA) as a **proof predicate**. If $\mathbf{Z}[x, y]$ is a proof predicate, the formula $(\exists x)\mathbf{Z}[x, y]$ is usually described as a **provability predicate**. Metaphorically speaking, $\mathbf{Z}[x, y]$ says that x is a PA proof of y and $(\exists x)\mathbf{Z}[x, y]$ says that y is provable from PA. In his proof of the First Incompleteness Theorem, Gödel constructed a specific proof predicate, which came to be known as "the standard proof predicate." The standard proof predicate permits us to "formalize" the proof of the First Incompleteness Theorem in Th(PA). We will give a rough explanation of this idea in this section, but we will prove the Second Incompleteness Theorem by invoking certain conditions that are called the "Provability Conditions."

Informally speaking, the Second Incompleteness Theorem says that if PA is consistent, then it is impossible to prove the consistency of PA in Th(PA) itself. It might sound strange to think that PA can prove its own consistency. However, during the first third of the twentieth century, there was a major school of thought in the philosophy of mathematics that attempted to show that the consistency of PA could be proved by some "computational" methods. The founder of this school was the great German mathematician David Hilbert (1862–1943). Hilbert argued that there are two types of mathematics: finitary and ideal. Although he never gave a precise definition of what finitary mathematics was, the idea is sufficiently intuitive. Hilbert seemed to mean by "finitary mathematics" computational procedures that are applied to finite collections of numbers. However, one must note that Hilbert included *potentially* infinite sets of numbers as part of finitary mathematics. Hence, roughly speaking, finitary mathematics consists of computational arithmetical properties, relations, and functions applied to well-defined finite or potentially infinite sets of natural numbers or of n-tuples of natural numbers. Ideal mathematics is the rest of mathematics. According to Hilbert, ideal mathematics has no "semantical significance"; one cannot attach a real meaning to an ideal mathematical statement. Rather, ideal mathematics is a *formal* enterprise that is an extremely powerful *instrument* for proving finitary theorems. These finitary theorems have finitary proofs. But most of these finitary proofs are very cumbersome, long, tedious, and inelegant. Ideal mathematical proofs, on the other hand, can be shorter, more manageable, and much more elegant. In principle, when it comes to finitary theorems, ideal mathematics is dispensable. But it would be foolish to dispense with ideal mathematics in favor of finitary mathematics. Many finitary truths might never be discovered without the employment of ideal mathematics. Hilbert's philosophy of mathematics was called "instrumentalism," "formalism," and "finitism." As we have seen, all of these terms have basis in Hilbert's thought.

Hilbert's Program was a program in the foundations of mathematics that was closely connected to his philosophy of mathematics. The program aimed at showing that ideal mathematics was a reliable instrument for arriving at finitary truths. The program had two parts: reduction and reliability. The first part of the program attempted to reduce almost all of mathematics to Peano Arithmetic. By the time Hilbert worked on his program, Peano Arithmetic had achieved such a foundational status. Analytic geometry reduced geometry to the theories of real and complex numbers; the complex numbers were reduced to ordered pairs of real numbers; the real numbers were reduced to certain sets (or sequences) of rational numbers; the rational numbers were reduced to classes of ordered pairs of integers; and the latter were reduced to classes of ordered pairs of natural numbers. Thus by the time of Hilbert, the project of reducing mathematics to the theory of natural numbers was accomplished. That was the reduction part of the program. The reliability part of the program was aimed at showing that Peano Arithmetic is a reliable instrument for proving finitary truths. Due to the efforts of Hilbert and some of his collaborators, such as the Swiss mathematician Paul Bernays (1888–1977), the task of demonstrating the reliability of Peano Arithmetic was narrowed down to the task of giving a *finitary* proof of the consistency of Peano Arithmetic.

To show that Peano Arithmetic is consistent, it is sufficient to show that the sentence **0 = 1**, where **1** is $s0$, is not a theorem of PA: PA is consistent iff PA $\nvdash$ **0 = 1**. Suppose that PA is consistent. It is clear that **0 ≠ 1** is a theorem of PA (by Universal Instantiation, Ax1, which is $(\forall x)0 \neq sx$, entails **0 ≠ s0**). Since PA is consistent, **0 = 1** cannot be a theorem of PA (otherwise, both **0 = 1** and **0 ≠ 1** would be theorems of PA). Now suppose that **0 = 1** is not a theorem of PA. It follows that no sentence and its negation are derivable from PA;

otherwise, if PA ⊢ X and PA ⊢ ¬X, then by Explosion PA ⊢ 0 = 1. Hence PA is consistent. Hilbert's and his collaborators' efforts were directed at finding a finitary proof of the statement that 0 = 1 is not a theorem of PA. Peano Arithmetic is a theory that contains both finitary and ideal mathematics. Almost all mathematicians and philosophers of mathematics believed that all of finitary mathematics was formalizable within Peano Arithmetic. This meant that if there was a finitary proof of the consistency of PA, then this proof must be formalizable as a PA proof. Gödel's Second Incompleteness Theorem establishes that a certain AV sentence that can be interpreted as affirming the consistency of PA is not derivable from the PA axioms. This result is almost universally held as establishing the impossibility of proving the consistency of PA within Th(PA). Since there is near consensus that every finitary proof can be formalized as a PA proof, the fact that there is no PA proof for the consistency of PA is seen as conclusive evidence that there is no finitary proof for the consistency of PA. Shortly after the publication of Gödel's Incompleteness Theorems, it was widely believed that the Second Incompleteness Theorem refuted Hilbert's program and with it Hilbert's instrumentalism. This view has been and remains the prevalent position.

There were some dissenters then and even more now. In fact, Gödel himself wrote in his concluding remarks after his sketch of the proof of the Second Incompleteness Theorem: "I wish to note expressly that Theorem XI (and the corresponding results for M and A) do not contradict Hilbert's formalistic viewpoint. For this viewpoint presupposes only the existence of a consistency proof in which nothing but finitary means of proof is used, and it is conceivable that there exist finitary proofs that *cannot* be expressed in the formalism of P (or of M or A)" (p. 195).[5] However, Gödel later changed his position, mostly due to Bernays's efforts, and joined the general agreement that all finitary proofs are formalizable as PA proofs. Most of those who dissent do so not because they think that there are finitary proofs beyond the reach of Peano Arithmetic but because they think that the formal sentence that Gödel constructed in order to express formally the statement that PA is consistent does not do justice to the metatheoretical statement. The standard consistency sentence incorporates several formulas, such as the standard proof predicate, which some find a grossly inadequate formalization of the concept of finitary proof. They argue that there are other proof predicates that are better candidates as representations of the types of proofs that are suitable for Hilbert's philosophy of mathematics. To be sure, there are other proof predicates, that is, AV formulas that represent PROOF in Th(PA), for which the Second Incompleteness Theorem fails. This is a striking feature of the Second Incompleteness Proof. The First Incompleteness Proof is totally insensitive to the choice of the proof predicate: the theorem holds for any AV formula that represents PROOF in Th(PA).

---

5   Kurt Gödel [1931], "On Formally Undecidable Propositions of *Principia Mathematica* and Related Systems I," in *Kurt Gödel Collected Works*, vol. I, edited by Solomon Feferman, John W. Dawson, Jr., Stephen C. Kleene, Gregory H. Moore, Robert M. Solovay, and Jean van Heijenoort, Oxford University Press, Oxford, 1986, pp. 145–95. (The German original of this paper was published in 1931.)

We will show later that the second Incompleteness Theorem holds for any proof predicate that satisfies certain "Provability Conditions." However, those who question the adequacy of the standard proof predicate also question the adequacy of some of these conditions as reasonable constraints on any acceptable proof predicate. For them, the fact that there are nonstandard proof predicates, which do not meet these conditions and for which the Second Incompleteness Theorem fails, demonstrates that one cannot accept at face value the common view that the Second Incompleteness Theorem implies that there can be no finitary proof for the consistency of PA. The construction of the standard proof predicate and the derivability conditions must be defended as correct representations of the concept of proof that is most relevant to Hilbert's instrumentalism. Perhaps the most famous nonstandard proof predicate is due to Rosser (1936). Informally stated, according to this proof predicate, a sequence D of AV sentences does not count as a PA proof if its last line contradicts the last line of a PA proof whose gödel number is less than the gödel number of D.[6] It is possible to define an AV sentence **Z** on the basis of this proof predicate such that **Z** can be interpreted as affirming the consistency of PA and PA ⊢ **Z**. Thus the Second Incompleteness Theorem fails for Rosser's proof predicate. We will not pursue this issue any further here. Our goal is only to create some awareness of the perceived impact of Gödel's Second Incompleteness Theorem on Hilbert's philosophy of mathematics and of some dissenting views. For a sustained defense of Hilbert's programs against the received view, see Detlefsen (1986).[7]

5.7.2 We said previously that Gödel proved the Second Incompleteness Theorem by using the standard proof predicate to formalize in Th(PA) the proof of the First Incompleteness Theorem. It is only the first part of the proof of the First Incompleteness Theorem that needs to be formalized in order to prove the Second Incompleteness Theorem. Let us rehearse this part here. We used the Diagonalization Lemma to prove that there exists an AV sentence G$_{PA}$ such that PA ⊢ G$_{PA}$ ↔ ¬**prov**[g], where g is the gödel number of G$_{PA}$, that is, g = [G$_{PA}$]. Recall that the formula **proof**[x, y] is any proof predicate; in other words, it is an AV formula that represents in Th(PA) the numerical relation PROOF. **prov**[y] is the provability predicate that is defined as ($\exists x$)**proof**[x, y]. As stated more than once, for every natural number m, if m = [**X**], where **X** is a theorem of PA, then PA ⊢ **prov**[m]. We present only the essential steps of the proof.

1. Assume that PA is consistent.
2. PA ⊢ G$_{PA}$ ↔ ¬**prov**[g], where g = [G$_{PA}$].
3. Reductio Assumption: PA ⊢ G$_{PA}$.
4. From 2 and 3 by Biconditional Modus Ponens: PA ⊢ ¬**prov**[g].

---

6 J.B. Rosser [1936], "Extensions of Some Theorems of Gödel and Church," *Journal of Symbolic Logic* 1: 87–91.
7 Michael Detlefsen [1986], *Hilbert's Program*, D. Reidel Publishing Company, Dordrecht.

5. From 3: since **prov**[y] is a provability predicate and G_{PA} is a theorem of PA, PA ⊢ **prov**[g].

6. From 4 and 5: PA is inconsistent, which contradicts 1.

7. From 3–6: PA ⊬ G_{PA}.

8. From 1–7: if PA is consistent, then PA ⊬ G_{PA}.

Gödel showed that, using the standard proof predicate, it is possible to prove from PA a formal sentence that represents the metatheoretical statement listed on line 8 above. Once such a theorem is proved, the Second Incompleteness Theorem easily follows. We begin by introducing some notation. Let **proof**s[x, y] be the standard proof predicate. It does not matter how this formula is actually constructed. What matters is that it can be used to formalize line 8 in Th(PA). The standard provability predicate **prov**s[y] is the formula ($\exists x$)**proof**s[x, y]. CONs is the standard consistency sentence; it is defined as the AV sentence ¬**prov**s[c], where c (for "contradiction") is the gödel number of the sentence $0 = 1$. Informally, CONs asserts that $0 = 1$ is not provable from PA, that is, $0 = 1 \notin$ Th(PA). In other words, CONs says that PA is consistent. To formalize line 8 above in Th(PA) is to show that the AV sentence CONs→¬**prov**s[g], where g = [G_{PA}], is a theorem of PA. Once we have this theorem of PA, the Second Incompleteness Theorem follows easily.

Gödel's Second Incompleteness Theorem (first proof): If PA is consistent, then PA ⊬ CONs.

Proof

1. Assume that PA is consistent.
2. By the Diagonalization Lemma: PA ⊢ G_{PA}↔¬**prov**s[g], where g = [G_{PA}].
3. Given: PA ⊢ CONS→¬**prov**s[g].
4. By the proof of the First Incompleteness Theorem: if PA is consistent, then PA ⊬ G_{PA}.
5. From 1 and 4: PA ⊬ G_{PA}.
6. Reductio Assumption: PA ⊢ CONs.
7. From 3 and 6 by Modus Ponens: PA ⊢ ¬**prov**s[g].
8. From 2 and 7 by Biconditional Modus Ponens: PA ⊢ G_{PA}, which contradicts 5.
9. From 6–8: the Reductio Assumption is false. We conclude that PA ⊬ CONs.
10. From 1–9: if PA is consistent, then PA ⊬ CONs.

Since at the beginning of this chapter we presupposed that PA has a model, by the Soundness Theorem, PA is consistent; and hence PA ⊬ CONs.

We will now establish the theorem via a different route. We will specify three conditions and stipulate that any "adequate" provability predicate must meet all of these conditions. We call these conditions the **Provability Conditions.** Said differently, we consider these three conditions as necessary conditions for any adequate provability predicate. We will continue to think of a provability predicate **prov**[y] as the formula $(\exists x)$**proof**[x, y], where **proof**[x, y] is any AV formula with two free variables that represents in Th(PA) the numerical relation PROOF. In fact our approach is quite unorthodox. As the reader will see from the proof we will present here, the fact that **prov**[y] is a provability predicate plays no role in the proof of the Second Incompleteness Theorem. The Provability Conditions on their own are sufficient for establishing the theorem. However, we decided to work with provability predicates rather than with any formula that satisfies these conditions because these conditions are quite permissive. Many strange formulas that have nothing to do with provability and proofs satisfy the Provability Conditions. For the sake of philosophical significance, we will work only with provability predicates and we will stipulate that the Provability Conditions are necessary conditions for the adequacy of the provability predicates.

Henceforth we will assume that every provability predicate **prov**[y] for PA meets the following three Provability Conditions. To improve the readability of these conditions and the proofs that are based on them, we will denote the AV numeral that represents the gödel number of an AV sentence **X** as $\langle X \rangle$. To reduce the number of brackets, we will write **prov**$\langle X \rangle$ instead of **prov**[$\langle X \rangle$]. Thus **prov**$\langle X \rangle$ is simply **prov**[k], where k is the gödel number of **X**; we replaced k with the new symbol $\langle X \rangle$ and we dropped the brackets from **prov**[k].

PC1   For every AV sentence **X**, if PA $\vdash$ **X**, then PA $\vdash$ **prov**$\langle X \rangle$.

PC2   For all AV sentences **X** and **Y**, PA $\vdash$ **prov**$\langle X \rightarrow Y \rangle \rightarrow$ (**prov**$\langle X \rangle \rightarrow$ **prov**$\langle Y \rangle$).

PC3   For every AV sentence **X**, PA $\vdash$ **prov**$\langle X \rangle \rightarrow$ **prov**$\langle$**prov**$\langle X \rangle\rangle$.

Every provability predicate satisfies PC1. We proved this previously. However, for the convenience of the reader, we repeat the proof here. Since **prov**[y] is the AV formula $(\exists x)$**proof**[x, y], where **proof**[x, y] represents PROOF in Th(PA), if PA $\vdash$ **X**, then there are natural numbers m and k such that k is the gödel number of **X**, m is the gödel number of a PL derivation of **X** from PA, and PA $\vdash$ **proof**[m, k]; by Existential Generalization, PA $\vdash$ $(\exists x)$**proof**[x, k]; and hence PA $\vdash$ **prov**[k]. Given that k is $\langle X \rangle$, it follows that if PA $\vdash$ **X**, then PA $\vdash$ **prov**$\langle X \rangle$, which is the first derivability condition. PC2 is motivated by the rule Modus Ponens. PC2 asserts that it is a theorem of PA that if the conditional **X**$\rightarrow$**Y** and its antecedent **X** are provable, then so is the consequent **Y**. Observe that by Exportation **prov**$\langle X \rightarrow Y \rangle \rightarrow$ (**prov**$\langle X \rangle \rightarrow$ **prov**$\langle Y \rangle$) is equivalent to (**prov**$\langle X \rightarrow Y \rangle \wedge$ **prov**$\langle X \rangle) \rightarrow$ **prov**$\langle Y \rangle$. Since Modus Ponens is an MDS rule, it seems very reasonable to require any provability predicate to represent this rule.

PC3 is contentious. It says that it is a theorem of PA that if **X** is provable, then the fact that **X** is provable is itself provable. In other words, PC3 asserts that it is a theorem of PA that if **X** is a theorem of PA, then its being a PA theorem is provable from PA. It is not clear why a provability predicate must satisfy this condition. The standard provability predicate **provs**[y] satisfies this condition. **provs**[y] has enough resources to formalize in Th(PA) the argument we used to show that if PA ⊢ **X**, then PA ⊢ **provs**⟨**X**⟩, which is PC1. This condition says that if **X** is a theorem of PA, then **provs**⟨**X**⟩ is also a theorem of PA. The formalization of this statement is the AV sentence **provs**⟨**X**⟩→**provs**⟨**provs**⟨**X**⟩⟩. If the proof we employed to establish PC1 is itself formalizable in PA, then there is a PA proof of the sentence **provs**⟨**X**⟩→**provs**⟨**provs**⟨**X**⟩⟩, that is, PA ⊢ **provs**⟨**X**⟩→**provs**⟨**provs**⟨**X**⟩⟩, which is PC3. It seems that PC3 is motivated by the thought that a provability predicate should be powerful enough to formalize in Th(PA) the argument used to prove PC1: if PA ⊢ **X**, then PA ⊢ **prov**⟨**X**⟩. We will not debate this condition here. Rather, we will accept these Provability Conditions as reasonable necessary conditions for any provability predicate (note that, according to our approach, PC1 is redundant since every provability predicate meets this condition).

We are now ready to prove the second Incompleteness Theorem on the basis of the Provability Conditions. Observe that the sentence ¬**prov**⟨0 = 1⟩ informally says that 0 = 1 is not provable from PA, which means that PA is consistent. Thus ¬**prov**⟨0 = 1⟩ is a formal representation of the metatheoretical assertion that PA is consistent. We will prove that ¬**prov**⟨0 = 1⟩ is not provable from PA.

Gödel's Second Incompleteness Theorem (second proof): Let **prov**[y] be any PA provability predicate that satisfies the three Provability Conditions. If PA is consistent, then PA ⊬ ¬**prov**⟨0 = 1⟩.

Proof

1. Assume that PA is consistent.
2. Reductio Assumption: PA ⊢ ¬**prov**⟨0 = 1⟩.
3. Consider the AV formula **prov**[y]→0 = 1. By the Diagonalization Lemma there is an AV sentence B such that PA ⊢ B↔(**prov**⟨B⟩→0 = 1).
4. From 2 and 3: there are PL derivations D_1 and D_2 of ¬**prov**⟨0 = 1⟩ and of B↔(**prov**⟨B⟩→0 = 1), respectively, from PA.
5. From 4: we can combine D_1 and D_2 to construct a PL derivation of 0 = 1 from PA. We will employ the three Provability Conditions as inference rules. PC2 and PC3 present no problem. However, care must be applied when PC1 is employed as an inference rule: the rule ensures that there is a PA derivation D of **prov**⟨**X**⟩ if there is a PA derivation E of **X**. So first we need to make sure that **X** is derived from PA axioms alone without the addition of any assumptions; and second we need to assume that any PA axioms that are needed for the derivation of **prov**⟨**X**⟩ are added to the PA proof. To simplify matters we will do two things: (1) we take PA^F to be the finite set of PA axioms that are invoked in the derivations D_1 and D_2 *as well as* any additional PA axioms that are needed to derive **prov**⟨**X**⟩ once a PA derivation of **X** is given, and (2) once **X** is obtained from PA^F we may freely enter **prov**⟨**X**⟩ in the

derivation without indicating that there is another derivation D of **prov**⟨X⟩ from PA^F.

[0	0	PA^F	Premises
	⋮	D$_1$	This PA derivation is given
	i	¬**prov**⟨0 = 1⟩	Conclusion of D$_1$
	⋮	D$_2$	This PA derivation is given
	j	B↔(**prov**⟨B⟩→0 = 1)	Conclusion of D$_2$
[1	j+1	**prov**⟨0 = 1⟩	CPA
[2	j+2	0 ≠ 1	RA
	j+3	**prov**⟨0 = 1⟩	j+1, Reit
2]	j+4	¬**prov**⟨0 = 1⟩	i, Reit
1]	j+5	0 = 1	(j+2)–(j+4), RAA
	j+6	**prov**⟨0 = 1⟩→0 = 1	(j+1)–(j+5), CP
	j+7	(B→(**prov**⟨B⟩→0 = 1))∧((**prov**⟨B⟩→0 = 1)→B)	
			j, Bc
	j+8	B→(**prov**⟨B⟩→0 = 1)	j+7, Simp
	j+9	**prov**⟨B→(**prov**⟨B⟩→0 = 1)⟩	j+8, PC1 (observe that line j+8 is derived from PA axioms alone)
	j+10	**prov**⟨B→(**prov**⟨B⟩→0 = 1)⟩→(**prov**⟨B⟩→**prov**⟨**prov**⟨B⟩→0 = 1⟩)	
			PC2 (substitute B for **X** and **prov**⟨B⟩→0 = 1 for **Y**)
	j+11	**prov**⟨B⟩→**prov**⟨**prov**⟨B⟩→0 = 1⟩	j+9, j+10, MP
	j+12	**prov**⟨**prov**⟨B⟩→0 = 1⟩→(**prov**⟨**prov**⟨B⟩⟩→**prov**⟨0 = 1⟩)	
			PC2 (substitute **prov**⟨B⟩ for **X** and 0 = 1 for **Y**)
	j+13	**prov**⟨B⟩→(**prov**⟨**prov**⟨B⟩⟩→**prov**⟨0 = 1⟩)	j+11, j+12, HS
	j+14	**prov**⟨B⟩→**prov**⟨**prov**⟨B⟩⟩	PC3 (substitute B for **X**)
[3	j+15	**prov**⟨B⟩	CPA
	j+16	**prov**⟨**prov**⟨B⟩⟩→**prov**⟨0 = 1⟩	j+13, j+15, MP
	j+17	**prov**⟨**prov**⟨B⟩⟩	j+14, j+15, MP
3]	j+18	**prov**⟨0 = 1⟩	j+16, j+17, MP
	j+19	**prov**⟨B⟩→**prov**⟨0 = 1⟩	j+15–j+18, CP
	j+20	**prov**⟨B⟩→0 = 1	j+6, j+19, HS
	j+21	B	j, j+20, BcMP
	j+22	**prov**⟨B⟩	j+21, PC1 (again, line j+21 is derived from PA axioms alone)
0]	j+23	0 = 1	j+20, j+22, MP

6. From 5: PA ⊢ **0 = 1**.
7. From Ax1 by Universal Instantiation: PA ⊢ **0 ≠ 1**.
8. From 6 and 7: PA is inconsistent, which contradicts 1.
9. From 2 through 8: PA ⊬ ¬**prov**⟨**0 = 1**⟩.
10. From 10 through 9: if PA is consistent, then PA ⊬ ¬**prov**⟨**0 = 1**⟩.

Since we originally presupposed that N is a model of PA, by the Soundness Theorem, PA is consistent; and hence PA ⊬ ¬**prov**⟨**0 = 1**⟩.

As indicated earlier, the proof above makes no use of the claim that **prov**[y] is a provability predicate. The fact that **prov**[y] is (∃x)**proof**[x, y], where **proof**[x, y] represents PROOF in Th(PA), is not invoked anywhere in the proof. The only resources needed are the Diagonalization Lemma, the three Provability Conditions, and the arithmetization of the metatheory of Th(PA). Hence any consistent theory T that meets the following three conditions fails to prove a sentence that expresses the consistency of T: (1) T is sufficiently powerful so that the Diagonalization Lemma is one of its metatheorems; (2) T has a one free-variable formula that satisfies the three Provability Conditions; and (3) a certain amount of the metatheory of T can be represented in T itself.

Observe that since PA is consistent, it is true that the sentence **0 = 1** is not provable in Th(PA). Hence the sentence ¬(∃x)**proofs**[x, c], where c is the gödel number of **0 = 1**, is true on the standard model N of Th(PA). Here is a rigorous proof of this claim. Let n be any natural number; the ordered pair ⟨n, c⟩ does not belong to PROOF. Thus PA ⊢ ¬**proofs**[**n**, **c**] (because **proofs**[x, y] represents PROOF in Th(PA)). By the Soundness Theorem, PA ⊨ ¬**proofs**[**n**, **c**]. Therefore ¬**proofs**[**n**, **c**] is true on N for every natural number n. By the truth conditions of the universal quantifier, (∀x)¬**proofs**[x, **c**] is true on N. It follows, by Negated Quantifier, that ¬(∃x)**proofs**[x, **c**] is true on N. In other words, ¬**provs**⟨**0 = 1**⟩ is a true sentence on N, but it is not a theorem of PA. Therefore, in addition to the sentence G$_{PA}$ of the First Incompleteness Theorem, ¬**provs**⟨**0 = 1**⟩ is another sentence that is true on N for which there is no PA proof. This is yet another demonstration that Th(PA) is a *proper* subset of Th$_{AV}$(N).

5.8 Exercises

Notes: All answers must be adequately justified. In all the exercises, we are assuming that the sets of all the sentences that are composed of the vocabularies indicated are decidable.

5.8.1* Let Σ be any set of PL sentences and τ be any PL sentence that is composed of Voc(Σ). Define Th(Σ, τ) to be the PL theory that consists of all the logical consequences of the set $\Sigma \cup \{\tau\}$ that are composed of Voc(Σ). (Observe that Voc($\Sigma \cup \{\tau\}$) = Voc(Σ).) In other words, Th(Σ, τ) = {**Z**: **Z** is a sentence composed of Voc(Σ) and $\Sigma \cup \{\tau\} \vDash$ **Z**}. Prove that if Th(Σ) is decidable, then Th(Σ, τ) is decidable.

5.8.2 Prove that for any $E \subseteq \mathbb{N}^n$, where n is a positive integer, if E is representable in Th(PA), then E is decidable.

5.8.3 RA is the set of Robinson's axioms (see Subsection 5.5.3). Let Σ be any PL theory whose vocabulary includes the standard arithmetical vocabulary, AV. Show that if $\Sigma \cup$ RA is consistent, then Σ is undecidable.

5.8.4* Let Σ be any decidable set of PL sentences whose vocabulary includes the standard arithmetical vocabulary, AV. Demonstrate that if $\Sigma \cup$ RA is consistent, then Th(Σ) is not a complete theory.

5.8.5 Give a different proof of Church's Undecidability Theorem.

5.8.6* Let G_{PA} be the Gödel sentence of PA (see the proof of the First Incompleteness Theorem). Suppose that we add to PA the sentence $\neg G_{PA}$. Denote the resulting set of axioms as PA⁺. Prove that PA⁺ is consistent but ω-inconsistent.

5.8.7 Formulate the ω-rule for Peano Arithmetic as follows: for every AV formula **X**[z] with one free variable, if **X**[n] is a theorem of PA for every natural number n, then $(\forall z)$**X**[z] is a theorem of PA. (See Subsection 5.6.4 for the definition of the ω-rule for Second-Order Logic.) Demonstrate that the ω-rule does not hold for Peano Arithmetic.

5.8.8 Let Γ be any PL theory whose vocabulary includes AV. Show that if Γ is consistent and ω-inconsistent, then the ω-rule for Γ (see Exercise 5.8.7) does not hold for Γ.

5.8.9* Prove that the standard interpretation of arithmetic, N, is not a model of any AV theory that is ω-inconsistent.

5.8.10 Demonstrate that there is an ω-consistent extension of Robinson Arithmetic, Th(RA), such that the standard interpretation of arithmetic, N, is not a model of it. (This is a difficult problem.)

5.8.11* The following statement is known as **Löb's Theorem** after the German mathematician and logician Martin Hugo Löb (1921–2006) who formulated this theorem in 1955: For every AV sentence **X**, if **prov⟨X⟩→X** is a theorem of PA, then **X** is also a theorem of PA. Use the Provability Conditions to give a proof of this theorem. (See Subsection 5.7.2 for the Provability Conditions, PC1–PC3, and for the notation **prov⟨X⟩**.)

5.8.12 Construct a PL sentence composed of standard arithmetical vocabulary that may be considered as asserting that it is a theorem of PA. What can you determine about the status of this sentence?

5.8.13 Construct an AV sentence that may be considered as asserting that either it or its negation is a theorem of PA. What can you determine about the status of this sentence and its negation?

5.8.14 Construct a sentence composed of AV that may be considered as asserting that neither it nor its negations is a theorem of PA. What can you determine about the status of this sentence and its negation?

5.8.15* A set D of n-tuples of natural numbers is said to be *arithmetically definable* if and only if there is an n-variable AV formula **D**[$x_1, x_2, x_3, ..., x_n$] such that for all natural numbers $k_1, k_2, k_3, ..., $ and k_n, ⟨$k_1, k_2, k_3, ..., k_n$⟩ ∈ D iff **D**[$k_1, k_2, k_3, ..., k_n$] is true on the standard interpretation of arithmetic, N. The following statement is sometimes described as *Tarski's Indefinability Theorem* instead of the version we stated and proved in Subsection 5.5.2: #Th$_{AV}$(N) is not arithmetically definable, where #Th$_{AV}$(N) is the set of the gödel numbers of all the sentences in Th$_{AV}$(N). Recall that Th$_{AV}$(N) is the theory that consists of all the AV sentences that are true on N. Since Arithmetic is standardly defined as the theory Th$_{AV}$(N), the previous theorem is commonly stated as "the notion of arithmetical truth is not arithmetically definable." Demonstrate the truth of this theorem.

5.8.16 Let Φ be a consistent theory composed of AV in which all recursive functions are representable. Prove that there is no decidable set E such that (1) E ⊆ Sent$_{AV}$, (2) Φ ⊆ E, and (3) Φ^{neg} ⊆ Sent$_{AV}$–E, where Sent$_{AV}$ is the set of all the AV sentences and Φ^{neg} = {**X**: **X** is an AV sentence and ¬**X** ∈ Φ}. Recall that Sent$_{AV}$–E is the complement of E in Sent$_{AV}$, that is, Sent$_{AV}$–E = {**X**: **X** ∈ Sent$_{AV}$ and **X** ∉ E}.

5.8.17 The goal of this exercise is to prove Gödel's First Incompleteness Theorem without the assumption that Peano Arithmetic is ω-consistent. Define the 2-place numerical relation DISPROOF as follows: for all natural numbers n and m, $\langle n, m \rangle \in$ DISPROOF iff $\langle n,$ NEG(m)$\rangle \in$ PROOF, where NEG is a 1-place numerical function that returns the gödel number of $\neg X$ when it is applied to the gödel number of the PL formula **X**, and PROOF is, as usual, a 2-place numerical relation that holds between the gödel number of a PL derivation D and the gödel number of an AV sentence **Z** such that D is a PA derivation of **Z**. We could have defined DISPROOF directly as follows: $\langle n, m \rangle \in$ DISPROOF iff n is the gödel number of a PL derivation from PA of the negation of an AV sentence whose gödel number is m. So while PROOF is the numerical counterpart of the proof relation in Th(PA), DISPROOF is the numerical counterpart of the disproof relation in Th(PA). Since DISPROOF is defined in terms of PROOF and NEG, and since both of these are recursive, DISPROOF is also a recursive relation. Hence it is representable in Th(PA). Let **proof** and **disproof** be 2-variable AV formulas that represent PROOF and DISPROOF in Th(PA), respectively. We use the standard symbol **<** (boldfaced) to denote the 2-variable AV formula that represents in Th(PA) the arithmetical binary relation "less than." The following AV formula contains only one free variable: $(\forall x)(\textbf{proof}[x, y] \rightarrow (\exists z)(z \mathbf{<} x \wedge \textbf{disproof}[z, y]))$. By the Diagonalization Lemma, there is a V sentence R_{PA} such that

$$PA \vdash R_{PA} \leftrightarrow (\forall x)(\textbf{proof}[x, r] \rightarrow (\exists z)(z \mathbf{<} x \wedge \textbf{disproof}[z, r]))$$

where r is the gödel number of R_{PA}. Metaphorically speaking, R_{PA} asserts: "if there is a PA derivation of me, then there is an earlier PA derivation of my negation." The language of "earlier" is warranted since PA derivations can be arranged in an infinite sequence $D_1, D_2, D_3, \ldots, D_i, \ldots$ according to the order of their gödel numbers, that is, D_i precedes D_j iff $[D_i]$ is less than $[D_j]$. R_{PA} is referred to as the *Rosser sentence* of PA. Prove without assuming that Th(PA) is ω-consistent that the Rosser sentence of PA is neither provable nor disprovable in Th(PA).

5.8.18* As usual, Arithmetic is the complete satisfiable theory $Th_{AV}(N)$, where N is the standard interpretation of arithmetic and AV is the standard arithmetical vocabulary. By the definition of $Th_{AV}(N)$, N is a model of $Th_{AV}(N)$. N is appropriately referred to as *the standard model of Arithmetic*. A *nonstandard model of Arithmetic* is a countable model of $Th_{AV}(N)$ that is not isomorphic to N. Prove the existence of a nonstandard model of Arithmetic.[8]

---

8   I am grateful to an anonymous reviewer of Broadview Press who suggested that I include a discussion of nonstandard models of Arithmetic.

Solutions to the Starred Exercises

SOLUTION TO 5.8.1

1. We will prove, first, that for every Z, $Z \in \text{Th}(\Sigma, \tau)$ iff $\tau \rightarrow Z \in \text{Th}(\Sigma)$.
2. Take Z to be any sentence in $\text{Th}(\Sigma, \tau)$. By definition, Z is composed of $\text{Voc}(\Sigma)$ and $\Sigma \cup \{\tau\} \vDash Z$.
3. Let M be any model of Σ. It is clear that M is relevant to τ, since τ is composed of $\text{Voc}(\Sigma)$.
4. From 3: either τ is true on M or it is false on M.
5. If τ is false on M, then, by the truth conditions of the conditional, $\tau \rightarrow Z$ is true on M.
6. From 3: if τ is true of M, then M is a model of $\Sigma \cup \{\tau\}$.
7. From 2 and 6: Z is true on M; and hence $\tau \rightarrow Z$ is true on M.
8. From 4, 5, and 7: $\tau \rightarrow Z$ is true on M.
9. From 3 and 8: $\tau \rightarrow Z$ is true on every model of Σ.
10. From 9 by the definition of logical consequence: $\Sigma \vDash \tau \rightarrow Z$.
11. From 10 by definition of $\text{Th}(\Sigma)$: $\tau \rightarrow Z \in \text{Th}(\Sigma)$. (Note that $\tau \rightarrow Z$ is composed of $\text{Voc}(\Sigma)$.)
12. From 2 through 11: for all Z, if $Z \in \text{Th}(\Sigma, \tau)$, then $\tau \rightarrow Z \in \text{Th}(\Sigma)$.
13. Now we take $\tau \rightarrow Z \in \text{Th}(\Sigma)$, that is, $\tau \rightarrow Z$ is composed of $\text{Voc}(\Sigma)$ and $\Sigma \vDash \tau \rightarrow Z$.
14. Let M be any model of $\Sigma \cup \{\tau\}$. Hence M is a model of Σ and τ is true on M.
15. From 13 and 14 by the definition of logical consequence: $\tau \rightarrow Z$ is true on M.
16. From 14 and 15 by Modus Ponens: Z is true on M.
17. From 14 through 16: Z is true on any model of $\Sigma \cup \{\tau\}$.
18. From 17 by the definition of logical consequence: $\Sigma \cup \{\tau\} \vDash Z$.
19. From 18 by the definition of $\text{Th}(\Sigma, \tau)$: $Z \in \text{Th}(\Sigma, \tau)$.
20. From 13 through 19: for all Z, if $\tau \rightarrow Z \in \text{Th}(\Sigma)$, then $Z \in \text{Th}(\Sigma, \tau)$.
21. From 12 and 20: for every Z, $Z \in \text{Th}(\Sigma, \tau)$ iff $\tau \rightarrow Z \in \text{Th}(\Sigma)$.
22. Assume now that $\text{Th}(\Sigma)$ is decidable. Thus there is an effective decision procedure P that determines whether any sentence composed of $\text{Voc}(\Sigma)$ is a member of $\text{Th}(\Sigma)$ or not.
23. From 21 and 22: we will describe an effective decision procedure for membership in $\text{Th}(\Sigma, \tau)$. We take any sentence X composed of $\text{Voc}(\Sigma)$. We form the sentence $\tau \rightarrow X$, and we apply the decision procedure P to it to see whether it is a member of $\text{Th}(\Sigma)$ or not. If it is not a member of $\text{Th}(\Sigma)$, we determine that X does not belong to $\text{Th}(\Sigma,$

τ). On the other hand, if it is a member of Th(Σ), we determine that **X** belongs to Th(Σ, τ).

24. From 23: since there is an effective decision procedure for Th(Σ, τ), it is decidable.
25. From 22 through 24: if Th(Σ) is decidable, Th(Σ, τ) is decidable as well.

The point of this exercise is to show that if a new sentence is added to the axioms of a decidable theory, then the resulting, possibly, larger theory would still be decidable, provided that the new axiom is composed of the vocabulary of the original theory. Since any finite number of sentences can be expressed as one sentence by taking their conjunction, we have the following result: adding any finite number of new axioms to a decidable theory produces a decidable theory, provided that the new axioms are composed of the vocabulary of the original theory.

SOLUTION TO 5.8.4

1. Σ is any decidable PL set whose vocabulary includes AV. RA is the set of Robinson's axioms. We assume that Σ∪RA is consistent. It follows that Th(Σ∪RA) is also consistent. (If Th(Σ∪RA) were inconsistent, a contradiction would be derivable from Th(Σ∪RA); since Th(Σ∪RA) is a theory, a contradiction would be a member of Th(Σ∪RA); and by the definition of Th(Σ∪RA), a contradiction would be derivable from Σ∪RA; hence Σ∪RA would be inconsistent.)
2. From 1: since Th(Σ∪RA) is consistent, and since it is obvious that Th(Σ)∪RA is a subset of Th(Σ∪RA), Th(Σ)∪RA is a consistent PL set whose vocabulary includes AV.
3. From 2 by Exercise 5.8.2: Th(Σ) is undecidable.
4. From 1 by the definition of axiomatizable set: since Σ is decidable, Th(Σ) is an axiomatizable theory.
5. From 4 by Theorem 3.5.1: if Th(Σ) is complete, then Th(Σ) is decidable.
6. From 3 and 5: Th(Σ) is incomplete.

SOLUTION TO 5.8.6

1. If PA⁺ is inconsistent, then PA∪{¬G_PA} is inconsistent, and hence by Lemma 3.2.1a, PA ⊢ G_PA, which contradicts the First Incompleteness Theorem. Thus PA⁺ is consistent. Now we show that PA⁺ is ω-inconsistent.
2. The two-variable AV formula **proof** represents the numerical relation PROOF in Th(PA), where PROOF consists of all ordered pairs ⟨n, m⟩ such that n is the gödel number of a PA proof of the AV sentence whose gödel number is m.
3. From 2 and the First Incompleteness Theorem: since PA ⊬ G_PA, for all natural numbers n, ⟨n, g⟩ ∉ PROOF, where g is the gödel number of G_PA.
4. From 2 and 3: for all natural numbers n, PA ⊢ ¬**proof**[n, g].

5. From 4: since PA ⊆ PA⁺, for all natural numbers n, PA⁺ ⊢ ¬**proof**[n, g].
6. From the proof of the First Incompleteness Theorem: PA ⊢ G$_{PA}$↔¬(∃x)**proof**[x, g]; and hence PA⁺ ⊢ G$_{PA}$↔¬(∃x)**proof**[x, g].
7. From 6: since PA⁺ ⊢ ¬G$_{PA}$, by Biconditional Modus Tollens and Double Negation, PA⁺ ⊢ (∃x)**proof**[x, g].
8. From 5 and 7: PA⁺ is ω-inconsistent.

This exercise demonstrates what we claimed previously that ω-consistency is a stronger notion than consistency. We explained before that every ω-consistent set is also consistent. However, the converse is not true: not every consistent set is ω-consistent. The exercise above shows that there are sets that are consistent but not ω-consistent.

SOLUTION TO 5.8.9

Let Γ be any ω-inconsistent theory whose vocabulary is AV. Suppose for reductio that N is a model of Γ. Since Γ is ω-inconsistent, there is a 1-variable AV formula **Z** such that Γ ⊢ ¬**Z**[**n**] for every natural number n and Γ ⊢ (∃x)**Z**[x]. Since Γ is a theory, for all natural numbers n, ¬**Z**[**n**] ∈ Γ and (∃x)**Z**[x] ∈ Γ. Hence (∃x)**Z**[x] is true on N, yet **Z**[**n**] is false on N for every natural number n. Given the truth condition of the existential quantifier, there is a name **t** in the list of names of N such that **Z**[**t**] is true on N. The referent of **t** on N is some natural number k. We know that the singular term **k** (which is s^k0) also refers on N to the number k. Thus on N the sentence **t** = **k** is true. By substitution, **Z**[**k**] is true on N, which contradicts the fact that **Z**[**n**] is false on N for every natural number n. Therefore we reject the Reductio Assumption, and conclude that N is not a model of Γ.

SOLUTION TO 5.8.11

We are given that PA ⊢ **prov**⟨**X**⟩→**X**, where **X** is an AV sentence. Thus there is a PL derivation D₁ of **prov**⟨**X**⟩→**X** from PA. We want to demonstrate that PA ⊢ **X**. We will give below a PL derivation of **X** from PA. But first we apply the Diagonalization Lemma to the one-variable formula **prov**[y]→**X**. By this lemma there is a sentence λ composed of AV such that PA ⊢ λ↔(**prov**⟨λ⟩→**X**). Hence there is a PL derivation D₂ of λ↔(**prov**⟨λ⟩→**X**) from PA. Let PA^F be the finite set of the PA axioms that are invoked in the derivations D₁ and D₂ as well as the PA axioms that are needed to derive **prov**⟨**Z**⟩ once a PA derivation of **Z** is given.

THE INCOMPLETENESS THEOREMS 287

[0	0	PA^F	Premises
	⋮	D$_1$	This derivation is given
	i	**prov**⟨X⟩→X	Conclusion of D$_1$
	⋮	D$_2$	This derivation is given
	j	λ↔(**prov**⟨λ⟩→X)	Conclusion of D$_2$
	j+1	λ→(**prov**⟨λ⟩→X)∧(**prov**⟨λ⟩→X)→λ	j, Bc
	j+2	λ→(**prov**⟨λ⟩→X)	j+1, Simp
	j+3	**prov**⟨λ→(**prov**⟨λ⟩→X)⟩	j+2, PC1 (observe that line j+2 is derived from PA axioms alone)
	j+4	**prov**⟨λ→(**prov**⟨λ⟩→X)⟩→(**prov**⟨λ⟩→**prov**⟨**prov**⟨λ⟩→X⟩) PC2	
	j+5	**prov**⟨λ⟩→**prov**⟨**prov**⟨λ⟩→X⟩	j+3, j+4, MP
	j+6	**prov**⟨**prov**⟨λ⟩→X⟩→(**prov**⟨**prov**⟨λ⟩⟩→**prov**⟨X⟩) PC2	
	j+7	**prov**⟨λ⟩→(**prov**⟨**prov**⟨λ⟩⟩→**prov**⟨X⟩)	j+5, j+6, HS
	j+8	**prov**⟨λ⟩→**prov**⟨**prov**⟨λ⟩⟩	PC3
[1	j+9	**prov**⟨λ⟩	CPA
	j+10	**prov**⟨**prov**⟨λ⟩⟩→**prov**⟨X⟩	j+7, j+9, MP
	j+11	**prov**⟨**prov**⟨λ⟩⟩	j+8, j+9, MP
1]	j+12	**prov**⟨X⟩	j+10, j+11, MP
	j+13	**prov**⟨λ⟩→**prov**⟨X⟩	j+9–j+12, CP
	j+14	**prov**⟨λ⟩→X	i, j+13, HS
	j+15	λ	j, j+14, BcMP
	j+16	**prov**⟨λ⟩	j+15, PC1 (again, line j+15 is derived from PA axioms alone)
0]	j+17	X	j+14, j+16, MP

Given the derivation above, it follows that PA ⊢ X.

SOLUTION TO 5.8.15

There are several proofs of this theorem. Here is perhaps the simplest one.

1. Since Th$_{AV}$(N) is a theory, for every AV sentence **X**, **X** ∈ Th$_{AV}$(N) iff Th$_{AV}$(N) ⊨ **X**.
2. From 1 by the Completeness Theorem: for every AV sentence **X**, **X** ∈ Th$_{AV}$(N) iff Th$_{AV}$(N) ⊢ **X**.
3. By definition: for every AV sentence **X**, **X** ∈ Th$_{AV}$(N) iff **X** is true on N.
4. From 2 and 3: for every AV sentence **X**, **X** is true on N iff Th$_{AV}$(N) ⊢ **X**.

5. Let D be any set of natural numbers, that is, D ⊆ ℕ. By the definition of arithmetically definable, D is arithmetically definable iff there is a 1-variable AV formula **D** such that for all natural numbers n, n ∈ D iff **D[n]** is true of N.

6. From 5 by the truth conditions of the negation: D is arithmetically definable iff there is a 1-variable AV formula **D** such that for all natural numbers n, if n ∈ D then **D[n]** is true on N, and if n ∉ D then ¬**D[n]** is true on N.

7. From 4 and 6: D is arithmetically definable iff there is a one-variable AV formula **D** such that for all natural numbers n, if n ∈ D then Th$_{AV}$(N) ⊢ **D[n]**, and if n ∉ D then Th$_{AV}$(N) ⊢ ¬**D[n]**.

8. From 7 by Definition 5.2.1a: D is arithmetically definable iff D is representable in Th$_{AV}$(N).

9. From 8: #Th$_{AV}$(N) is arithmetically definable iff #Th$_{AV}$(N) is representable in Th$_{AV}$(N).

10. Since Th(PA) ⊆ Th$_{AV}$(N), and since all recursive functions are representable in Th(PA), all recursive functions are representable Th$_{AV}$(N).

11. By definition: Th$_{AV}$(N) is satisfiable. Hence by the Soundness Theorem, Th$_{AV}$(N) is consistent.

12. From 10 and 11 by Theorem 5.5.2: #Th$_{AV}$(N) is not representable in Th$_{AV}$(N).

13. From 9 and 12: #Th$_{AV}$(N) is not arithmetically definable.

Solution to 5.8.18

The standard arithmetical vocabulary, AV, consists of the logical vocabulary of PL and the following extra-logical vocabulary: one name **0**, one 1-place function symbol s, and two 2-place function symbols + and •. On the basis of this vocabulary infinite collections of terms and formulas can be constructed. We previously defined the AV numerals as the infinite sequence **0**, **1**, **2**, **3**, ..., **n**, ..., where **n** is an abbreviation for $s^n\mathbf{0}$; and the latter notation is defined recursively as follows: $s^0\mathbf{0}$ is **0**, and for every natural number k, $s^{k+1}\mathbf{0}$ is $ss^k\mathbf{0}$. We previously proved that the standard interpretation N assigns to every AV numeral **n** the natural number n as its referent.

We add a new name c to AV. We denote the resulting vocabulary as AV⁺. It is clear that every AV expression is an AV⁺ expression, but the converse is not true. For instance, the expression $x = c$ is an AV⁺ formula but not an AV formula. Let Σ be the set whose members are all the sentences of the form $c \neq \mathbf{n}$, where **n** is any AV numeral. Symbolically, Σ = {$c \neq \mathbf{n}$: n is a natural number}. It is clear that Σ is countably infinite. Let Ω be the union of Th$_{AV}$(N) and Σ, that is, Ω = Th$_{AV}$(N)∪Σ. We first prove that Ω is finitely satisfiable. Take any finite subset Γ of Ω. If Γ contains no members of Σ, then it is clearly satisfiable since N is a model of Th$_{AV}$(N) and Γ ⊆ Th$_{AV}$(N). The general case is when Γ contains members of Σ. Since Γ is finite, it can only contain finitely many sentences of the form $c \neq \mathbf{n}$. Let **j** be the greatest numeral such that the sentence $c \neq \mathbf{j}$ belongs to Γ. We take N$_j$ to be exactly like N except that the list of names of N_j contains the name c and N_j assigns the number j+1 as the referent of c. It is clear that N_j is a model of Γ: N_j is a model of $\text{Th}_{\text{AV}}(N)$, since it agrees with N on the interpretations of all the extra-logical vocabulary of AV; and every sentence of the form $c \neq \mathbf{n}$ that is in Γ is true on N_j, because the referent $N_j(c)$ is greater than all the referents of the numerals used in the sentences that Γ shares with Σ. Hence Ω is finitely satisfiable. By the Finite-Satisfiability Theorem (see Subsection 3.3.2), Ω is satisfiable. By the Löwenheim-Skolem Theorem (see Subsection 3.6.1), Ω has a countable model. Let this model be M.

Since M is a model of Ω, and $\text{Th}_{\text{AV}}(N)$ is a subset of Ω, M is also a model of $\text{Th}_{\text{AV}}(N)$. We said above that for each natural number k, $N(\mathbf{k}) = k$. It follows from this that for all natural numbers n and m, if n ≠ m, the AV sentence $\mathbf{n} \neq \mathbf{m}$ is true on N, and hence it belongs to $\text{Th}_{\text{AV}}(N)$; since M is a model of $\text{Th}_{\text{AV}}(N)$, the sentence $\mathbf{n} \neq \mathbf{m}$ is true on M as well. This entails that the referents that M assigns to the AV numerals are all distinct, which implies that M is an infinite model. Therefore M is countably infinite.

Furthermore, we can also prove that N and M are elementarily equivalent with respect to AV. Let **X** be any AV sentence that is true on N. By definition, $\text{Th}_{\text{AV}}(N)$ consists of all the AV sentences that are true on N. Thus $\mathbf{X} \in \text{Th}_{\text{AV}}(N)$. Since M is a model of $\text{Th}_{\text{AV}}(N)$, **X** is true on M. On the other hand, if we suppose that **X** is false on N, then ¬**X** is true on N; and hence ¬$\mathbf{X} \in \text{Th}_{\text{AV}}(N)$. Again, since M is a model of $\text{Th}_{\text{AV}}(N)$, ¬**X** is true on M, which means that **X** is false on M. So we proved that for every AV sentence **X**, **X** is true on N iff it is true on M. This establishes that N and M are elementarily equivalent with respect to AV.

However, M and N are not isomorphic with respect to AV. To see this let h be any one-to-one correspondence between ℕ (the universe of discourse of N) and **M** (the universe of discourse of M). Let β be the individual in **M** that is the referent of c on M (c is in the list of names of M, since M is a model of Ω). Since h is an onto-function, there is a number m in ℕ such that h(m) = β. Suppose that h(N(**m**)) = M(**m**). We know that the standard model N assigns to every numeral **n** the referent n. So N(**m**) = m. It follows that h(N(**m**)) = h(m) = M(**m**). Thus the value of h at the argument m is both M(**m**) and β. A function cannot have two distinct values at the same argument. It follows that M(**m**) and β are the same individual in **M**. However, we said that β is the referent of c on M. By the truth conditions of the identity predicate, the AV⁺ sentence $c = \mathbf{m}$ is true on M. But this is impossible. M is a model of Ω, $\Sigma \subseteq \Omega$, and $\Sigma = \{c \neq \mathbf{n}:$ n is a natural number$\}$. Therefore the AV⁺ sentence $c \neq \mathbf{m}$ is true on M. M, as any PL interpretation, cannot make both $c = \mathbf{m}$ and $c \neq \mathbf{m}$ true. We conclude that our earlier supposition "h(N(**m**)) = M(**m**)" is false.

It can easily be verified that if h is an isomorphism between N and M, then for every VA numeral **n**, h(N(**n**)) = M(**n**). Since we proved above that there is an AV numeral **m** such that h(N(**m**)) ≠ M(**m**), h cannot be an isomorphism between N and M, and hence N and M are not isomorphic. Given that M is a countably infinite model of $\text{Th}_{\text{AV}}(N)$ that is not isomorphic to the standard model N, it follows that M is a nonstandard model of Arithmetic. The existence of nonstandard models of Arithmetic and, consequently, of Peano Arithmetic demonstrates that these theories are not only non-categorical, they are also not $\aleph_0$-categorical, since these nonstandard models have cardinality $\aleph_0$.

Index

NOTE: A much more elaborate and detailed index is available for free download at http://sites.broadviewpress.com/metalogicindex/.

Addition (+), 12, 90, 189, 213
  function (A), 90–91, 100, 189, 200–03, 213, 215, 218, 225
    Inductive definition of, 215, 225
    Precise recursive definition of, 218
    T_A is a Turing machine that computes, 200–03
Addition (rule of inference; Add), 53, 61–62
Algorithm, 200, 218–20
Anti-diagonal number, 107
Arbitrary object, 32
Arbitrary-object proof, 32–33, 37
Argument, 31–34, 42, 165, 185, 257–59, 269
  Demonstrative, 205
  Invalid, xiv, 33–34, 41, 165, 185, 257
  Second-order, 257, 259
  Semantical, 269
  Valid, xiv, 31–34, 41–42, 67, 84, 165, 185, 257, 259
Arithmetic Th$_{AV}$(N), 90, 103, 209, 224, 227, 242–43, 247, 249–50, 263–64, 271–72, 282–83, 289
  is not $\aleph_0$-categorical, 243, 289
  is not axiomatizable, xiv, 227, 249–50
  is the PL theory that consists of all AV sentences that are true on N, 227, 243, 249–50, 282–83, 289
  is undecidable, 227, 271
Arithmetical truth, 250–52, 282
  is an AV sentence that is true on N, 250
  a formula **true**[x] defines arithmetical truth in Peano Arithmetic only if it satisfies TS: for every AV sentence **X**, PA ⊢ **true**[**k**]↔**X**, where k = [**X**], 250–252
Arithmetization of effective procedures, 190
  arithmetization of the effective procedure that generates SL sentences, 190, 192
  a prime number is decoded, and a composite number is prime-factored, 191
  SENT$_{SL}$ consists of the numerical codes of all SL sentences, 192
  SENT$_{SL}$ is decidable, 192
Arithmetization of the metatheory of Peano Arithmetic Th(PA), 231, 232, 235, 280
  AFORM, 233–34
    consists of the gödel numbers of all AV atomic formulas, 233
  AFORM is a recursive set, 233
  AFORM is represented in Th(PA) by an AV formula **aform**[x], 233
  arithmetization of PA proofs, 235
    a PA proof is a PL derivation D of an AV sentence **X** from PA, 235–36, 272, 278, 285–86
    ⟨m, k⟩ ∈ PROOF iff m is the gödel number of an AV sentence **X** and k is the gödel number of a PA proof of **X**, 235, 242–44, 272
  PROOF is a recursive relation, 235–36
  PROOF is represented in Th(PA) by an AV formula **proof**[x, y], 237, 242, 272, 274–75, 280, 285
  arithmetization of PA provability, 237–38

PROV consists of the gödel numbers of all theorems of PA, 237–38
k ∈ PROV iff there is m such that ⟨m, k⟩ ∈ PROOF, 237
PROV's definition is not an effective decision procedure for membership in PROV, 237–38
  PROV is not a recursive set but only recursively enumerable, 237–38, 246
  PROV is not representable in Th(PA), 238, 246
prov[y] is defined as $(\exists x)$**proof**[x, y], 238, 242, 244, 272, 275, 277, 280
arithmetization of the syntax of Peano Arithmetic, 232–35
  1st stage is the arithmetization of the category of AV terms, 232–33
  2nd stage is the arithmetization of the category of AV atomic formulas, 233
  3rd stage is the arithmetization of the category of AV formulas, 233–34
  4th stage is the arithmetization of the category of AV sentences, 234–35
  5th stage is the arithmetization of the category of PA proofs, 235
encoding the metatheory in N, 231–32
  is based on encoding the syntax of Th(PA) into numerical codes, 232
FORM consists of the gödel numbers of all AV formulas, 234
FORM is a recursive set, 234
FORM is represented in Th(PA) by an AV formula **form**[x], 234
SENT consists of the gödel numbers of all AV sentences, 234–35
SENT is a recursive set, 234
SENT is represented in Th(PA) by an AV formula **sent**[x], 234–35
TERM consists of the numerical codes of all AV terms, 232–33
TERM is a recursive set, 233

TERM is represented in Th(PA) by an AV formula **term**[x], 233
Arithmetization of PL metatheory, 174
Artificial intelligence, 193
Association (rule of inference; Assoc), 58
Assumption, 34, 42, 51, 56–58, 121, 139, 141–42, 147, 278
Asymmetry, 74, 76, 133
Axiom(s), xii, 172–73, 235–36, 249, 252, 260, 281, 285
  Arithmetical, 90, 223–24, 249
  of Euclidean geometry, 172
  of First-Order Peano Arithmetic (see also *Peano Axioms*; PA), 223–24, 235–36, 249, 252–53, 260, 274, 278, 286
  schema, 224–25
  of Second-Order Peano Arithmetic (PA²), 260, 264, 266–71
  Set of, 172–73, 223
    Decidable, 173, 236
    Finite, 172, 188
    Infinite, 172, 223
  of set theory, 178–79
Axiom of Choice, 94, 178
Axiomatic method, 249, 265
  Limitations of, 249, 265
Axiomatic system, 172, 227
  is a PL theory whose members are logical consequences (or, theorems) of a set of axioms, 172
Axiomatizability, xv, 175, 250
  for a consistent PL theory, the triad of completeness, axiomatizability, and undecidability is inconsistent, 175, 250

Basic number, 232–34
Bernays, Paul 273–74
Biconditional (↔, ≡), 5, 8, 18–19, 24, 45, 88, 93, 112, 114, 251
  Truth conditions of, 22, 88, 244
  Truth table for, 24, 112

Biconditional (rule of inference; Bc), 59, 241, 279, 287
Biconditional Elimination (↔E), 63, 116
Biconditional Introduction (↔I), 63, 116
Biconditional Modus Ponens (BcMP), 54, 63, 242, 246, 275–76, 287
Biconditional Modus Tollens (BcMT), 54, 243, 246, 286
Bivalence, 12
Blank, 2–4, 92
Block(s), 49, 51–52, 57, 139, 141, 147, 269
  0- (Main), 51, 82, 139
    All non-0-blocks are subblocks of, 51
    closes at the conclusion of the derivation, 51
  Closed, 49, 140
  can be closed only if all its subblocks are closed, 51, 139
  Conditional Proof (CP), 51, 141, 236
  Nested, 51
    are stacks, 51
  Open, 49, 140–43, 269
  Reductio Ad Absurdum (RAA), 141

Cantor, Georg, 102–06
Cantor's Diagonal Argument, xiv, 106
  establishes that $\mathbb{R}$ is uncountable, 106–07
Cantor's Theorem, xiv, 104–05
  asserts: card($\mathcal{P}A$) > card(A), 104
Cardinal numbers, 101
Cardinalities, xiv, 21–22, 103, 137, 181–82
  Countable, 182
    are less than or equal to $\aleph_0$, 182
  Infinite, 21, 104–05, 182
    are greater than or equal to $\aleph_0$, 103
    Theory of, 103
  Relation of greater-than (>) between, 102, 104–05, 107, 181
  Relation of greater-than or equal-to (≥) between, 102, 104, 106
  Relation of identity (=) between, 101

Relation of less-than (<) between, 101–02
Relation of less-than or equal-to (≤) between, 102, 104
Transfinite, 103
Uncountable, 105
  are greater than $\aleph_0$, 105
Cartesian product A×B, 96, 98, 125, 132–33
Categorical PL set, 176
  all its models are isomorphic with respect to its vocabulary, 176
  $\aleph_0$-, 176, 289
  κ-, 176
  Satisfiable, 176, 262
    defines the structure of its models relative to its vocabulary, 176, 262
Categoricity, xv, 176, 260–62
  Second-Order Peano Arithmetic Th(PA²) is categorical, xv, 260, 262
Characteristic function, 205–06, 208–09, 217, 219, 222, 231, 233
  K is decidable iff its characteristic function is computable, 206, 208–09
  of the relation of identity (=) between natural numbers ($\chi_=$) is a recursive function, 219
  of the relation "less-than or equal-to" (≤) between natural numbers ($\chi_\leq$) is a recursive function, 219
Choice set, 124, 128–29, 178
Church, Alonzo, 204
Church's Thesis, xiv, 204–06, 209, 211, 213, 235–37, 247
  asserts: the class of computable numerical functions is identical with the class of Turing-computable functions, 204–05
  There is no known counterexample to, 205
  Turing-computable functions are equivalent to partial recursive functions, 205
  is unprovable, 205

Church's Undecidability Theorem, xi, xiv, 174, 254–55, 281
　the finite set of Robinson's Axioms RA is adequate for representing all recursive functions, 252
　Proof of, 253–54
　states that Th(∅) is undecidable, 174, 254–55
　Th(∅) is the set of all valid PL sentences, 254–55
　Th(RA) is undecidable, 254
　if Th(∅) is decidable, then Th(RA) is decidable, 255
　therefore, Th(∅) is undecidable, 255
　Th(∅) is semidecidable, 255–56
Circular reasoning, 89
　Benign, 89
Classical logic, 24, 60–61, 174, 251
　Laws of, 251
　Systems of, 88
Colon (:), 16
　indicates identity in the metalanguage, 16
Commutation (rule of inference; Com), 58
Compactness Theorem, xi, xiv, 164–67, 182, 186–87, 265–68
　asserts: if **X** is a logical consequence of Γ, **X** is a logical consequence of a finite subset of Γ, 164, 167, 265–66
　is a corollary of the Soundness and Completeness Theorems, 164, 265–66
　is equivalent to the Finite-Satisfiability Theorem, 164
　fails for PL², 265–66
　Generic proof of, 164, 186–87, 265–66
　Philosophical implication of, 165
　Proof of, 164–67, 265–66
Complement of B in A (A–B), 96, 134–35, 181, 208, 282
　consists of the elements contained in A but not in B, 96, 208

Complete PL set Σ, 169–72, 174–76, 179, 182, 223, 227, 243, 245, 248–50, 264–65, 271, 281, 283
　Arithmetic Th$_{AV}$(N) is, 227, 249–50, 283
　for a consistent PL theory, the triad of completeness, axiomatizability, and undecidability is inconsistent, 174–75, 248–50
　Σ is a complete PL set iff all its models are elementarily equivalent with respect to Voc(Σ), 175–76, 179
　Σ is a complete PL set iff for every **X** composed of Voc(Σ), either Σ ⊨ **X** (Σ ⊢ **X**) or Σ ⊨ ¬**X** (Σ ⊢ ¬**X**), 169–72, 223, 227, 242–43
Completeness Theorem, xi, xiii-xiv, 43–45, 59, 88, 109, 115, 139, 144, 147–48, 157, 164, 169, 177, 182
　asserts: all logical consequences of a PL set are derivable from it, 143, 146
　asserts: every consistent PL set has a model, 147–48, 161
　Henkin's proof of, 144, 157, 177, 182
Complex numbers, 178, 273
　The set of all (ℂ), 178
　is uncountable, 178
Complexity of a sentence, 92–93, 127–28, 135–36, 159–61, 169, 182–83
Composite numbers, 191–92
　Numerical codes of SL compound sentences are, 191
Comprehension scheme, 97–98
Computability, xii-xiii, 189, 193, 205
　Formalization(s) of, 205
　are all equivalent, 205
Computable function, xii-xiv, 189–90, 193, 204–09
　There is an effective procedure that computes the values of, 204, 209

It is a philosophical question whether
   Turing-computable function captures
   the notion of, 204
Two formalizations of, 190, 209
Computable numerical function, xiv, 189–90, 192–93, 206–07, 209, 213, 247
   when applied to its argument it returns its value via a mechanical computational procedure, 190
   Partial, 189–90, 203, 206
      assigns to every argument *at most one* value, 189, 203
   Total, 189–90, 203, 206–07
      assigns to every argument *exactly one* value, 189, 203, 206
Computation time, 190, 193, 207
Computational theory of the mind, 193, 204
Computing machine, 190, 193
Concept, 41, 97, 189, 208, 223, 256
   of actual infinity, 178
   Decidable, xii, 40, 173, 189, 256
   Existential, 41
      its denial is a universal concept, 41
   Extension of, 97
   Semidecidable, 41, 173, 189, 256
   Set-theoretic, 94
   Universal, 41–42
      its denial is an existential concept, 41
   Well-defined predicate designates, 97
Conditional Elimination (→E), 62–63, 116
Conditional Introduction (→I), 62–63, 115
Conditional Proof (CP), 50–52, 57, 62–63, 115–16, 141, 236, 254, 287
   assumption (CPA), 50–51, 57, 141, 236, 254, 287
Conjunct(s), 8, 145, 187
Conjunction (∧, &, •), 5, 8, 88, 93, 111, 113, 137, 178, 184, 187, 266–67, 285
   Infinite, 126
   Repeated, 184, 187
   Truth conditions of, 22, 59, 88
   Truth table for, 24, 111

Conjunction (rule of inference; Conj), 52, 62
   Soundness of, 59
Conjunction Elimination (∧E), 62, 115–16
Conjunction Introduction (∧I), 62, 115–16
Connective. See *sentential connective*.
Connexity, 133
Consistency, 17, 36, 151, 153, 241–42, 272–74, 280–82, 286
   ω-, 241–43, 281–82, 285–86
      there is an ω-consistent extension of Th(RA) of which N is not a model, 282
      there are sets that are consistent but not ω-consistent, 242, 281, 285,-286
   of PA, 242, 243, 246, 272–75, 278
      a finitary proof for, 273–75
   the Second Incompleteness Theorem asserts informally that if PA is consistent, it is impossible to prove its consistency from PA, 272
   CON$_s$ is the standard PA consistency sentence, 274, 276
   the Second Incompleteness Theorem's first proof establishes: If PA is consistent, PA ⊬ **CON**$_s$, 275–76
   the Second Incompleteness Theorem's second proof establishes: If PA is consistent, PA ⊬ ¬**prov**⟨0 = 1⟩, 276–20
   the structure of the natural numbers N is not a model of any AV theory that is ω-inconsistent, 281, 286
Consistent set (proof-theoretically), xiii, 38–39, 45, 65, 72, 83, 144, 147–57, 161, 171, 174, 177, 180–81, 226–27, 237, 241–42, 248–49, 252, 280–82, 286
   PA∪{¬G$_{PA}$} is consistent but ω-inconsistent, 281, 285–86
   is proof-theoretically a set from which no contradiction is derivable, 144, 147–57, 161, 180, 226–27, 242

is semantically a set that has a model, 144, 148, 171, 237
Constructive Dilemma (CD), 48–49, 55
Contingent PL sentence(s), xiv, 36–37, 41, 65, 77, 256
 The set of all, 256
  is neither decidable nor semidecidable, 256
Contradiction, 35–36, 40, 44, 60, 144, 151–52, 226, 242, 273, 285
 is a sentence and its negation, 144, 226, 273
Contradiction (sentential connective; $\bot$), 110, 113–14
 Truth table for, 110, 113
Contradictory PL sentence(s), xiv, 35–36, 41, 45, 66, 81, 174, 180–81, 185, 251–52, 255, 256
 The set of all, 256
  is undecidable but semidecidable, 256
Contraposition (Cont), 59, 180, 242
 is sound, 180
Convention-T, 251
 asserts: any adequate definition of truth must entail all instances of the Tarskian Schema, 251
Converse implication ($\leftarrow$, $\subset$), 112, 114
 Truth table for, 112
Converse nonimplication ($\not\leftarrow$, $\not\subset$), 112, 114
 Truth table for, 112
Corollary, 45, 88, 164, 170, 172, 179, 231
Creativity, 190

Davidson, Donald, 30
De Morgan, Augustus, 50
De Morgan's Laws (DeM), 50, 58
Decidability, xiv, 236
Decidable relation, 208, 217
 is equivalent to recursive relation, 217
Decidable set, xiv, 41, 173–75, 182, 189–90, 192, 205–06, 208, 217, 225, 247–50, 255, 257, 264, 270–71, 281–82, 284–85

Arithmetic is not, 227
For $E \subseteq \mathbb{N}^n$, if E is representable in Th(PA), E is, 281
There is an effective decision procedure for determining membership in, 173–75, 189, 225
is equivalent to recursive set, 208, 217
Every complete axiomatizable PL theory is, 174, 227, 248–50
for $K \subseteq \mathbb{N}^n$, K is decidable iff its characteristic function χ_K is computable, 205–06
Th($\emptyset$) is not, 255–56
Th(PA) is not, 248
Th$_{AV}$(N) is not, 248
Dedekind, Richard, 223
Deduction system (DS), 126, 185, 263
 any sound deduction system for PL^2 is incomplete, 263
Deductively closed PL set, 149, 153, 156–57, 161, 163–64, 169–70, 172
 contains all its theorems, 149
Definability in Peano Arithmetic, 250–51
 a formula **true**[x] defines arithmetical truth in Peano Arithmetic only if it satisfies TS: for every AV sentence **X**, PA $\vdash$ **true**[k]$\leftrightarrow$**X**, where k = [**X**], 250-251
 Tarski's Indefinability Theorem, xi, xiv, xv, 252, 282
  asserts in general that the notion of arithmetical truth is not definable in Peano Arithmetic, 252
The definite article (*the*), 3
Definite description, 3–4, 13
Definition, 94, 173, 176, 204–09, 213–17, 219, 226, 228, 230, 237–38, 251–52, 254, 263, 272
 Inductive, 151, 214–16
Demonstration, 42, 180, 280

INDEX 297

Derivability, 44–45, 122–23, 139, 145, 152, 164, 172, 263, 256–66, 268, 270–71
   is equivalent to logical consequence in PL, 44
   is a formal notion, 44
   in MDS, 122–23
   in NDS, 122
   in PL², 263, 265–66, 268, 270–71
      cannot be sound, complete, and finite, 265–66, 268
Derivable, 38, 40, 43–46, 61, 65, 76, 81–84, 115, 121–22, 144, 154, 161, 172–74, 185, 226, 242, 253, 264–65, 270, 273–74, 285
   the conclusion of a valid argument is derivable from its premises, 84
   every PL sentence is derivable from an inconsistent set, 174, 242
   no MDS rule is derivable from other MDS rules, 115
   a sentence that is derivable from ∅ is a logical theorem, 45, 82, 161, 173
Derivation D, xii, xiv, 42–44, 48–49, 52, 66, 88, 145–47, 152–53, 155, 157, 163–64, 187, 226–27, 254–55, 265, 268–69
   Conclusion of, 147, 227, 235, 248–49, 254, 257, 269
   Formal, 88, 94, 187, 265, 269
   is formal representation of demonstrative proof, 42, 268
   Infinite, 268–70
   Main, 51
   PA, 235–38, 240, 244, 248, 253, 272, 277–78, 283, 286
   PL, 42, 43, 66, 88, 89, 139–40, 145–47, 154, 163, 174–75, 226, 235–36, 238, 240, 242–43, 248–49, 253–56, 272, 277–78, 283, 286
   that is without premises, 82, 254–55
   PL², 45, 265–66, 268, 270–71
   PL⁺, 155, 157
   The set of premises invoked in (Σ_D), 139, 145–47, 152, 155–56, 164, 266
   Sound, 89
   Zero stage of, 43, 51, 82
Deterministic steps, 40–41, 173, 190, 204, 271
Detlefsen, Michael, 275
Diagonal, xiii, 107, 212–13
Diagonalization, xiv, 106, 126, 212, 238–39, 240–42, 245
   is the 3rd major component of the proof of the First Incompleteness Theorem, 238–42
   of AV formula X[z] is the AV sentence X[k] where k is the gödel number of X[z], 239
   the Diagonalization Lemma, 240–41, 246, 286
      asserts: if W[z] is an AV formula, there is an AV sentence G such that PA ⊢ G↔W[g] where g = [G], 240, 242, 275, 283, 286
   the proof of the Diagonalization Lemma, 240–41, 244
   the proof of the First Incompleteness Theorem, 242–43, 245
      the theorem states that there is a sentence G$_{PA}$ such that if Th(PA) is ω-consistent, Th(PA) ⊬ G$_{PA}$ and Th(PA) ⊬ ¬G$_{PA}$, 227, 242
   a sufficient condition for incompleteness, 248
      if Σ is a consistent axiomatizable PL theory in which all recursive functions are representable, Σ is incomplete, 248
   a sufficient condition for undecidability, 246–47
      if Σ is a consistent PL theory in which all recursive functions are representable, Σ is undecidable, 246–47
Diagonalization Lemma, xiii, 238, 240–42, 245–46, 278, 280, 283, 286

asserts: for any AV formula **W**[z], there is an AV sentence G such that
PA ⊢ G↔**W**[g], where g = [G], 240, 242, 275, 283, 286
Proof of, 240–41
Dictionary ordering, 132
Disjoint sets (A∩B = ∅), 96, 163
Disjunct(s), 8, 34, 49, 187
Disjunction, 5, 8, 49–50, 88, 93, 109, 133, 187
  Exclusive (⊕, +), 111, 113
    its expansion in terms of {¬, →}, 114
    Truth table for, 111
  Inclusive (∨), 111, 113
    Truth conditions of, 22, 88
    Truth table for, 24, 111
  Repeated, 187
Disjunction Elimination (∨E), 62, 116
Disjunction Introduction (∨I), 62, 116
Disjunctive Syllogism (DS), 48–49, 53, 61–62, 180
  is sound, 61, 180
  is truth-preserving, 61
Distribution (rule of inference; Dist), 58
Double Negation (DN), 58, 246, 286
Doubling function (D(n) = 2n), 218–19, 222
  A precise recursive definition of, 218, 222
  A Turing machine algorithm for computing, 218–20
  A Turing machine T_D that computes, 219
  Diagram of, 218, 221
  Instruction set of, 218, 220

Economical version of PL, xiv, 59, 109, 115, 123, 126, 148, 180, 183, 229
  is equivalent to the full version of PL, 109
  Logical symbols of, 115, 123
  Semantics of, 115
  Syntax of, 115, 148, 183, 232–35
Effective decision procedure, 40–41, 173–75, 189–90, 192, 206, 209, 225, 227, 233, 235–38, 247–49, 255–57, 270–72, 284–85

is an effective procedure that can fully decide a Yes-No question, 173, 189
Procedure C is, 192
two formalizations of, 189, 193, 213
Effective No-procedure, 41, 173
Effective procedure(s), xii, xiv, 40–41, 173–74, 189–92, 207, 219, 232, 235–38, 247, 249, 256, 270
  Arithmetical, 190, 192
  Decoding, 191, 206, 209–11, 219, 232–33, 235–36, 247
  Encoding, 174, 190–92, 206–07, 209, 211, 219, 232–35
Effective Yes-procedure, 173–74, 189, 206–07, 249, 256–57
Effectively enumerable set, xiv, 173, 182, 206–08, 271
  is equivalent to semidecidable set, 173, 206, 208
  K ⊆ ℕⁿ is effectively enumerable iff there is a computable numerical total function F such that ran(F) = K, 206
  F defines the computable listing function λ_K of K, 207
  there is a Turing machine T_F that computes F, 207
Elementarily equivalent, 167, 169, 175–76, 179, 182, 289
  all finite elementarily equivalent models of {(∃x)Px} are isomorphic, 182
  PL interpretations with respect to V (I≡vJ), 167, 169, 175–76, 179, 187
Elementary equivalence, xi, xiv, 167, 169, 175–76, 179
  the completeness of a PL set and the elementary equivalence of all its models are equivalent notions, 175, 178
  Isomorphism is a much stronger relation than, 169, 176, 179
The empty set (∅), 21, 45, 82, 95–96, 129–30, 140, 174, 185–86
  All PL interpretations are models of, 140

Logical consequence of, 34, 140, 185–86
  is unique, 95, 129
English, 2–3, 9, 12, 24–25, 165, 251
  conditional, 25, 60–61, 88
  sentences, 181, 184
  the sentence 'Most individuals are P' is not expressible by any PL sentence, 181
  the sentence 'There are at least n individuals that are not P' is expressible by a PL sentence, 181
  the sentences 'There are at least n individuals', for each n, are expressible by a set of PL sentences, 181, 184
Equinumerosity, 102, 104, 106, 124–25
  is the condition for identity among cardinalities, 102
Equinumerous sets (A ≈ B), 101, 103, 125
  There is a one-to-one correspondence between, 101
  have the same cardinality, 101–02
Equivalence class(es), 162–64
  Intersection of two, 163
  of PL+ singular terms, 162–64
Euclidean geometry, 172
Even natural numbers, 90, 102–04
  The set of all (E), 102–04
    Cardinality of ($\aleph_0$), 103–04
Evening Star, 257–59
Existential Elimination (∃E), 62, 116
Existential Generalization (EG), 56, 59, 62, 180, 238, 277
  is sound, 180
Existential Instantiation (EI), 58, 62, 180
  assumption (EIA), 58
  is sound, 180
Existential Introduction (∃I), 62, 116
Explosion (rule of inference; Expl), 55, 60–61, 174, 180, 242, 274
  Justification for, 60–61
  Lewis's Argument for, 61
  is sound, 60, 180
  is truth-preserving, 61

Exponentiation m^n, 218
  Precise recursive definition of, 218
Exportation (Expr), 59, 277
Expressive Completeness, 109–10, 113–14, 124
  Definition of, 110
  of {¬, →}, {¬, ∧}, {↑}, and {↓}, 113–14, 124
Extendibility, 74, 76
Extension, 15, 23, 72, 74, 259
Extension of a set, 150, 152, 159, 227, 248, 250, 282
  A is an extension of B iff B ⊆ A, 150
  Arithmetic is a consistent extension of Peano Arithmetic, 227, 250
  every consistent extension of Peano Arithmetic is undecidable, 227
  Maximal consistent, 152
Extensional system, 15, 23, 74, 259
Extensionality of PL, 23

Factorial function n!, 218
Fact(s), 5, 13–15, 21, 27, 44–45, 51, 70, 73, 90, 115, 125–26, 148, 174, 185, 205, 231, 235, 247, 268
  Arithmetical, 90, 233–35
  Semantical, 5, 13–15, 27, 70, 126
  Physical, 73
  Proof-theoretic, 115
  Set-theoretic, 125
Family of sets (F), 96, 124–26, 128–29, 131, 151–52, 162–63, 177–78
  Cardinality of, 128–29, 177–78
  Countably infinite, 124
  Finite, 124, 152
  Inductive, 124, 131
  Intersection of the members of (∩F), 96
  Pairwise disjoint, 96, 124, 126, 128–29, 162–63, 177
  Transitive, 125
  Union of the members of (∪F), 96, 125, 151–52, 162–63

Finite-Satisfiability Theorem, 165–66, 182, 185, 289
  asserts: a finitely satisfiable PL set is satisfiable, 165
  is equivalent to the Compactness Theorem, 165
Finitely satisfiable set, 165–66, 185, 288–89
  each of its finite subsets has a model, 165, 289
First Incompleteness Theorem, xiii-xiv, 223, 227, 238–45, 250, 272, 274–76, 280–83, 285–86
  asserts: there is an AV sentence G_{PA} such that if Th(PA) is ω-consistent, Th(PA) $\nvdash$ G_{PA} and Th(PA) $\nvdash \neg G_{PA}$, 227, 242
  Th(PA) is not categorical, 249
  Th(PA) is not $\aleph_0$-categorical, 249
  Proof of, 227, 238–45, 250, 281–82, 286
  A detailed outline of, 227–43
  if Σ is a consistent axiomatizable PL theory in which all recursive functions are representable, it is incomplete, 248
  Th(PA) is undecidable, 248
  under the supposition that N is a model of Th(PA), G_{PA} is true on N, 243–45, 280
First Incompleteness Theorem (proof without the assumption of ω-consistency), 282–83
  $R[y]$ is the AV formula: $(\forall x)(\mathbf{proof}[x, y] \rightarrow (\exists z)(z < x \wedge \mathbf{disproof}[z, y]))$, 283
  by the Diagonalization Lemma, there is a sentence R_{PA} such that PA $\vdash$ $R_{PA} \leftrightarrow R[\mathbf{r}]$, where r = $[R_{PA}]$, 283
  R_{PA} asserts metaphorically: if there is a PA derivation of me, then there is an earlier PA derivation of my negation, 283
  R_{PA} is neither provable nor disprovable in Th(PA), even without the assumption of ω-consistency, 283
  R_{PA} is "the Rosser sentence of PA," 283

First-Order Logic, 1, 257–58, 265
First-Order Predicate Logic (PL), xi-xii, 1, 43–44, 108–09, 136, 144, 157–58, 161, 164, 172, 178, 180–81, 185, 229, 257–58, 265
  first-order predicate, 257–58
  first-order quantifier, 257–58
  first-order sentences, 258
  first-order variable, 258, 261, 269
  Language of, 2, 28, 30, 88, 177–78, 182
  NDS is the deduction system of, 123
  Standard version of, 161
FORM, 234, 237
  consists of the gödel numbers of all AV formulas, 234
  Construction of, 234
  is a recursive set, 234
  is represented in Th(PA) by an AV formula $\mathbf{form}[x]$, 234
Form, 43–44, 192, 215
  of a formula, 43, 180, 182
  of an inference's antecedent, 43, 49, 140, 142–43
  of an inference's conclusion, 43, 140, 141–43
  of a sentence, 43, 49, 185–86, 192
Formal logic, 44
Formalizable semantical notion, 44
Formalization(s), 189–90, 193, 209, 213, 216
Formation rule, 7–8, 29, 115, 190, 233–34
  Biconditional, 7–8
  Conditional, 7–8, 234
  Conjunction, 7–8
  Disjunction, 7–8
  Existential Quantifier, 7–8
  Negation, 7–8, 234
  Universal Quantifier, 7, 234
Formula, 7–11, 19–20, 25, 27, 29–30, 52, 56–58, 94, 109, 120–21, 136, 142, 174, 177, 180, 182–83, 187, 213–14, 224–25, 229–31, 233–35, 237–40, 242, 246, 251–53, 258–60, 269, 272, 274–78, 280–83, 285–86, 288.
  Arithmetical, 94, 151, 224–25, 260

Atomic, 7–8, 233
AV, 223–25, 229–31, 233–35, 237–39, 240–41, 252, 260, 272, 282, 288
Components of, 8, 43, 59
Compound, 7–8, 124
First-order, 258
Main connective of, 43
Main operator of, 8, 43
Metalinguistic, 92, 260
One-variable, 233–34, 238–42, 246, 250–52, 260, 286, 288
Open, 9–10, 50, 258
Second-order, 258–59
Formula formation rules, 233–34
Full version of PL, 115, 123–24, 180, 229
Function(s), xii, 2, 11–15, 90–91, 98, 100–01, 104, 106, 137, 168–69, 177–78, 189–90, 203–19, 229–31, 261, 288–89
Algebraic, 209
Argument(s) of, 12–13, 99–101, 105, 168, 189, 203–05, 208, 211, 213–17, 222, 289
Arithmetical, 88, 90, 209, 215
Bijective, 100
Composition of (G∘F), 100–01, 214
Domain of (dom(F)), 99, 101, 203–04, 213, 217
Enumerating, 208
Existence condition of, 98–99, 101, 203, 217–18
Extension of, 13, 15
Identity (I, J_1^1), 101, 214–15
Injective, 100
Inverse (F^{-1}), 101
Maximum (Max), 219
One-to-one, 100–01, 178
Onto-, 100–01, 105, 210, 289
Partial, 91, 99–100, 189–90, 203, 205–06, 213, 217–18
Range of (ran(F)), 99–101, 178, 206–07, 213, 217
Set-theoretic, 88
Surjective, 100
Total, 99–100, 189–90, 203, 205–06, 210, 213, 217–19, 230–31
Uniqueness condition of, 98–99, 101, 203, 217–18, 230
Value of, 12–13, 99–101, 104–05, 137, 168, 189, 203–04, 206, 208, 213–14, 216, 289
Function symbol(s), 1–6, 10–11, 13–16, 20, 27, 29, 88, 126, 148, 158, 163, 167, 177, 224
Extension of, 163
Places of, 3, 148, 158, 163, 167, 224
Functional description, 3–5, 13, 15

GDS rules of inference (see also *standard rules of inference*), 62–63, 115–16
Elimination, xiv, 62–63, 115–16
Introduction, xiv, 62–63, 115–16
Gentzen, Gerhard, 62
Gentzen Deduction System (GDS; see also *Standard Deduction System*), 62–63
is equivalent to the Natural Deduction System NDS, 63
Gödel, Kurt, 44, 193, 227, 272, 274–75
gödel numbers, xiii, 232–42, 245–52, 256, 272, 275–77, 280, 282–83, 285
The Gödel Sentence G$_{PA}$, 227, 242–45, 275–76, 280–81, 285–86
cannot be proved or disproved from the Peano Axioms PA, 226, 242–43
is true on the standard model of Arithmetic, 226, 243–44, 280
Gödel's First Incompleteness Theorem (see also *First Incompleteness Theorem*), xiii-xiv, 223, 227, 238–43, 245, 274–75, 281–83.
asserts: there is a sentence G$_{PA}$ that is neither provable nor disprovable from the Peano Axioms PA, if PA is ω-consistent, 227, 242–43, 245
Gödel's Proof of, 227–28, 231, 238–43, 245, 250, 252, 272, 274, 276, 281, 286
its proof is formalizable in Peano Arithmetic Th(PA), 272, 275

the Rosser sentence R_PA is neither
  provable nor disprovable from PA,
  even without the assumption of ω-
  consistency, 283
Gödel's Second Incompleteness Theorem
  (see also *Second Incompleteness Theorem*),
  271–80
  fails for Rosser's proof predicate, 275
  holds for any proof predicate that
    satisfies the Provability Conditions,
    272, 275–78, 282
  Original proof of, 275–76
  Second proof of, 275–80
Greater-than, 90, 102
  or equal-to (≥), 90–92
  Strictly (>), 90
Greatest common divisor, 13

Half-open unit interval **I**, 106
The Halting Problem, xi, xiv, 209, 211, 213
  assets the Halting Function H is not
    Turing-computable, 211, 213
  asserts informally there is no effective
    decision procedure that can determine
    for any Turing machine T and any
    number m whether T($\bar{m}$) will halt or
    not, 209
  Proof of, 211–13
Hardware, 190, 205
Henkin, Leon, 44, 144
Henkin Interpretation H$^\Sigma$, 157–59, 161, 180, 184
  assigns a singular term as its own
    referent, 158
  for a Henkin set is a model of it, 157, 159, 161
  List of Names of, 158–62, 184
  Modified, 163–64
    assigns to a singular term its
      equivalence class as its referent, 162, 164

  is countable, 177
  for a Henkin set is a model of it, 163–64
  List of Names of, 162–63
  Universe of Discourse of ($\mathcal{U}$), 162–63
  Universe of Discourse of (UD), 158, 161–63, 177, 184
Henkin Model of a Henkin set (H$^\Pi$), 157, 159, 164, 177
  is countable, 177
Henkin set Π, 152–53, 156–57, 159–64, 177, 180
  is deductively closed, 161, 163
  Every consistent PL set could be
    extended into, 161, 177
  is maximal consistent, 156, 161
Hilbert, David, 272–75
Hilbert's philosophy of mathematics, 272–75
  posits two types of mathematics, 272
    finitary mathematics, 272–74
      is formalizable in Th(PA), 274
    ideal mathematics, 272–73
      has no semantical significance, 274
  Th(PA) contains both finitary and ideal
    mathematics, 273
Hilbert's Program, xv, 273–75
  consisted of two parts: reduction and
    reliability, 273
  reduction was aimed at reducing almost
    all of mathematics to Th(PA), 273
  reliability was reduced to giving a
    finitary proof of the consistency of
    Th(PA), 274
  the Second Incompleteness Theorem
    establishes that an AV sentence
    expressing the consistency of PA is not
    a PA theorem, 274–75
  Rosser's proof predicate can be used to
    construct a PA consistency sentence
    that is a PA theorem, 274

the Second Incompleteness Theorem holds for any proof predicate that satisfies the Provability Conditions, 275
Hubble Telescope, 257, 259
Hume's Principle, 102, 104, 106, 137
   card(A) = card(B) iff A ≈ B, 102
Hypothetical rule(s) of inference, 47, 50–52, 57, 59, 140, 236
   Assumption of, 51, 57, 139–40, 147, 236
   Discharged, 51, 121, 139, 140, 147
   Undischarged, 51, 56, 58, 121, 139–40
Hypothetical Syllogism (HS), 55, 252, 287

Idempotence (rule of inference; Idem), 58
Identity (rule of inference; Id), 56, 62, 115–16, 140, 236
Identity Elimination (=E), 62, 116
Identity Introduction (=I), 62, 116
Identity predicate =, 2, 5, 7, 10–11, 62, 115–16, 126, 157, 159, 161, 163–64, 177, 180, 222
   Truth conditions of, 22, 36, 136, 163, 289
Identity relation = on ℕ, 219
   Characteristic function of ($\chi_=$), 219, 222
      There is a Turing machine $T_=$ that computes, 222
      instruction set of $T_=$, 222
   is a recursive relation, 219
Identity sentence **s** = **t**, 136, 140, 161, 163–64
Incomplete PL set, 223, 243, 245, 247–48, 250, 254, 281, 285
   every incomplete consistent PL theory is not categorical, 243
   Peano Arithmetic is incomplete, 242–43, 245, 247, 250
   if Σ is a consistent axiomatizable PL theory in which all recursive functions are representable, it is incomplete, 247–48
Incompleteness Theorems, xi, xiii, xiv, 115, 223, 225, 265, 272, 274–75, 280

Inconsistent set (proof-theoretically), 40, 45, 144–45, 148–53, 155–56, 174–76, 242–43
   is proof-theoretically a set from which a contradiction is derivable, 144–45, 148–53, 155–56, 174, 242
   is semantically a set that has no model, 144, 171, 176
Indirect proof, 34–35, 40
Individual(s), 2–3, 5, 10–16, 19–20, 22–23, 25–31, 75–76, 129, 136–37, 142–43, 161, 167, 178, 181, 187, 257–62, 267, 269, 289
   are first-order objects, 257
   Philosophical sense of, 2
   Unnamed, 14, 26
Inductive set, 124–25, 130–31
   Smallest (ω), 124–25, 130–31
      ∈ is asymmetric, connex, irreflexive, transitive, and has a minimal element on, 125, 131
      ∈ is well-founded on, 125, 131
      ∈ is a well-ordering of, 125
   is a subset of all inductive sets, 125, 130
Inference, 59–60, 109
   Truth-preserving, 59–60
   Unsound, 59
Inference's antecedent(s), 42–43, 48–49, 51, 56, 60, 140, 142–43, 269–70
   of a hypothetical rule is the derivation enclosed in its block, 51
   The ω-rule of inference requires infinitely many, 269–70, 281
Inference's conclusion, 42–43, 48–49, 60, 140, 141–43
   Intermediate, 42
   of the ω-rule of inference is a PL² sentence of the form $(\forall z)X$, 269
Inferential license, 42, 49, 52
Inferential step, 42
Infinite decimal expansion, 106–07
Infinity, 102–03, 178
   Actual, 102–03, 178
      Properties and relations of, 103

Potential, 103
Integer(s), 100, 222, 273
   Positive, 13, 137–39, 144, 148, 214, 216, 218, 222, 234, 238–39, 266–67, 269
   The set of all ($\mathbb{Z}$), 100
Interderivable sentences, 45–46, 65, 76
Intersection of two sets (A∩B), 26, 96, 124, 163
Invalidity (of arguments), 41, 257
   is neither decidable nor semidecidable, 41, 257
Isomorphic, 167, 176, 179, 182, 289
   PL interpretations with respect to V (I~vJ), 167, 169, 176, 179, 187, 289
   models are elementarily equivalent, 168, 176
Isomorphism, xi, xiv, 167–69, 175, 179, 187, 289

Justification, 60–61, 63, 89, 102, 239
   Demonstrative, 89
   for Explosion, 60–61

Kleene's Theorem, 208
   asserts: for K ⊆ $\mathbb{N}^n$, K is decidable iff K and $\mathbb{N}^n$–K are effectively enumerable, 208

Language, 27–30, 177, 178, 232–33, 235, 251
   Bivalent, 12
   Formal, 87
   Formal arithmetical, 90, 223–24
   Natural, xiii, 1, 3
      argument, 164–65
   Object, 1–2, 16, 228, 232, 251
   of Peano Arithmetic, 233, 235
   of a PL interpretation, 14, 29
   of a set of PL sentences, 28–29, 182
   Symbolic, 87
Lemma, xiii, 144–45, 148–50, 152, 157, 226, 238–41, 245–46, 252, 254, 275, 280, 283, 286

Less-than, 90, 101–02, 283
   or equal-to (≤), 90–91, 130–31
   Strictly (<), 90–92, 140, 217
Lewis, C.I., 60
   his argument for Explosion, 60–62
Lexicographical ordering, 132–33
Liar Paradox, 251
Liar Sentence, 251–52
   is any sentence that affirms of itself that it is not true, 251–52
Lindenbaum, Adolf, 148
Lindenbaum's Lemma, 148, 150, 152, 154, 156–57
   asserts: every consistent set can be extended into a maximal consistent set, 150, 156
Lindström, Per, 87
Lindström's Theorem, 87
Linear ordering of a set, 125, 132–34
List of names (LN), 11, 14–15, 19–20, 23, 26–29, 142–43, 153, 162, 184, 186, 224, 244, 266–68, 289
   of Henkin Interpretation, 158–62, 184
   of modified Henkin Interpretation, 162
   Name in, 184, 224
   of the structure of the natural numbers, 224, 244, 286
   Uncountable, 29
Listing function (λ_K), 206–09, 217
   K is semidecidable iff λ_K is computable, 206–09, 217
Löb, Martin Hugo, 282
Löb's Theorem, xv, 282, 286–87
   asserts: for every **X**, if **prov**⟨**X**⟩→**X** is a PA theorem, **X** is also a PA theorem, 282
Logic fiction, 60–61
Logical concept(s), xi, xiv, 31–41, 45, 87, 181, 257
   of contingent sentence is neither decidable nor semidecidable, 256
   of contradictory sentence is undecidable but semidecidable, 256

of invalid argument is neither decidable nor semidecidable, 257
non of the 8 logical concepts is decidable in PL, 254–57
of satisfiable set is neither decidable nor semidecidable, 257
of unsatisfiable set is undecidable but semidecidable, 257
of valid argument is undecidable but semidecidable, 257
of valid sentence is undecidable but semidecidable, 254–56
Logical consequence, xii, 31, 34, 43–44, 64, 87–88, 122, 140, 143, 148, 164–67, 169–71, 181, 185–86, 223, 226, 254, 263–65, 268, 270–71
is equivalent to derivability in PL, 44
is equivalent to validity (of arguments), 31
is formalizable in PL, 44
in PL^2 ($\vDash^2$), 44, 263, 265–68, 270, 271
and derivability in PL^2 ($\vdash^2$), 263, 266, 268–69, 270–71.
is not formalizable, 44
Logical derivation(s), 255–56
is a PL derivation without premises (from $\emptyset$), 255–56
The set of all (LD), 255–56
is decidable, 255
Logical identity (sentential connective), 110, 113
Truth table for, 110
Logical operators, 5, 121–22
Logical possibility, xii, 12, 73
Logical symbols, 1–2, 5, 28, 62–63, 115, 123
Logical system, 44, 61, 87–88, 126, 164–65, 181, 186, 265
Incomplete, 265
Symbolic, 87
Logical theorem(s), 45, 82, 161, 163, 173, 185, 254–55

is derivable from the empty set $\emptyset$, 82, 174, 185
The set of all (Th($\emptyset$)), 174, 254–55
is semidecidable, 174
is undecidable, 254–55
Logically equivalent sentences, xiv, 37–38, 45, 66, 76, 92–93, 109, 122, 127, 180, 182–83, 257
Logically false sentence, 35, 174
Logically true sentence, 34, 174
Loop, 74–76, 211–12
Löwenheim, Leopold, 177
Löwenheim-Skolem Theorem, xi, 177–79, 289
asserts: every satisfiable PL set has a countable model, 177, 179
Proof of, 177

Material conditional ($\rightarrow$, $\supset$), 5, 8, 17, 20, 25, 42, 49–52, 63, 88, 93, 108, 112–14, 116, 140–41, 152–53, 160, 182–83, 190, 234
Antecedent of, 8, 17, 20, 25, 33, 42, 51, 183
Consequent of, 8, 20, 25, 33, 50–51, 183
Truth conditions of, 22, 60, 88, 141, 153, 160
Truth table for, 24–25, 60, 112
Material Conditional (rule of inference; MC), 50, 58, 159
Material implication. See *material conditional*.
Material nonimplication ($\nrightarrow$, $\not\supset$), 112, 114
Truth table for, 112
Mathematical induction, xiv, 94, 127, 139, 151, 159, 169, 261–62, 267
Mathematics, xi–xii, xv, 172, 272–73
Foundations of, 275
Hilbert's philosophy of, 272–75
Maximal consistent PL set, 149–50, 152–54, 156–57, 160–61, 180
Maximal PL set, 149–50, 152–53, 161, 169
Γ is maximal iff for each **X**, either **X** ∈ Γ or ¬**X** ∈ Γ, 149, 169

MDS rules of inference, 115–16, 140, 235–36, 248–49, 255, 277
   Deriving the GDS rules from, 116–21
   Deriving Modus Tollens from, 116
Membership ($\in$), 94, 125, 133–34, 153, 156, 189–90, 192, 206, 225, 230
   $\in$-minimal element, 131, 134–35
   Circular, 97
Memory, 190
Metalanguage, 1–2, 16, 228
Metalogic, xii, xiv-xvi, 34–35, 38, 40, 45, 87, 89, 144, 174, 238
Metatheorems, xiv, 31, 87–90, 92, 94, 109, 121–23, 174, 179
Metatheory, 3, 60, 87–89, 94, 174, 231–32, 235, 252, 280
   of classical logic, 60, 88
   contains counterparts of the NDS rules of inference, 89
   of First-Order Predicate Logic, xi-xiii, 87–88, 90, 94, 174
   Language of, 3, 88–89
   of Peano Arithmetic, 231–32, 235
Mini Deduction System (MDS), 115, 122–23, 126
   is complete, 126
   is decidable, 236
   is equivalent to the Natural Deduction System (NDS), 121
   is sound, 139
Minimality of an element in a set, 131, 134
Model, 23, 31, 33, 38, 40, 60, 64–65, 76, 126, 137, 140, 143–44, 157, 161–62, 165, 171, 176–79, 181–82, 184–87, 225–26, 250, 254, 260–62, 264, 280, 283, 288–89
   all models of PA² are isomorphic to N², 260–64, 269
   of a collection of sentences, 88, 139, 141–44, 157, 166, 171, 175–76, 178, 180, 182, 184–86, 225, 262–63
   Countable, 177–79
   Countably infinite, 176, 178, 182
   Finite, 137, 181–82, 187
   Infinite, 65, 76, 178, 181, 185
   N is not a model of any ω-inconsistent AV theory, 281, 286
   Nonstandard, 263, 283, 289
   Second-order, 260, 263, 265, 266–67
   Size of, 137, 181–82, 184–85
   standard model N² of Th(PA²), 260, 263, 266, 268–69
   Uncountable, 178–79
Model theory, xiv, 88
Modus Ponens (MP), 53, 62–63, 115–16, 140, 141, 153, 156, 254, 276–77, 284, 287
   is sound, 140–41
Modus Tollens (MT), 54, 116, 180, 231
   is sound, 180
Morning Star, 257–58
Multiplication, 90, 191, 218, 225
   function (M), 90–91, 218
   the product of d and k (Prod(d, k)), 215–16, 218–19
   Recursive definition of, 225

n-tuple(s), 12–13, 15, 23, 96, 99, 131–32, 158, 162–64, 167, 190, 203, 206, 208, 230, 273, 282
   Coordinates of, 12–13, 95–96, 99, 132, 203
   of objects, 230
   Relations and functions are sets of, 230
   is a set-theoretic object in which order and repetition matter, 95
   Sets of, 98, 158, 208
Name(s), 1, 3–6, 11, 13–16, 19–20, 23, 26–30, 56, 58, 121, 136, 142–43, 145, 147–48, 153–54, 161, 163, 168, 177–78, 180, 184–87, 224, 251, 257–58, 286
   Ambiguous, 14
   Arbitrary, 56, 121, 142, 147, 154–56
   Metalinguistic, 154–56, 232
   Non-PL, 11, 28–29, 154, 181
   Non-referring, 14

Natural Deduction System (NDS), xi, xiv, 43–44, 52, 61–63, 109, 115, 122–23, 185
 is equivalent to the Gentzen Deduction System (GDS), 63
 is equivalent to the Mini Deduction System (MDS), 121
 is sound and complete in PL, 43–44, 59, 63, 122–23, 185
Natural number(s), xiv, 12–13, 22, 90–94, 96, 99–101, 103, 106–07, 124–25, 127–31, 136, 151–52, 174, 181–82, 185–86, 189–90, 192, 206–07, 214, 224, 228–29, 230–34, 236–39, 241, 244, 247, 260, 263, 266–67, 272–73, 282, 288
 n-tuples of, 189–90, 193, 203, 206, 208–09
 Ordered pairs of, 99, 189, 236, 272–73
 Properties and relations of, 229
 The set of all ($\mathbb{N}$), 21, 37, 71, 90, 96, 99–104, 106, 108, 125, 131, 150, 179, 181, 185–86, 205–08, 213, 217, 224, 229
  Cardinality of ($\aleph_0$), 21–22, 64, 101, 103–06, 177, 182
  is discrete, 106
NDS rules of inference, 43–45, 49, 59, 63, 89, 109, 115, 185
 are derivable from the MDS rules of inference, 115
 are derivable from the standard rules of inference (GDS), 61, 63, 115
 are truth-preserving, 43, 44
Negated Biconditional (rule of inference; NBc), 59
Negated Conditional (rule of inference; NC), 58
Negated Quantifier (rule of inference; NQ), 59, 280
Negation ($\neg$, ~, –), 5, 8, 35, 45, 49, 81, 83, 88, 93, 95, 108, 110, 113–14, 116, 153, 159, 161, 169, 183, 190–91, 234, 263–65, 273, 282–83, 288
 Truth conditions of, 22, 88, 153, 159
 Truth table for, 24, 110

Negation Elimination ($\neg$E), 62, 116
Negation Introduction ($\neg$I), 62, 116
No-procedure, 41, 173
Non-zero stages, 43
Nonempty set, 31, 101, 125, 128–29, 131, 133, 156, 162, 180, 186
Nonstandard models of Arithmetic, 283, 289
 are models of $Th_{AV}(N)$ that are not isomorphic to N and whose cardinality is $\aleph_0$, 283
 N is the standard model of Arithmetic, 280–83, 288–89
Normal rules of inference, 47–50, 52, 59–60, 140, 143
 are truth-preserving, 60
Notation, 183, 203, 222, 228–29
 Defined, 108–09, 228–29
 Set-theoretic, 139
 Standard arithmetical, 215, 224
 Vector, 203, 206
  $\vec{m} = \langle m_1, m_2, \ldots, m_n \rangle$, 203, 206
  $\bar{m} = \langle m, m, \ldots, m \rangle$, 203, 209
Number, 2, 4, 7, 21, 37, 40, 92, 94, 101–02, 107, 129, 135, 156, 189, 191–92, 206
 The anti-diagonal, 107
Number Logic NL, 181, 185–87
 The Compactness Theorem fails for, 186
 derivations, 181
 interpretation, 181, 185–86
 Logical concepts in, 181
 Names of, 181
 NDS is the deduction system of, 185
 Proof theory of, 181, 185–87
 Sentences of, 185–86
 Valid sentences of, 185–86
 Vocabulary of, 181
Number theory, xi, 172, 250, 254
 Finitely axiomatizable, 254
 Infinite set of axioms of, 172
 no number theory is complete, 250

Numerical code(s), 190–92, 211, 231, 231-232, 235–36
  of AV expressions, 231–32
  Basic, 190–91
  basic numbers, 232–34
  gödel numbers, 232–38
  SENT$_{SL}$ consists of the numerical codes of all SL sentences, 192
  TERM consists of the numerical codes of all AV terms, 232–33
Numerical function, 203, 205–07, 209–10, 213
  Partial, 203, 205–06, 213
    F from $\mathbb{N}^n$ into $\mathbb{N}$ is partial iff dom(F) $\subseteq$ $\mathbb{N}^n$, 203, 205–06, 213
  Strictly partial, 203, 213
    F from $\mathbb{N}^n$ into $\mathbb{N}$ is strictly partial iff dom(F) $\subset$ $\mathbb{N}^n$, 203, 213
  Total, 203, 205–07, 210, 213
    F from $\mathbb{N}^n$ into $\mathbb{N}$ is total iff dom(F) = $\mathbb{N}^n$, 203, 206, 213

Object(s), 2, 32, 41, 88, 94–95, 97–98, 101, 103–04, 151, 173, 178, 230, 256–57
  Arbitrary, 32
  First-order, 257
  Proof-theoretic, 88
  Second-order, 257
  Set-theoretic, 94–95, 97–98, 101, 103–04, 230
Object language. See *language*.
Odd natural numbers, 104, 232–34
  The set of all (**O**), 104
    Cardinality of ($\aleph_0$), 104
The ω-rule of inference, 269–70, 281
  Antecedents of, 269–70
  Conclusion of, 268–69
  is truth-preserving, 269
One-to-one correspondence, 100–01, 103–06, 108, 137, 150, 167, 179, 187, 262, 289

One-variable formula, xiii, 224, 233–34, 238–39, 240, 242, 246, 250–52, 260, 280–81, 283, 286, 288
  Diagonal instance of, xiii, 239–40
  Substitutional instance of, 239
Operations, xiii, 88, 90, 96, 100–01, 105, 124, 129, 194–95, 213–17, 232, 234–35
  Arithmetical, 88, 90, 232, 234–35
  Recursive, 213–17
  Set-theoretic, 88, 96, 100–01, 105, 124, 129
  Turing machine, 194–95
Ordered pairs, 12–13, 15, 23, 95–96, 98, 132–33, 168, 184, 235–37, 272–73, 280, 285
  Coordinates of, 96, 98, 132–33, 183
  Sets of, 98
Ordered triples, 12–14, 95, 183
  Coordinates of, 13, 183
  Sets of, 98
Ordering relations, 91, 102
Ordinal(s), 125, 133–34
  The class of all (**Ord**), 125
  is a transitive set that is well-ordered by $\in$, 125, 133–34

Paradox, 97–98, 178, 251
Parentheses, 1, 10, 28, 124, 148, 223, 261
Partition, 96, 137, 162, 178
  of $\mathbb{N}$, 96
  of the set of all PL⁺ singular terms, 162
Peano, Giuseppe, 223
Peano Arithmetic Th(PA), xiii-xiv, 90, 223, 226, 227–32, 234–35, 237, 239, 242–46, 428–50, 252, 254, 257, 373–76
  AFORM consists of the gödel numbers of all AV atomic formulas, 233
  AFORM is a recursive set, 233
  AFORM is represented in Th(PA) by an AV formula **aform**[x], 233
  All recursive functions are representable in, 228–32, 248, 250, 252–54
  Consistent extension of, 248

consists of all logical consequences (or, theorems) of PA that are composed of Voc(PA), 223, 226, 237, 248
Language of, 233, 235
Metatheory of, 231–32, 235–38, 242, 248, 272, 274–75, 278, 285–86
Proof(s) of (PA proofs), 235, 271, 274, 278, 283, 285–86
   is a PL derivation D of an AV sentence **X** from PA, 235–37, 242, 248, 253–55, 272, 277–78, 283, 285–86
      ⟨m, k⟩ ∈ PROOF iff m is the gödel number of an AV sentence **X** and k is the gödel number of a PA proof of **X**, 235, 272, 285
PROOF is a recursive relation, 235–37
PROOF is represented in Th(PA) by an AV formula **proof**[x, y], 237, 244, 272, 277, 280, 283, 285
Proof theory of, 232
Provability in, 237–38, 272, 76–278
   PROV consists of the gödel numbers of all theorems of PA, 237, 246
   b ∈ PROV iff there is k such that ⟨k, b⟩ ∈ PROOF, 237–38
   PROV is not recursive but only recursively enumerable, 237–38
   PROV is not representable in Th(PA), 238, 246
   prov[y] is defined as (∃x)**proof**[x, y], 238, 242, 272, 275–77
      Gödel proved that there is a sentence G$_{PA}$ such that PA ⊬ G$_{PA}$ and PA ⊬ ¬G$_{PA}$, 227, 242–43, 282
      G$_{PA}$ is true on N, 227, 243–45, 280
      in PL, PA ⊬ G$_{PA}$ and PA ⊬ ¬G$_{PA}$, 243
      Th(PA) is not categorical, 243
Representability in, 228–31, 233–35, 237–40, 248, 250, 252–54, 276, 278, 283
The set of the axioms of, 223–27
Standard consistency of, 272–76, 278

the standard PA consistency sentence is **CON**$_s$, 274, 276
   the Second Incompleteness Theorem's first proof establishes: If PA is consistent, PA ⊬ **CON**$_s$, 275–76
   the Second Incompleteness Theorem's second proof establishes: If PA is consistent, PA ⊬ ¬**prov**⟨0 = 1⟩, 277–80
Syntax of, 231–38
   Arithmetization of, 232–35
   FORM consists of the gödel numbers of all AV formulas, 233–34
   FORM is a recursive set, 234
   FORM is represented in Th(PA) by an AV formula **form**[x], 234
   SENT consists of the gödel numbers of all AV sentences, 234
   SENT is a recursive set, 234
   SENT is represented in Th(PA) by an AV formula **sent**[x], 234–35
   TERM consists of the gödel numbers of all AV terms, 232
   TERM is a recursive set, 233
   TERM is represented in Th(PA) by an AV formula **term**[x], 233
is undecidable, xiv, 248
   every consistent extension of Th(PA) is undecidable, 227
Vocabulary of (AV), 223–25, 228, 230–35, 243, 245, 248–49, 260, 264, 271–72, 281–83, 286, 288
   AV atomic formulas, 233
   AV formulas, 223–25, 229–30, 233–34, 237–40, 242, 252–53, 260, 272, 275, 277–78
   AV numerals, 228–30, 234–35, 239, 244, 277, 288–89
   AV sentences, 223, 227, 234–40, 242–46, 248–50, 252, 254–56, 264, 271, 273–78, 282–83, 285–87, 289

AV singular terms, 228
AV terms, 229, 232–33, 235
Peano Axioms, 224–27, 230, 235–37, 240, 244, 248–49, 252, 272–82
are the following:
Ax1: $(\forall x)0 \neq sx$, 224–25, 260–61, 273, 280
Ax2: $(\forall x)(\forall y)(sx = sy \rightarrow x = y)$, 114–15, 228, 260–61
Ax3: $(\forall x)(x + 0) = x$, 224–25, 262, 262
Ax4: $(\forall x)(\forall y)(x + sy) = s(x + y)$, 224–25, 262, 262
Ax5: $(\forall x)(x \bullet 0) = 0$, 224–25, 262, 262
Ax6: $(\forall x)(\forall y)(x \bullet sy) = ((x \bullet y) + x)$, 224–25, 262, 262
IS: If **X[z]** is an AV formula in which z is free, the following is a Peano Axiom: $X[0] \rightarrow ((\forall v)(X[v] \rightarrow X[sv]) \rightarrow (\forall y)X[y])$, 224
IS (the Induction Schema) is the PL representation of the Principle of Mathematical Induction PMI, 224–26
Gödel constructed an AV sentence that affirms the consistency of, 272
His Second Incompleteness Theorem established that this sentence is not provable from PA, 272
Gödel proved that there is an AV sentence G$_{PA}$ such that PA ⊬ G$_{PA}$ and PA ⊬ ¬G$_{PA}$, 227, 243
representing PA proofs in Th(PA), 235–36
a PA proof is a PL derivation D of an AV sentence **X** from PA, 235–36, 248, 272, 277–78, 283, 286
⟨m, k⟩ ∈ PROOF iff m is the gödel number of an AV sentence **X** and k is the gödel number of a PA proof of **X**, 235–36, 272, 283
PROOF is a recursive relation, 235–37

PROOF is represented in Th(PA) by an AV formula **proof**[x, y], 237–38, 242, 244, 272, 274–80, 283, 285
representing provability in Th(PA), 237–38, 272
PROV consists of the gödel numbers of all PA theorems, 237
PROV is not a recursive set but only recursively enumerable, 237, 246
PROV is not representable in Th(PA), 238, 246
prov[y] is defined as $(\exists x)$**proof**[x, y], 238, 242–44, 272, 275, 280
The set of (PA), 224–25, 227, 236, 274, 278, 281, 285–86
An AV sentence asserting that it is a theorem of, 282
An AV sentence asserting that either it or its negation is a theorem of, 282
An AV sentence asserting that neither it nor its negation is a theorem of, 282
is decidable, 224, 236
Theorem(s) of, 230, 237, 240, 242, 246, 248, 253, 272–78, 280–82
The union PA∪{¬G$_{PA}$} is consistent but ω-inconsistent, 281, 285–86
Peirce's Arrow (↓), 111, 113–14
its expansion in terms of {¬, →}, 114
Expansions in terms of, 114
Truth table for, 111
PL interpretation(s), xiv, 4–5, 10–17, 19–20, 22, 24, 26–43, 59–61, 64, 73, 88, 94, 126, 135, 137, 140, 143, 148, 153–54, 166–67, 169–70, 180, 184–85, 224, 228, 244, 267, 289
Constituents of, 11, 167, 169, 176
Finite, 21, 64, 137, 178
Infinite, 21–22, 64, 74
Language of, 14, 29, 154
for (relevant to) a PL sentence, 11, 141–42, 144, 171

for (relevant to) a set of PL sentences, 11, 28–29, 142, 167, 171, 180, 183
  Size of, 21–22, 64, 71, 184–85
  that is a standard interpretation of arithmetic (N), 224, 241, 243, 250, 254, 260, 262–63, 266, 269, 280–83, 288–89
  Structurally identical, 167, 169, 187
  Uncountable, 22, 28–29
PL set(s), 169–70, 176–82, 184–85, 188, 223, 226, 281–82, 285–86, 288
  Finite, 184, 188, 227
  ω-consistent, 241–43, 282–83, 286
    PA∪{¬G_PA} is consistent but ω-inconsistent, 281, 285
  Properties of, 169, 226
PL theory, xi, xiv, 169–75, 179, 182, 223, 243, 245–50, 252, 254–55, 280–83, 285–87
  in which all recursive functions are representable, 245–48, 250, 252–55, 282, 288
  AV, 248–50, 271, 282–85, 286–87, 289
    N is not a model of any AV theory that is ω-inconsistent, 281, 286
  Axiomatic, 94, 172, 227
  Axiomatizable, 173–75, 182, 223, 285
  Consistent, 175, 237, 241–43, 245–49, 252, 255, 280–82, 288
  contains all its logical consequences (or, theorems) that are composed of its vocabulary, 169–70, 172, 174, 223, 245–46, 287
  Every deductively or semantically closed PL set is, 170
  Finitely axiomatizable, 173, 182, 188, 252–54, 263
  Non-axiomatizable, 227, 250
  of a PL interpretation J with respect to V (Th_V(J)), 170–72, 182, 227, 243, 245
  of PL interpretation J with respect to Voc(J) (Th(J)), 170
  of PL set Σ (Th(Σ)), 170–72, 182, 188, 223, 227
  consists of all logical consequences (or, theorems) of Σ that are composed of Voc(Σ), 170–72, 223, 227
  Scope and limitations of, 223
  Set of axioms of, 172–73, 223
    is decidable, 173
    the PL theory whose set of axioms is ∅ consists of all valid sentences (or, logical theorems), 173
  if Σ is a consistent PL theory in which all recursive functions are representable, it is undecidable, 245–47
PL⁺, 154–55, 157–58, 177, 184
  Expanded language of, 177
  Names of, 155–58, 177
  Sentences of, 154–59, 161, 163
  Singular terms of, 157–64, 177
    Equivalence classes of, 162–63
    Partition of the set of all, 162
  Vocabulary of, 154, 157–58, 177
PL² interpretation, 45, 259–60, 262, 266–69
  allows quantification over individuals and over properties and relations of individuals, 257–60
  if a PL² interpretation is a model of PA², it is isomorphic to N², 260–63
PL-English, 2–3, 9, 13, 15, 19, 23, 88, 137
Placeholders, 2, 9
Powerset of a set ($\mathcal{P}$A), 95–96, 104–05, 125, 131–32
  Cardinality of, 104–05, 125
Predecessor, 91–92
Predicate, 1–3, 6–7, 9–11, 14–16, 22–23, 29–30, 88, 97, 126, 136, 148, 158–59, 162, 167, 177, 181, 185, 257–59
  Extension of, 12, 64, 74, 97, 137, 158, 162–64, 169, 181, 184, 185–87
  First-order, 257–58
  Monadic, 2, 88
  Order, 229

Places of, 2–3, 9–10, 148, 158–59, 162, 167, 181
Second-order, 257–58
Unary, 2, 88
Well-defined, 97–98
  designates a concept, 97
Premise(s), 42–43, 56, 58, 121, 140–41, 241, 254–55, 257, 268–70, 279, 287
Second-order, 257, 269–70
Set of, 31, 34, 42, 44, 139, 147
Prime-factorization, 191–92, 233–35
is unique, 191
Prime numbers, 90, 191–92, 233–35
  Numerical codes of the basic SL vocabulary are all, 191
  Powers of, 191–92, 235
  Sequences of, 235
Prime powers, 191–92
Prime-powers of a number n (PP(n)), 234
Principle, 92, 95, 97–98
  Arithmetical, 89–90
  Consistent, 98
  Contradictory, 98
  Set-theoretic, 89, 94
Principle of Complete Induction PCI, 91–93, 124, 127–28, 130, 135–36, 139, 144, 161, 182–83
  Base Step of, 91–93, 127, 130, 136, 140, 159, 183
  is equivalent to PMI, 91
  Induction Hypothesis of, 93, 127–28, 130, 136, 140–43, 159–60, 183
  Inductive Step of, 91, 93, 127–28, 130, 136, 140, 144, 159, 161, 180, 183
Principle of Extensionality, 89, 95, 129, 134–35, 163, 188, 250, 264
Principle of Mathematical Induction PMI, xiv, 14, 89, 91–92, 94, 124, 127–30, 151, 214, 226, 228–29, 260–62, 267
  Base Step of, 91, 94, 127–30, 151, 228
  is equivalent to PCI, 91

Induction Hypothesis of, 94, 127–30, 151, 228–29
Inductive Step of, 91, 94, 127–30, 228–29
is represented in PL by the Induction Schema (IS) of PA, 224, 226
is represented in PL^2 by the Induction Axiom (IA) of PA^2, 260
Principle of Ordered Tuples, 95, 133
Procedure(s), 189–92, 207, 209, 232–38, 247–49, 255–57, 270–72, 284–85
  Arithmetical, 190–92
  Coding, 232
  Computational, 190, 207–08
  Decoding, 191, 206, 211, 232–33, 235–36, 247
  Encoding, 191, 206, 209, 232–35
  Mechanical, 204
  Mechanical computational, 190–92
  Numerical, 190
Process, 103, 151, 162, 175, 204, 207–08
  Computational, xii, 204, 207–08
  Effective, 204
  Counting, 124
  Inductive, 151
  Infinite counting, 108
  Numerical, 190
  Potentially infinite, 103
Product, 90, 215
  function Prod(d, k), 215–16
Projection functions J_i^n, 213–18, 222
PROOF, 235–38, 242–44, 272–72, 276–77, 280, 283, 285
  Construction of, 243–44
  ⟨m, k⟩ ∈ PROOF iff k is the gödel number of an AV sentence **X** and m is the gödel number of a PA proof of **X**, 235–36, 277, 283, 285
  is a recursive binary relation, 237, 242, 272
  is represented in Th(PA) by an AV formula **proof**[x, y], 237–38, 242, 244, 272, 274–75, 277, 280, 283, 285

Proof(s), 42, 89, 106, 109, 173, 186–87, 206, 211, 231–32, 235–39, 244–45, 247–48, 250, 252, 254, 260, 256, 268, 270, 272–75, 277–78, 280
    Demonstrative, xi–xii, 42, 44, 268
        Formal derivations represent, 42, 268
    Finitary, 273–75
    Infinitely long, 268–70
    Peano Arithmetic (PA proofs), 235–38, 242, 248, 253, 272, 274–75, 278, 280, 285
Proof$_{PA}$, 248–49
    is the set of all PA proofs, 248
Proof predicate, 272, 274–76
    Any AV formula that represents PROOF in Th(PA) is, 272, 274
    Nonstandard, 274–75
    proof[x, y] is, 272, 275, 277, 280, 285–86
    Rosser's, 275
    Standard (**proof**$_s$[x, y]), 272, 274–76
Proof-theoretic concepts, 87–88
Proof-theoretically consistent set. See *consistent set*.
Proof-theoretically inconsistent set. See *inconsistent set*.
Proof theory, xii–xiv, 39, 41–44, 48, 87–88, 164, 172, 181, 185–87, 265–66, 268, 270
    Complete, 44, 164, 172, 181, 185–87, 265–66, 268, 270
    Finite, 265–66, 268
    in PL^2, 265–66, 268, 270, 272
    Incomplete, 181, 265, 268, 270, 272
    in NL, 181
    Sound, 43–44, 164, 172, 181, 185–87, 265–66, 268
    Sound and complete, 265–66, 268
    Sound, complete, and finite, 265–66, 268
Property, 2, 10, 12, 14–16, 22–23, 29–31, 167, 229, 257–59
    Arithmetical, 90
    Bivalent, 11–12, 15
    Extension of, 12, 15–16, 23, 26, 259
    of individuals is a second-order object, 257
    Non-bivalent, 12
Proposition, xii, 227, 229, 274
PROV, 237–38, 246
    consists of the gödel numbers of all PA theorems, 237, 246
    b ∈ PROV iff there is k such that ⟨k, b⟩ ∈ PROOF, 237
    is not a recursive set but is recursively enumerable, 237–38, 246
    is not representable in Th(PA), 238, 246
    PROV's formula **prov**[y] is defined as ($\exists x$)**proof**[x, y], 238, 242, 272, 275–77, 280
Provability, 87–88, 230, 272, 275, 277–78, 280, 282
Provability Conditions, xiii, 272, 275, 277–78, 280, 282
    a "proof predicate" is any AV formula **proof**[x, y] that represents PROOF in Th(PA), 272, 274–76
    prov[y] is a "provability predicate" that is defined as ($\exists x$)**proof**[x, y], 272, 275–78, 280
    prov⟨X⟩ is **prov**[k] where k = [X], 277–78, 282, 286
    prov⟨X⟩ satisfies the following Provability Conditions, 276–78
    PC1: if PA ⊢ X, PA ⊢ **prov**⟨X⟩, 277–78, 282, 287
    PC2: PA ⊢ **prov**⟨X→Y⟩→(**prov**⟨X⟩→**prov**⟨Y⟩), 277–78, 282, 287
    PC3: PA ⊢ **prov**⟨X⟩→**prov**⟨**prov**⟨X⟩⟩, 277–78, 282, 287
    The Second Incompleteness Theorem holds for any provability predicate that satisfies, 275, 277–78
Provability predicate, 272, 275–78, 280

any AV formula **proof**[x, y] that
represents PROOF in Th(PA) is called
"proof predicate," 272, 275–77
prov[y] is a "provability predicate" that
is defined as ($\exists x$)**proof**[x, y], 272, 275–
77, 280
prov⟨**X**⟩ is **prov**[k], where k = [**X**], 277,
282, 286
Standard (**prov**$_s$[y]), xiii, 275–77
satisfies the Provability Conditions,
275–78
Punctuation marks, 177

Quantification, 8–9, 11, 14, 19–21, 25–31,
257–60
First-order, 257–58
Multiple, 8
Objectual, 14, 19, 21, 25–26, 30–31
Second-order, 257–60
Substitutional, 11, 14, 19–21, 25–31
Vacuous, 8
Quantified clauses, 10
Quantified sentence, 19, 21, 25–27, 29–31,
153–54, 156–57, 159, 259
Basic substitutional instance of, 19–20, 23,
27, 30, 136, 142–43, 244, 259
Objectual interpretation of, 20, 25–26, 30
Substitutional instance of, 19–20, 26–27,
29–30, 156–57, 244, 259
Substitutional interpretation of, 20, 25–
26, 30, 156
Universally, 153–57, 160–61, 183
Quantifier(s), 2, 5, 7–10, 14, 19–20, 56, 58,
88, 109, 115, 121, 126, 135, 159, 177, 182–
83, 223, 229, 234, 239, 257–58, 269
Existential ($\exists x$), 5, 9, 109, 180, 229
Truth conditions of, 244, 286
First-order, 257–58
range over the universe of discourse, 16,
37
rules, 8
Second-order, 257–58

Scope of, 8–10
Truth conditions of, 14, 19–20, 23, 25, 30–
31, 88, 136, 142–43
Universal ($\forall x$), 5, 9, 19–20, 109, 116, 142–
43, 153, 156, 159–60, 183, 234
Truth conditions of, 153, 156, 160–61,
183, 280

Rational numbers, 100, 106, 108, 273
Sequences of, 273
The set of all ($\mathbb{Q}$), 100, 107
Cardinality of ($\aleph_0$), 106–07
is dense, 106
Set of non-negative ($\mathbb{Q}^+$), 107–08
Set of non-positive ($\mathbb{Q}^-$), 108
Sets of, 273
Real numbers, 21, 29, 106–07, 273
Ordered pairs of, 273
The set of all ($\mathbb{R}$), 21, 28–29, 105–06, 178
Cardinality of (C), 21–22, 105
is uncountable, 22, 105–06, 178
Recursion theory, 217
Recursive definition, 215, 218, 222, 225, 228
of addition, 225
of AV numerals, 228, 288
of the doubling function D, 218, 222
of the factorial function n!, 218
of multiplication, 225
of the total subtraction function G, 222
Recursive function(s), xiii, 205, 208, 213–14
Basic, 213–14, 216–19, 245–48, 250, 252-
255, 282, 288
Projection functions J_i^n are, 213–14, 216–
18, 222
The successor function S is, 213, 216,
222
are total functions, 217
The zero function Z is, 213, 215–16, 222
Partial, xi, xiii–xiv, 205, 208, 213, 216, 217,
219
are equivalent to Turing-computable
functions, xiii–xiv, 205, 208, 213, 217

Primitive, 217
  are total functions, 217
  are representable in Peano Arithmetic, xiii-xiv, 228–31, 245–48, 250, 252–55, 282
  Total, 217, 219
Recursive operation, 213–15, 217
  Composition is, 213–18
  Minimization is, 213, 216–17
  Primitive recursion is, 213, 215, 217–18
Recursive relation, 217, 230, 235–37, 242, 247, 272
Recursive set, 208, 217, 230–31, 233–34, 236–37, 246–47
Recursively enumerable relation, 217
Recursively enumerable set, 208, 217, 237–38, 270–71
Reductio Ad Absurdum (RAA), 34, 57, 62, 115–16, 141, 145, 236
Reductio assumption (RA), 34–35, 40, 57, 141
Referent, 3–4, 15, 17, 27, 161–64, 181, 228, 244, 259–62, 266–69, 286, 288, 289
  of an AV numeral, 244, 266–68, 286, 288, 289
  of a name, 11, 14, 143, 163, 181, 185–86, 244, 261, 266–68, 286, 288, 289
  of a singular term, 14, 22–23, 126, 136, 163–64, 228, 244, 266, 269, 286
Reiteration (rule of inference; Reit), 52, 62, 115, 140, 145, 172
Relation, 2, 12–15, 22–23, 30–31, 98–99, 167, 208, 207, 229–31, 236–37, 247, 257, 259, 272, 285
  Antisymmetric, 98, 130
  Arithmetical, 90
  Asymmetric, 74–75, 90, 98, 125, 132–34
  Binary, 3, 12, 23, 76, 98, 124–25, 132, 235, 237, 272, 283, 285
  Bivalent, 11–12, 15
  Connex, 90, 99, 125, 133–34
    satisfies trichotomy, 99
  Equivalence, 124
  Extendible, 74–75, 90, 98
  Extension of, 12, 15, 23, 76, 208
  Injective, 99
  Irreflexive, 90, 98, 125
    has a minimal element, 90, 99, 125
    has a maximal element, 90, 99
  Non-bivalent, 12
  Places of, 2–3
  Reflexive, 98, 124
    on a set, 98, 125
  Symmetric, 98, 124
  Total, 98
    satisfies dichotomy, 98
  Transitive, 74–75, 90, 98, 124–25, 132, 134
  Well-founded, 125, 133–34
Relation of token identity, 5, 11
Relational predicate, 3, 88
Replacement rules of inference, 47, 49–50, 52, 58–59
Representability in Peano Arithmetic Th(PA), 228–31, 233–35, 237–39, 245–48, 250, 252, 254–55, 272, 274, 278, 281, 283
  all recursive functions are representable in Th(PA), 228–31, 245–46, 248, 250, 252, 254, 288
  definition of the representability in Th(PA) of an n-place relation on ℕ, 230
  definition of the representability in Th(PA) of a subset of ℕ, 230, 233
  definition of the representability in Th(PA) of a total function from ℕⁿ into ℕ, 230
  FORM consists of the gödel numbers of all AV formulas, 234
  FORM is a recursive set, 234
  FORM is represented in Th(PA) by an AV formula **form**[x], 234
  representing PA proofs in Th(PA), 235–38, 242

a PA proof is a PL derivation D of an
AV sentence **X** from PA, 235–38, 248,
271, 278, 283, 285–86
⟨m, k⟩ ∈ PROOF iff k is the gödel
number of an AV sentence **X** and
m is the gödel number of a PA
proof of **X**, 235–37, 272, 285
PROOF is a recursive relation, 236–38,
242, 272
PROOF is represented in Th(PA) by an
AV formula **proof**[x, y], 237–38, 242,
272, 275, 277, 280, 285
representing provability in Th(PA), 237–38
PROV consists of the gödel numbers of
all theorems of PA, 237–38, 246
b ∈ PROV iff there is k such that
⟨k, b⟩ ∈ PROOF, 237–38
PROV is not a recursive set but only
recursively enumerable, 237–38, 246
PROV is not representable in Th(PA),
238, 246
prov[y] is defined as ($\exists x$)**proof**[x, y],
238, 242, 272, 277, 280
representing the syntax of Th(PA) within
Th(PA), 232–35
SENT consists of the gödel numbers of all
AV sentences, 234
the construction of SENT, 234
SENT is a recursive set, 234
SENT is represented in Th(PA) by an AV
formula **sent**[x], 234–35
Representability Theorem, 231, 252–53
asserts: every recursive function is
representable in Th(PA), 231, 252–53
the set of Robinson's Axioms (RA) is
finite and is adequate for representing
all recursive functions, 252–53
Representative, 32, 177,
Resources of the metatheory, xiv, 87–91, 94
Arithmetical, xiv, 88–91

Linguistic, xiv, 87
Logical, xiv, 87, 89
Set-theoretic, xiv, 88–89, 94
Robinson Arithmetic Th(RA), 252–54, 257,
281–82
All recursive functions are representable
in, 252
there is an ω-consistent extension of
Th(RA) of which N is not a model, 282
RA is much weaker than PA, 253
Robinson Arithmetic set of axioms RA
consists of the following 7 axioms, 253,
285
RA1: ($\forall x$)**0** ≠ sx, 253
RA2: ($\forall x$)($\forall y$)($sx = sy \rightarrow x = y$), 253
RA3: ($\forall x$)($x \neq$ **0** $\rightarrow (\exists y) x = sy$), 253
RA4: ($\forall x$)($x +$ **0**) = x, 253
RA5: ($\forall x$)($\forall y$)($x + sy$) = $s(x + y)$, 253
RA6: ($\forall x$)($x \bullet$ **0**) = **0**, 253
RA7: ($\forall x$)($\forall y$)($x \bullet sy$) = (($x \bullet y$) + x), 253
Th(RA) is incomplete, 254
Th(RA) is undecidable, 254
Rosser, John Barkley, Sr, 275
Rosser's proof predicate, 275
The Second Incompleteness Theorem
fails for, 275
Rosser Sentence R$_{PA}$ of PA, xv, 283
Rule(s) of inference, xii, 42–43, 48, 87, 89,
109, 115, 236, 242, 248–49, 255–56, 265,
268–70, 277–78
Classical, 115
Complete, 87, 115
Formal, 42–43
Independent, 115
are invoked in the metatheory, 89
The ω-, 269–70
PL, 88, 229
Set of, 87, 115
Sound, 43, 59, 87, 115, 270
Traditional, 61

Truth-preserving, 43, 269
Unsound, 44
Russell's Paradox, 97–98

Satisfaction, 20, 25, 30, 87
Satisfiability, 41, 164, 257
   is neither a decidable nor semidecidable concept in PL, 41, 257
Satisfiable set, xiv, 38, 41, 45, 65, 72, 147, 161, 165–66, 171–72, 177–80, 182, 185, 226, 252, 262, 266–67, 271, 283, 288–89
   The concept of satisfiable set is neither decidable nor semidecidable in PL, 257
   has a model, 144, 147–48, 161, 165, 171, 178, 226, 237, 241, 243–45, 254, 260–64, 266–69, 276, 280–83, 288–89
Second Incompleteness Theorem, xiii, xv, 272, 274–78, 280
   asserts informally that if PA is consistent, it is impossible to prove its consistency from PA, 272
   an AV formula **proof**$[x, y]$ that represents PROOF in Th(PA) is called "proof predicate," 272, 274–77, 280, 285
     prov$[y]$ is a provability predicate and is defined as $(\exists x)$**proof**$[x, y]$, 272, 275–78, 280
     the Second Incompleteness Theorem fails for Rosser's proof predicate, 274–75
   First proof of, 272, 275–76
     the standard PA consistency sentence is **CON**$_s$, 274, 276
     first proof establishes: If PA is consistent, PA $\nvdash$ **CON**$_s$, 276–77
   Second proof of, 275–80
     prov$\langle X \rangle$ is **prov**$[k]$, where k = $[X]$, 277–78, 286
     $\neg$**prov**$\langle 0 = 1 \rangle$ is a PA consistency sentence, 276
     second proof establishes: If PA is consistent, PA $\nvdash \neg$**prov**$\langle 0 = 1 \rangle$, 276–80
Second-Order Arithmetic Th$_{AW}$(N²), xv, 263–66, 271
   Axioms of, 260, 264
   is finitely axiomatizable, xv, 263–64
   is semantically complete, 263–64, 271
   The standard interpretation of (N²), 254, 260, 262, 266, 269–83, 288–89
   Th(PA²) = Th$_{AW}$(N²), 263–64, 271
Second-Order Peano Arithmetic Th(PA²), xv, 260, 263–66, 271
   The axioms of (PA²), 260–71
   every model of PA² is isomorphic to N² with respect to AW, 260–64, 269
   PA² is categorical, 260–62
   The standard model of (N²), 260, 262, 266, 269
   The vocabulary of, 260, 262–63, 266
Second-Order Predicate Logic PL², xi, xv, 87, 257–59, 263, 265–66, 268, 270, 272, 281
   Finite proof theory of, xi, xv, 44, 265–66, 268
   is an incomplete logical system, 263, 265, 270, 272
   Infinite proof theory of, 268–71
     the ω-rule of inference, 269–70, 281
   Logical consequence in, 44, 263–68, 270–71
   PL² theory, 260, 263–64, 271
   proof of the incompleteness of PL² infinite proof theory, 269–72
   second-order argument, 257–60
   second-order predicate, 257–58
   second-order predication, 257–58
   second-order quantification, 257–59
   second-order quantifier, 257–58
   second-order sentence, 257–59
   second-order variable, 258–61
   Semantics of, 258–60
   Syntax of, 258

Semantical assignments (SA), 11, 14–16, 25, 161–62, 164, 166, 184, 224
Semantical concepts, xii, 87–88, 223, 256–57, 263–64
Semantically consistent set. See *satisfiable set*.
Semantically inconsistent set. See *unsatisfiable set*.
Semantics, xi-xiv, 10, 14–15, 43, 60, 87–89, 94, 115, 181, 251, 258
  of PL, xi-xiv, 10, 15, 21, 43, 60, 94, 115, 181, 252, 258
Semidecidable relation, 208
  is equivalent to recursively enumerable relation, 217
Semidecidable set, 173–74, 189, 205–08, 217, 248, 255–57
  is equivalent to recursively enumerable set, 208, 217
  K is semidecidable iff its listing function is computable, 206–07
  Th($\emptyset$) is, 255–57
  Th(PA) is, 248–49
SENT, 234–35
  consists of the gödel numbers of all AV sentences, 234
  Construction of, 234
  is a recursive set, 234
  is represented in Th(PA) by an AV formula **sent**[x], 234–35
Sentence(s), 3, 7–12, 14–16, 20, 22–23, 25–27, 35–36, 41–45, 49–50, 52, 55–58, 60–61, 64, 81, 83, 88, 92–94, 109–10, 113, 115, 120–22, 126–27, 135–45, 148–50, 152, 165, 169, 174, 177–78, 180–81, 235–36, 242, 245–47, 251, 254–56, 265–68, 270, 274, 281–85, 288–89
  Atomic, 23, 136, 158–59, 164, 184
    Truth conditions of, 23, 136, 158, 163
  AV, 223, 227, 234–37, 239–40, 242–46, 248–53, 255, 257, 264–65, 271–78, 280–83, 285–86, 289
  AW, 263–64, 266, 271
  Bivalent, 12
  Components of, 59, 109, 113, 183
  Compound, 23–24, 109–10, 113–14, 136, 159
    Truth conditions of, 23
  Declarative, 1–2, 12, 92, 251
  Main connective of, 113–14
  Meaning of, 30, 87
  Natural reading of, 14, 15, 19–20, 137
  Non-bivalent, 12
  PL², 257–60, 265–66, 268–71
  Quantifier-free, 92–93, 124, 127, 190
  Sentential components of, 92–93
  Set of, 14, 16, 20, 28, 39–40, 41–44, 64, 66, 83, 88, 122, 139, 144, 147–52, 155–57, 167, 171, 178, 180–81, 184
  The set of all PL (Sent$_{PL}$), 149–50, 247
  Symbolic, 42
  Truth conditions of, 22, 30, 32, 34–35, 37, 87, 93, 110, 113–14, 152
  V, 170–72, 174–76, 187–88
  Well-formed, 126
Sentence letters, 190–91
Sentence Logic SL, xi-xii, 190–92
  arithmetization of SL "sentencehood," 192
  Basic vocabulary of, 191
  Compound sentences of, 191
  Expressions of, 191–92
  Formation rules of, 190
  Sentences of, 190–92
    The set of all (Sent$_{SL}$), 190, 192
    The set of all numerical codes of (SENT$_{SL}$), 192
  Sentence letters of, 190–91
  Symbols of, 190–91
Sentential connectives, 5, 10, 24, 62, 88, 92–93, 108–10, 114, 126–27, 135, 159, 177, 182–83, 190, 223, 229
  Binary, 5, 63, 108–10, 113, 115, 190, 229
    Truth conditions of, 110, 113
  Constant, 113

Expressively complete sets of, 108–10, 113
Natural-language, 109
Set of, 109–10
Traditional set of, 109
Truth-functional, 24, 108–10, 113–14
  Truth conditions of, 110, 113
Unary, 5, 8, 109–10, 113, 115, 190
  Truth conditions of, 110, 113
Sequence (also, *list*), 151, 187, 255, 270–71, 273
  of AV sentences, 235, 236, 275
  Finite, 173, 187, 195, 206, 248–49, 252-253, 255, 265–66
  Infinite, 151, 174, 206
  of objects, 151
  of PL sets, 151–52
  of prime numbers, 235
  of sentences, 42–43, 88, 154, 174, 187
Set(s), 13, 21, 41, 45, 94–98, 101, 103, 105–06, 125–26, 129, 133–35, 179, 189–90, 208, 226–27, 229–32, 241, 248, 256, 265, 268, 273, 280–82, 285–86, 288
  Cardinality of, 21, 101, 103–05
  can consist of sets, 94
  Consistent. See *consistent set*.
  Countable, 21–22, 103, 106, 126, 178–79, 182
  Countably infinite, 22, 103
  are determined solely by their members, 95
  Elements of, 94–96, 102–05
  Empty. See *empty set*.
  Finite, xiv, 21–22, 101–02–103, 125, 131, 156, 178, 227
  Identity condition for, 95
  Inconsistent. See *inconsistent set*.
  Inductive. See *inductive set*.
  Infinite, xii, xiv, 19, 21–22, 74, 101–04, 106, 178
    is definable in PL, 74–76, 178
  Members of, 95–96, 98, 102–04, 106, 122, 124, 129, 131–34, 141–42, 144–45, 147, 152–53, 155–56, 184, 186–87, 189–90, 208, 226–27, 267, 270, 288
  Nonempty. See *nonempty set*.
  Satisfiable. See *satisfiable set*.
  Size of, 101, 182
  Transfinite, 103
  Transitive. See *transitive set*.
  Uncountable, xii, 21–22, 103, 105, 126, 178–79
  Universal, 97–98
  Unsatisfiable. See *unsatisfiable set*.
Set theory, xi, xiv, 94–95, 97–98, 101, 178–79, 230
  Infinite set of axioms of, 94, 172
  Set of standard axioms of, 178–79
  Standard, 95
    Language of, 178
Sheffer Stroke (↑, |), 111, 113–14
  Truth table for, 111
Simplification (rule of inference; Simp), 52, 59, 61–62, 287
  is sound, 61
  Soundness of, 59
  is truth-preserving, 61
Singleton of an object {x}, 95, 104, 124, 128
  is unique, 95
Skolem, Thoralf, 177
Skolem's Paradox, 178–79
Soundness and Completeness Theorems, 44–45, 88, 139, 149, 169–71, 223, 226, 254, 265
  Corollaries of, 45, 88, 164, 170
Soundness Theorem, xi, xiii-xiv, 43–44, 63, 89, 115, 122, 139, 144, 149, 164, 169–70, 180, 223, 226, 252, 254, 266, 271, 276, 280, 288
  asserts: every satisfiable PL set is consistent, 180, 226
  asserts: if **X** is derivable from Γ, it is a logical consequence of Γ, 43–44, 139

Proof of, 89, 115, 122, 139–44, 180
Standard arithmetical vocabulary AV, 223, 228, 260, 263, 266, 271–72, 281–83, 285–86, 288–89
  Second-order (AW), 260–64, 271
Standard Deduction System (see also *Gentzen Deduction System*), xiv, 61–62
Standard interpretation of arithmetic N, 250, 254, 260, 263, 266, 269, 280–83, 288–89
  is not a model of any AV theory that is ω-inconsistent, 281, 286
  Second-order (N^2), 260, 263, 266
    All models of PA^2 are isomorphic to, 260–64, 269
Standard rules of inference (see also *GDS rules of inference*), 61–62
Statement, 25, 45, 91, 94, 140, 204, 240, 242, 252, 273, 276, 278, 282
  Arithmetical, 91, 94
  "Hereditary," 91
  Identity, 140
Structure, 167, 169, 176, 187, 224
Structure of the natural numbers N, 90, 224–29, 231–32, 237, 241, 243–45, 250, 254, 260, 263, 280–83, 288–89
  FORM consists of the gödel numbers of all AV formulas, 234
  FORM construction, 234
  FORM is a recursive set, 234
  FORM is represented in Th(PA) by an AV formula **form**[x], 234
  The Gödel Sentence G$_{PA}$ is true on N but not a theorem of PA, 227, 242–45, 280
  is the intended interpretation of PA, 224–25, 241, 250, 281
  List of names of, 224, 244, 286
  is a model of Th(PA), 226
  is not a model of any AV theory that is ω-inconsistent, 281, 286
  the referent of the AV numeral s^k0 on N is the natural number k, 228, 269, 286

Second-order (N^2), 260, 263–64, 266–69
  All models of PA^2 are isomorphic to, 260–61, 263–64
Semantical assignments of, 224
SENT consists of the gödel numbers of all AV sentences, 234–35
SENT construction, 234
SENT is a recursive set, 234
SENT is represented in Th(PA) by an AV formula **sent**[x], 234–35
Th$_{AV}$(N) consists of all AV sentences that are true on N, 227, 243, 245, 249–50, 282, 287, 289
Universe of discourse of, 224
Subblock(s), 51–52, 139–41
Subderivation, 51, 147
Subformula, 8, 234
  Proper, 8
Subject-predicate sentence, 136, 164
Subscripts, 1, 6, 28–29
  Numeric, 1, 5–6, 155
Subset(s) ($\subseteq$), 21, 95–96, 98–100, 102, 104, 106, 113, 125, 131–32, 141, 143, 145, 151–52, 157, 162–64, 173, 182, 184, 186, 188, 213, 280, 285, 288–89
  Finite, 139, 152, 155, 164–67, 182, 184, 186, 188, 240–41, 265–68, 288
  Nonempty, 90, 96, 125–26, 131, 133–34, 155, 162, 163, 186
  Proper ($\subset$), 95, 102–03, 106, 108, 125, 134, 140, 150, 213
Substitution, xiii, 3–4, 19, 122, 132, 135, 251
Substitution (rule of inference; Sub), 57, 62, 105, 115–16, 143, 163–64
Subtraction, 91, 96, 189, 222
  function, 189, 222
    Partial (Sub), 189, 218, 219
    Total (G), 218, 222
Successor, 90–92, 94, 127, 213, 215, 225, 227
  function Sn, 90, 194, 199, 213, 215–16, 225, 227, 229
    is a basic recursive function, 213

T$_S$ is a Turing machine that computes, 194, 199
   of a set A (SA), 124–25, 129–30
Sum, 90, 99–100, 189, 215
   function Sum(d, k), 215–16, 218–19, 222
Superscripts, 1, 3, 6, 10
Symbol(s), 88, 108–10, 113, 115, 174, 190–93, 209, 228
   Basic Turing machine, 193, 209–10
   Metalinguistic, 228
   PL, 150, 174
   SL, 190–92
   Standard, 110–12
Symbolic logic, xi–xii, xv, 144
Symbolic system, xii, 1, 42, 87
Syntactical categories, 1, 5, 7–8, 232–34
Syntactical structure of a sentence, 43
Syntactically identical, 143
Syntax, xiii, 10, 87, 115, 232–35, 258
   of PL, xi–xiii, 1, 8, 94, 108, 115, 181, 258
Systems of non-classical logic, 61, 63, 174

Tarski, Alfred, 251–52
Tarskian Schema, 251
   can generate a contradictory biconditional by being instantiated for a Liar Sentence, 251
   a Liar Sentence is any sentence that asserts of itself that it is not true, 251–52
   the Liar Paradox, 251
Tarski's Indefinability Theorem (first version), xi, xiv, xv, 251–52
   any formula **true**[*x*] defines arithmetical truth in Peano Arithmetic only if it satisfies TS: for every AV sentence **X**, PA ⊢ **true[k]**↔**X**, where k = [**X**], 250–52
   asserts: the notion of arithmetical truth is not definable in Peano Arithmetic, 251–52
   Convention-T, 251

Tarskian Schema, 251
   truth, 250–52, 282
Tarski's Indefinability Theorem (second version), 282, 287–88
   asserts: #Th$_{AV}$(N) is not arithmetically definable, 282
   #Th$_{AV}$(N) is the set of the gödel numbers of all members of Th$_{AV}$(N), 282, 288
Tautology (sentential connective; T), 110, 113–14
   Truth table for, 110, 113
Term(s), 4–7, 10, 13, 23, 29, 88, 173, 177
   AV, 228–29, 232–33, 235
   Complex, 3, 6, 11
   Functional, 5–7, 11
   Singular, 2–7, 11, 13–14, 19–20, 22–23, 27, 56–57, 88, 120, 126, 135–36, 140, 142–43, 158, 184
Term-Formation Rule, 5
Theorem(s), xi, xiii–xv, 14, 21, 27, 43–45, 59, 63, 84, 87–89, 93, 104–07, 115, 122–23, 139, 144, 147, 149–50, 155, 157, 159, 163–65, 169, 174–75, 177–78, 180, 207–09, 211, 217, 223, 226–32, 238–50, 252–25, 261, 263–66, 268, 270, 272, 274–78, 280–82, 385–87, 289
   Mathematical, 21, 209
   of PA, 226, 230, 237, 240, 242, 246, 248, 253, 272–78, 280–82
   of a set of sentences, 43, 170–72, 185, 226, 227, 230, 237, 240, 242, 245–48, 250, 253, 272–78, 280–82
Theory, 87
Transfinite arithmetic, 103
Transitive set, 125, 133–35
   is a family of sets that contains the members of its members, 125
   an ordinal is a transitive set that is well-ordered by ∈, 125
Transitivity, 74–75, 90, 132, 135
   of a set, 134
True-on-an-interpretation, xiv, 87
Truth, 45, 157, 250–52, 273, 282

Arithmetical, 250–52, 282
Convention-T, 251
Definition of, 250–52
Tarskian Schema, 251
  the Liar Paradox, 251
Tarski's Indefinability Theorem, xi, xiv, xv, 252, 282
  asserts: the notion of arithmetical truth is not definable in Peano Arithmetic, 252
Truth conditions, 15, 22–23, 59, 63, 87, 110, 280, 284, 286, 288–89
  of an atomic sentence, 23, 136, 159, 164
  of biconditional, 22, 88, 244
  of binary connectives, 110, 113
  of a compound sentence, 23
  of conjunction, 22, 59, 88
  of disjunction, 22, 88
  of the existential quantifier, 244, 286
  of the identity predicate, 22, 36, 136, 163, 289
  of material conditional, 22, 60, 88, 141, 153, 160
  of negation, 22, 88, 153, 159
  of PL sentences, 30, 32, 34–35, 37, 87, 93, 110, 113–14, 152
  of the quantifiers, 14, 19–20, 23, 25, 30–31, 88, 136, 142–43
  of truth-functional connectives, 110, 113
  of unary connectives, 110, 113
  of the universal quantifier, 153, 156, 160–61, 183, 280
Truth table, 23, 110, 113–14, 183, 190
Truth values, 15, 22–27, 31, 37, 109, 113, 126, 135–36, 142–44, 169, 183, 190
Truth-functionality, 24
Truth-Membership Theorem, 157, 159, 161, 163–64
  asserts: a PL⁺ sentence belongs to a Henkin set iff it is true on its Henkin Interpretation, 157, 159
  Modified, 163

Turing, Alan, 193, 209
Turing computability, 193, 204–05
Turing machine(s), xiv, 193–204, 206–07, 209–13, 218–19, 222
  Components of, 193
  that computes the addition function (T_A), 200–03
  that computes the successor function (T_S), 194, 199
  that computes the zero function (T_Z), 197–99, 210
  Diagram of, 196, 218, 221
  may enter into a loop, 211
  Examples of, 197–202
  Halting position of, 194–95, 208–09, 211–12, 220
  Inputs of, 193–94, 204, 206–07, 219–20
  Instruction lines of, 193–97, 209–10, 212, 222
  Instruction set of, 193–95, 211–12, 218, 220, 222
  Outputs of, 193–94, 204, 206–07, 212, 219–20
  Pointer of, 193–99, 202–03, 212, 219–20
  Terminal internal state of, 193, 195–96, 199, 203, 209, 211–12, 221
  Turing machines are extremely powerful computing devises, 193, 205
    every known computable function is Turing-computable, 193, 205
Turing-computable functions, xi, xiii-xiv, 193, 204–09, 213, 217–18, 247
  are equivalent to partial recursive functions, xiii-xiv, 205, 208, 213, 217, 247

Undecidability, 175
  for a consistent PL theory, the triad of completeness, axiomatizability, and undecidability is inconsistent, 175, 250
Undecidable set, 174–75, 227, 237, 245–48, 250, 252, 254–55, 257, 270

Arithmetic is, 227, 248
  if Σ is a consistent PL theory in which all recursive functions are representable, it is undecidable, 246–48, 252, 254, 255
  Th(∅) is, 255–56
  Th(PA) is, 246, 248, 257
Union of two sets (A∪B), 96, 125, 152, 155–56, 166
  consists of the members of the two sets, 96
Universal Elimination (∀E), 62, 116
Universal Generalization (UG), 32, 56, 62, 115–16, 142–43, 147
Universal Instantiation (UI), 56, 59, 62, 115–16, 142, 153, 157, 231, 242, 273, 280
Universal Introduction (∀I), 62, 116
Universal set. See *set*.
Universe of discourse (UD), 11, 14–15, 19–21, 23, 25–31, 137, 142–43, 153, 161, 178–79, 181, 184, 186–87, 259, 266, 269
  Countable, 179
  Countably infinite, 178
  Finite, 137
  Infinite, 18
  is nonempty, 11, 74
  of a Henkin Interpretation, 158, 161, 163, 184
  of a modified Henkin Interpretation, 162–63
  of a PL² interpretation, 259, 269
  of the structure of the natural numbers N, 224
  Uncountable, 179
Unordered pair {x, y}, 95
Unsatisfiability, 41, 166, 257
  is not a decidable concept but only semidecidable, 41, 257
Unsatisfiable set, xiv, 39–42, 45, 66, 83, 148, 166, 171
  The concept of, 257
    is not decidable but only semidecidable, 257

  has no model, 144, 148, 171

Valid PL sentence(s), xiv, 34–35, 41, 45, 66, 82, 140, 170, 173–74, 181, 185, 254–56
  are logical consequences of ∅, 34, 140, 186
  The set of all (Th(∅)), 174, 254–55
    is semidecidable, 174, 255–56
    is undecidable, 174, 254–55
Validity (of arguments), 31, 41, 165, 185, 256–57
  The concept of, 41, 257
    is not decidable in PL but only semidecidable, 41, 256–57
  is equivalent to logical consequence, 31
Variable(s), 1–3, 5–10, 13, 19, 27–28, 56, 58, 88, 120–21, 142, 145, 147–48, 157, 180, 223–24, 234, 238, 257–58
  Bound occurrences of, 8–9, 234
  First-order, 257–58
  Free, 92, 224, 234
  Free occurrences of, 8–10, 19, 234, 258
  Metalinguistic, 2, 5, 7, 49, 88
  Numerical, 92, 94
  Second-order, 257–59
Vocabulary, 28–29, 144, 148, 170–71, 181–82, 190, 223–24
  Basic, 1, 30, 190–91, 223–24, 254, 259, 271
  Countable, 29, 182
  Extra-logical, 1, 10, 28, 29, 142, 157, 167, 170–71, 223–24
    of Peano Arithmetic, 223–24
  of the language of the metatheory, 88
  Logical, 1, 10, 28–29, 89, 109, 126, 158, 167–68, 170–71, 223
  of a PL interpretation Voc(I), 11, 14, 29, 141–43, 166, 168–71, 176, 226
  of the set {(∃x)Px}, 182, 187
  of a set of PL sentences Voc(Γ), 10, 14, 28–29, 141–42, 167–70, 172, 174, 176, 180, 182, 184, 187–88, 223, 245
  of a set of PL² sentences Voc(Δ), 262

of a set of PL⁺ sentences Voc(Σ), 158
Uncountable, 29, 182

W-conditional(s) θ_n, 155–57
    The set of all (Θ), 155, 157
Well-ordering of a set, 125, 132–35
Whole-Part Principle, 102–03

X-negation, 113–14
    Truth table for, 113
X-projection, 113–14
    Truth table for, 113

Y-negation, 113–14
    Truth table for, 113
Y-projection, 113–14
    Truth table for, 113
Yes-procedure, 41, 173, 206–07, 238, 248 5–57

Zermelo-Fraenkel Axioms, 94
Zero (0), 90, 197, 210, 213, 216, 225
    function Z, 197, 210, 213, 215, 216
ZFC Set Theory, 94, 97

from the publisher

A name never says it all, but the word "broadview" expresses a good deal of the philosophy behind our company. We are open to a broad range of academic approaches and political viewpoints. We pay attention to the broad impact book publishing and book printing has in the wider world; we began using recycled stock more than a decade ago, and for some years now we have used 100% recycled paper for most titles. As a Canadian-based company we naturally publish a number of titles with a Canadian emphasis, but our publishing program overall is internationally oriented and broad-ranging. Our individual titles often appeal to a broad readership too; many are of interest as much to general readers as to academics and students.

Founded in 1985, Broadview remains a fully independent company owned by its shareholders—not an imprint or subsidiary of a larger multinational.

If you would like to find out more about Broadview and about the books we publish, please visit us at
www.broadviewpress.com.
And if you'd like to place an order through the site, we'd like to show our appreciation by extending a special discount to you: by entering the code below you will receive a 20% discount on purchases made through the Broadview website.

Discount code: **broadview20%**

Thank you for choosing Broadview.

Please note: this offer applies only to sales of bound books within the United States or Canada

The interior of this book is printed
on 100% recycled paper.